	$(f/g)x$	$f(x)/g(x)$
	$(f \circ g)x$	$f(g(x))$
[5.1]	a^n	the nth power of a, or a to the nth power; $a \cdot a \cdot a \cdot \cdots \cdot a$ (n factors)
	a^0	$1 \quad (a \neq 0)$
	a^{-n}	$1/a^n \quad (a \neq 0)$
	$a^{m/n}$	$(a^{1/n})^m$
[5.3]	$\log_b x$	the logarithm to the base b of x, or the logarithm of x to the base b
[5.4]	$\text{antilog}_b x$	the antilogarithm to the base of b of x, or the antilogarithm of x to the base b
	e	an irrational number, approximately equal to 2.7182818
[6.1]	$\cos x$	an element in the range of the cosine function
	$\sin x$	an element in the range of the sine function
[6.3]	$\bar{s}$ or $\bar{x}$	the length of reference arc corresponding to the arc of length s or x
[6.7]	$\tan x$	an element in the range of the tangent function
	$\sec x$	an element in the range of the secant function
	$\csc x$	an element in the range of the cosecant function
	$\cot x$	an element in the range of the cotangent function
[7.1]	$\alpha \cong \beta$	α is congruent to β
	$\overrightarrow{AB}$	the ray AB
	$m^\circ(\alpha)$	the measure of α in degree units
	$m^R(\alpha)$	the measure of α in radian units
	x°	x degrees
	x^R	x radians
[7.2]	$T(\alpha)$	any one of the trigonometric functions of α
	$C(x)$	any of the circular function values of x
[7.3]	$\mathscr{A}$	area
	$\overline{AB}$	the line segment with endpoints A and B, or the line containing points A and B
	$l(\overline{AB})$	length of line segment AB
[8.4]	$\text{Sin}^{-1}x$, $\text{Cos}^{-1}x$, etc.	elements in the range of the principle-valued inverse circular or trigonometric functions; $\text{Arcsin } x$, $\text{Arccos } x$, etc.
[8.5]	$\text{cis } \theta$	$\cos \theta + i \sin \theta$
	z^0	$1 + 0i$
	z^{-n}	$1/z^n \quad (z \neq 0 + 0i)$
[9.2]	$s(n)$, or s_n	the nth term of a sequence
[9.3]	S_n	the sum of the first n terms in a sequence
	S_∞	the sum of an infinite sequence
[9.4]	$n!$	n factorial, or factorial n
	$0!$	zero factor
	$\binom{n}{r}$	the number of combinations of n things taken r at a time, also the binomial coefficient
[9.5]	$n(A)$	the number of elements in the set A
	$P_{n,n}$	the number of permutations of n things taken n at a time
	$P_{n,r}$	the number of permutations of n things taken r at a time
[10.1]	(x, y, z)	the ordered triple of numbers whose first component is x, second component is y, and third component is z
	$(R \times R) \times R$, or R^3	the Cartesian product of $(R \times R)$ and R

Elementary Functions

Second Edition

William Wooton

Irving Drooyan
Los Angeles Pierce College

Wadsworth Publishing Company, Inc., Belmont, California

© 1975 by Wadsworth
Publishing Company, Inc.
© 1971 by Wadsworth
Publishing Company, Inc.,
Belmont, California 94002. All
rights reserved. No part of this
book may be reproduced, stored
in a retrieval system, or tran-
scribed, in any form or by any
means, electronic, mechanical,
photocopying, recording, or
otherwise, without the prior
written permission of the
publisher.

ISBN-0-534-00339-7
L. C. Cat. Card No. 73-89412
Printed in the United States
of America

2 3 4 5 6 7 8 9 10—
79 78 77 76

Preface

This text is essentially an expansion of the first edition for the purpose of improving the teachability of the elementary functions course. In addition to the inclusion of six new sections, we have added two new chapters, one an introductory chapter reviewing the properties of real numbers and complex numbers, and the other a chapter covering natural-number functions. We have also undertaken an extensive revision of many of the other sections in the first edition. In particular, the material on logarithms has been expanded to include a separate section on applications of logarithms; the first two sections of the chapter on circular functions have been rewritten to provide a simpler introduction to the topic; and the sections on analytic geometry in space have been revised and now constitute a separate chapter. All exercise sets and the chapter reviews in the original book have been reviewed, with many being completely revised.

This book is designed for students who have completed the equivalent of one and one-half to two years of high school algebra and one year of high school geometry but, nevertheless, need additional preparation before undertaking more advanced courses in mathematics. Upon successful completion of this course, the student can proceed directly to the study of calculus.

The topics covered, as well as the spirit in which they are presented, reflect the recommendations of various mathematics curriculum study groups. In particular all of those topics recommended by the Committee on the Undergraduate Program of the Mathematical Association of America for the Level O course are covered.

The function concept plays a heavy role in the development, starting with linear functions in Chapter 2. The study of the properties of various kinds of functions proceeds through quadratic, polynomial, rational, exponential, and logarithmic functions; heavy emphasis is placed on sketching graphs of these functions. Chapters 6 to 8 offer a concise treatment of circular and trigonometric functions, culminating in a discussion of polar coordinates. Chapters 9 and 10 are new, Chapter 9 dealing with natural-number functions including sections on mathematical induction, sequences, series, the binomial theorem, and simple combinatorics, and Chapter 10 dealing with the analytic geometry of three space. A three-section appendix offers coverage of parametric equations and simple transformations of the plane.

The second color in the text is used functionally to highlight key procedures in routine manipulations and to stress basic concepts. Color in graphs focuses attention on particular parts of the figure under discussion. Marginal annotations direct the reader's attention to important ideas. Exercises are provided in each section, and answers for the odd-numbered exercises, including graphs, are given at the end of the book. Answers for all chapter review problems are also given.

We sincerely thank Dennis G. Zill of Loyola University of Los Angeles for his careful reading of this new edition and his suggestions for improving it. Our thanks go also to David Roselle of Louisiana State University for his suggestions and to our editor, Don Dellen, for his guidance and encouragement.

William Wooton
Irving Drooyan

Contents

Numbers and Their Properties

In earlier mathematics courses, you explored the basic properties of the set of real numbers and saw how some of their properties apply to real-life situations. Moreover, you probably have encountered the set of complex numbers in some form and have seen how the set of real numbers is related to this set. The purpose of this first chapter is to review, briefly, some facts about these sets of numbers, facts we shall need in our examination of the elementary functions during the remainder of this course.

1.1   Preliminary Concepts

A **set** is simply a collection of some kind. In algebra, we are interested in sets of numbers of various sorts and in their relations to sets of points or lines in a plane or in space. Any one of the collection of things in a set is called a **member** or **element** of the set, and is said to be **contained in** or **included in** (or, sometimes, just **in**) the set. For example, the counting numbers 1, 2, 3, . . . (where the three dots indicate that the sequence continues indefinitely) are the elements of the set we call the set of **natural numbers.**

Set notation   Sets are usually designated by means of capital letters, A, B, C, etc. They are identified by means of **braces,** $\{\ \}$, with the members either listed or described. For example, the set N of *natural numbers* is $\{1, 2, 3, \ldots\}$.

Using the undefined notion of set membership, we can be more specific about some other terms we shall be using.

Definition 1.1   *Two sets A and B are **equal**, $A = B$, if and only if they have the same members.*

The order in which the members of a set are named is of no import in determining its membership, nor is the fact that a member might be named more than once. Thus, if A denotes $\{1, 2, 3\}$, B denotes $\{3, 2, 1\}$, C denotes $\{2, 3, 4\}$, and D denotes $\{$natural numbers between 1 and 5$\}$, then $A = B$ and $C = D$. The phrase "if and only if" used in this definition is simply the mathematician's way of making two statements at once. Definition 1.1 means: "Two sets are equal if they have the same members. Two sets are equal only if they have the same members." The second of these statements is logically equivalent to: "Two sets have the same members if they are equal."

Definition 1.2   *If every member of a set A is a member of a set B, then A is a **subset** of B.*

Subset symbol   The symbol $\subset$ (read "is a subset of" or "is contained in") will be used to denote the subset relationship. Thus

$$\{1, 2, 3\} \subset \{1, 2, 3, 4\} \quad \text{and} \quad \{1, 2, 3\} \subset \{1, 2, 3\}.$$

Notice that, by definition, every set is a subset of itself.

The set that contains no elements is called the **empty set,** or **null set,** and is denoted by the symbol $\emptyset$ (read "the empty set" or "the null set"); $\emptyset$ is a subset of every set. If a set S is the null set or contains exactly n elements for some fixed natural number n, then S is said to be **finite.** A set that is not finite is said to be **infinite.** For example, the set of *all* natural numbers, $\{1, 2, 3, \ldots\}$, is an infinite set.

Definition 1.3   *The **union** of two sets A and B is the set of all elements that belong either to A or to B or to both.*

Set-union symbol   The symbol $\cup$ is used to denote the union of sets. Thus $A \cup B$ (read "the union of A and B") is the set of all elements that are in A or B or both.

For example, if $A = \{1, 2, 3, 4\}$ and $B = \{3, 4, 5\}$, then

$$A \cup B = \{1, 2, 3, 4, 5\}.$$

Notice that each element in $A \cup B$ is listed only once in this example, since repetition would be redundant.

Definition 1.4   *The **intersection** of two sets A and B is the set of all elements common to both A and B.*

Set-intersection symbol   The symbol $\cap$ is used to denote the intersection of sets. Thus $A \cap B$ (read "the intersection of A and B") is the set of all elements that are in both A and B.

For example, if $A = \{1, 2, 3, 4\}$ and $B = \{3, 4, 5\}$, then

$$A \cap B = \{3, 4\}.$$

Definition 1.5   *Two sets A and B are **disjoint** if and only if A and B contain no member in common.*

Thus A and B are disjoint if and only if $A \cap B = \emptyset$. For example, if $A = \{1, 2, 3\}$ and $B = \{5, 6, 7\}$, then A and B are disjoint.

Set-member-ship notation   The symbol $\in$ (read "is a member of" or "is an element of") is used to denote membership in a set. Thus,

$$2 \in \{1, 2, 3\}.$$

Note that we write

$$\{2\} \subset \{1, 2, 3\} \quad \text{and} \quad 2 \in \{1, 2, 3\},$$

since $\{2\}$ is a *subset*, whereas 2 is an *element*, of $\{1, 2, 3\}$.

When discussing an individual but unspecified element of a set containing more than one member, we usually denote the element by a lowercase italic letter (for example, a, d, s, x), or sometimes by a letter from the Greek alphabet: α (alpha), β (beta), γ (gamma), and so on. Symbols used in this way are called *variables*.

Definition 1.6   *A **variable** is a symbol representing an unspecified element of a given set containing more than one element.*

The given set is called the **replacement set,** or **domain,** of the variable. For example,

$$x \in A$$

means that the variable x represents an (unspecified) element of the set A.

The members of the replacement set are called the **values** of the variable. If the set has just one member, this one value is called a **constant.**

Negation symbol   The slant bar, $/$, drawn through certain symbols of relation, is used to indicate negation. Thus $\neq$ is read "is not equal to," $\not\subset$ is read "is not a subset of," and $\notin$ is read "is not an element of." For example,

$$\{1, 2\} \neq \{1, 2, 3\}, \quad \{1, 2, 3\} \not\subset \{1, 2\}, \quad \text{and} \quad 3 \notin \{1, 2\}.$$

Set-builder notation   Another symbolism useful in discussing sets is $\{x \mid x \text{ has a certain property}\}$; for example,

$$\{x \mid x \in A \quad \text{and} \quad x \notin B\}$$

(read "the set of all x such that x is a member of A and is not a member of B"). This symbolism, called **set-builder notation,** is used extensively in this book. What it does is specify a variable (in this case, x) and, at the same time, state a condition on the variable (in this case, that x is contained in the set A and is not contained in the set B).

Exercise 1.1

Designate the given set by using braces and listing the members.

Example   {natural numbers between 8 and 12}

Solution $\{9, 10, 11\}$

1. {natural numbers between 2 and 7}
2. {natural numbers between 13 and 20}
3. {natural numbers between 88 and 89}
4. {natural numbers between 100 and 102}
5. {days in the week}
6. {months in the year}

Replace the colored comma with either $=$ *or* $\neq$ *to make a true statement.*

7. {natural numbers less than 3}, {1, 2}
8. {natural numbers between 1 and 3}, {1, 2}
9. {2}, {−2}   10. {4}, {7}
11. $\emptyset$, {0}   12. {5, 7, 9}, {7, 9, 5}

Replace the colored comma with either $\subset$ *or* $\not\subset$ *to make a true statement.*

13. {5}, {4, 5, 6}   14. {5, 4, 6}, {4, 5, 6}
15. $\emptyset$, {4, 5, 6}   16. {3, 4}, {4, 5, 6}

Replace the colored comma with either $\in$ *or* $\notin$ *to make a true statement.*

17. 3, {2, 3, 4}   18. 15, {2, 4, 6, . . .}

19. {2}, {2, 3, 4}   20. $\emptyset$, {2, 3, 4}

21. Let $U = \{5, 6, 7\}$. List the subsets of U that contain

   a. three members   b. two members

   c. one member   d. no members

22. Let $U = \{1, 2, 3, 4\}$. List the subsets of U that contain

   a. four members   b. three members   c. two members

   d. one member   e. no members   f. 2 and one other member

23. Let $A = \{2, 4, 6, 8\}$, $B = \{1, 2, 3, 4\}$,   and $C = \{1, 3, 5, 7\}$. List the members of the following sets.

   a. $A \cap B$ b. $A \cup B$ c. $A \cup \emptyset$

   d. $A \cap \emptyset$ e. $(A \cap B) \cup C$ f. $(A \cup B) \cap C$

24. Let $A = \{6, 8, 10, 12\}$, $B = \{5, 6, 7, 8\}$,   and $C = \{5, 7, 9, 11\}$. List the members of the following sets.

   a. $B \cup C$ b. $B \cap C$ c. $(A \cap C) \cup B$

   d. $(A \cup B) \cap C$ e. $A \cap (B \cup \emptyset)$ f. $A \cup (B \cap \emptyset)$

Designate the given set by using set-builder notation.

Example   {even natural numbers}

Solution $\{x \mid x = 2n,\ n$ a natural number$\}$

25. {odd natural numbers}   26. {natural numbers}
27. {solutions of $2^x = 5$}   28. {solutions of $x^x = 5$}
29. {natural-number multiples of 3}   30. {elements in set B}

1.2   Real Numbers

We shall frequently refer to the following five sets of numbers.

1. The set N of **natural numbers,** whose elements are the counting numbers:

$$N = \{1, 2, 3, \ldots\}.$$

2. The set J of **integers,** whose elements are the counting numbers, their negatives, and zero:

$$J = \{\ldots, -2, -1, 0, 1, 2, \ldots\}.$$

3. The set Q of **rational numbers,** whose elements are all those numbers that can be represented as the quotient of two integers $\dfrac{a}{b}$ (or a/b, or $a \div b$), where b is not 0. Among the elements of Q are such numbers as $-3/4$, $18/27$, $3/1$, and $-6/1$. In symbols,

$$Q = \left\{x \mid x = \frac{a}{b},\quad a, b \in J,\quad b \neq 0\right\}.$$

Every rational number can be represented by a terminating or repeating decimal numeral.

4. The set H of **irrational numbers,** whose elements are the numbers with decimal representations that are nonterminating and nonrepeating. Among the elements of this set are such numbers as $\sqrt{2}$, $\sqrt{3}$, $\sqrt{5}$, $\sqrt{7}$, π, and $-\sqrt{7}$. An irrational number cannot be represented in the form a/b, where a and b are integers.

5. The set R of **real numbers,** which contains all of the members of the set of rational numbers and all of the members of the set of irrational numbers:

$$R = Q \cup H.$$

*Equality
postulates*
The foregoing sets of numbers are related as indicated in Figure 1.1. Thus we have

$$H \subset R, \quad Q \subset R, \quad \text{and} \quad N \subset J \subset Q.$$

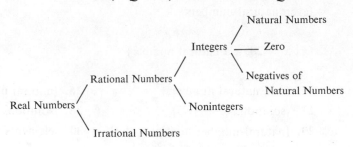

Figure 1.1

We characterize each set by stating properties that we assume it has. In mathematics, when we make formal assumptions about members of a set or about their properties, we call the assumptions **axioms,** or **postulates.** The words **property, law,** and **principle** are sometimes used in referring to assumptions, although these words may also be applied to certain consequences thereof.

The first assumptions to be considered here have to do with *equality*. An **equality,** or an "is equal to" assertion, is simply a mathematical statement that two symbols or words, or two groups of symbols or words, are names for the same thing.

We postulate that, for any members a, b, and c of any set S, the equality ($=$) relationship satisfies the following laws:

E-1 $a = a$ *Reflexive law for equality.*

E-2   If $a = b$, then $b = a$.   *Symmetric law for equality.*

E-3   If $a = b$ and $b = c$, then $a = c$.   *Transitive law for equality.*

E-4   If $a = b$, then a may be replaced by b *Substitution law for equality.*
and b by a in any mathematical statement without altering the truth or
falsity of the statement.†

The equality axioms indicate how we use symbols and warn that we must neither change the meaning of a symbol in the middle of a discussion nor use the same symbol for two different things in the same context. Other axioms, somewhat different conceptually, are used to characterize mathematical systems. They are assumptions made about the behavior of elements of sets under *binary operations*.

Definition 1.7 *A **binary operation** in a set A is an operation that assigns to each pair a and b of elements of A, taken in a definite order (a first and b second), a corresponding element of A.*

We express the fact that the assigned element is also in A by saying that A is **closed** under the operation.

† We have adopted a very powerful postulate in E-4, one that includes E-2 and E-3 as special cases. By wording E-4 as we have, we can eliminate a great deal of detail in later arguments, and, because E-2 and E-3 are fundamental properties of the equality relation, we have elected to retain them as postulates.

Addition and multiplication postulates Since the set R of real numbers is the set of greatest interest to us at the moment, we postulate the basic laws for the elements of R in relation to the binary operations of addition and multiplication. You are probably familiar with these laws from your earlier work. They are listed on this page for reference. Parentheses are used in stating some of the relations to indicate that symbols within parentheses are to be viewed as representing a single entity.

Field postulates of the set **R** *of real numbers*

Let a, b, c be arbitrary elements of R.

F-1 $a + b$ is a unique element of R. *Closure law for addition.*

F-2 $(a + b) + c = a + (b + c)$. *Associative law for addition.*

F-3 There exists an element $0 \in R$ (called the **identity element for addition**) with the property *Additive-identity law.*

$$a + 0 = a \quad \text{and} \quad 0 + a = a$$

for each $a \in R$.

F-4 For each $a \in R$, there exists an element $-a \in R$ (called the **additive inverse**, or **negative**, of a) with the property *Additive-inverse law.*

$$a + (-a) = 0 \quad \text{and} \quad (-a) + a = 0.$$

F-5 $a + b = b + a$. *Commutative law for addition.*

F-6 $a \times b$ is a unique element of R. *Closure law for multiplication.*

F-7 $(a \times b) \times c = a \times (b \times c)$. *Associative law for multiplication.*

F-8 $a \times (b + c) = (a \times b) + (a \times c)$ and $(b + c) \times a = (b \times a) + (c \times a)$. *Distributive law.*

F-9 There exists an element $1 \in R$ (called the **identity element for multiplication**), $1 \neq 0$, with the property *Multiplicative-identity law.*

$$a \times 1 = a \quad \text{and} \quad 1 \times a = a$$

for each $a \in R$.

F-10 $a \times b = b \times a$. *Commutative law for multiplication.*

F-11 For each element $a \in R$, $a \neq 0$, there exists an element $1/a \in R$ (called the **multiplicative inverse**, or **reciprocal**, of a) with the property *Multiplicative-inverse law.*

$$a \times \frac{1}{a} = 1 \quad \text{and} \quad \frac{1}{a} \times a = 1.$$

If, for a given set and a given pair of binary operations (not necessarily the set of real numbers or ordinary addition and multiplication), Postulates F-1 through F-11 are satisfied by the elements of the set, then the set is called a **field** under these operations. Thus, we speak of the **field R of real numbers.**

To give meaning to expressions $a + b + c$, $a \times b \times c$, $a + b + c + d$, and so on, let us make the following agreement.

Definition 1.8   *If $a, b, c, d, \ldots \in R$, then*

$$a + b + c = a + (b + c), \quad a + b + c + d = a + (b + c + d), \ldots,$$

and

$$a \times b \times c = a \times (b \times c), \quad a \times b \times c \times d = a \times (b \times c \times d), \ldots.$$

Inverse operations

In terms of the operations of addition and multiplication, we can define two additional operations on real numbers: **subtraction** (finding a *difference*) and **division** (finding a *quotient*).

Definition 1.9   *The **difference** of elements $a \in R$ and $b \in R$, denoted by $a - b$, is given by*

$$a - b = a + (-b).$$

Definition 1.10   *The **quotient** of elements $a \in R$ and $b \in R$, $b \neq 0$, denoted by $\dfrac{a}{b}$, a/b, or $a \div b$, is given by*

$$\frac{a}{b} = a \times \frac{1}{b}.$$

In all that follows, we shall adopt the customary practice of writing ab or $a \cdot b$ for $a \times b$.

Exercise 1.2

Let $N = \{natural\ numbers\}$, $J = \{integers\}$, $Q = \{rational\ numbers\}$, $H = \{irrational\ numbers\}$, $R = \{real\ numbers\}$. State whether each statement is true or false.

1. $-5 \in N$ 2. $0 \in Q$ 3. $\sqrt{3} \in R$
4. $-1 \in Q$ 5. $-3 \in R$ 6. $-7 \in H$
7. $\{-1, 1\} \subset N$ 8. $\{-1, 1\} \subset J$ 9. $\{-1, 1\} \subset Q$
10. $\{-1, 1\} \subset H$ 11. $\{-1, 1\} \subset R$ 12. $\{0\} \subset R$

Represent the given set by listing members if finite, and using three dots, . . . , if infinite.

Examples   a.   {natural numbers between 5 and 9}   b.   {integers greater than 3}

Solutions   a.   {6, 7, 8}   b.   {4, 5, 6, . . .}

13.   {first five natural numbers}   **14.**   {integers between −4 and 5}
15.   {natural numbers less than 7}   **16.**   {integers greater than −5}
17.   {integers between −10 and −5}   **18.**   {nonnegative integers}

Use set-builder notation, $\{x \mid \text{condition on } x\}$, to represent the given set.

Example   {natural numbers greater than 12}

Solution $\{x \mid x \in N \text{ and } x \text{ is greater than } 12\}$

19.   {natural numbers}   **20.**   {integers}
21.   {real numbers}   **22.**   {integers less than −6}
23.   {real numbers between −4 and 3}   **24.**   {nonnegative real numbers}

25.   Let $A = \{4, -2, 2/5, 0, -3/4, \sqrt{2}, \sqrt{7}\}$. Designate each of the following sets by using braces and listing the members.

    a.   {natural numbers in A}   **b.**   {integers in A}

    c.   {rational numbers in A}   **d.**   {real numbers in A}

26.   Let $B = \{-6, 3, \sqrt{5}, -3/4, 0, 5, -1, \sqrt{3}\}$. Designate each set by using braces and listing the members.

    a.   {natural numbers in B}   **b.**   {integers in B}

    c.   {irrational numbers in B}   **d.**   {real numbers in B}

Each variable in Exercises 27–50 denotes a real number.

Each of the statements 27–32 is an application of one of the postulates E-1 *through* E-4. *Justify the statement by citing an appropriate postulate. (There may be more than one correct justification.)*

Example   If $3 = a$,   then $a = 3$.

Solution   Symmetric law for equality, E-2

27.   If $a + 3 = b$ and $b = 7$,   then $a + 3 = 7$.
28.   If $x = 5$ and $y = x + 2$,   then $y = 5 + 2$.

29. If $2 + y = 6$, then $6 = 2 + y$.
30. If $a = 2c$ and $c = 6$, then $a = 2 \cdot 6$.
31. If $y = 7 + x$ and $7 + x = z$, then $y = z$.
32. $x + 8 = x + 8$.

Each of the statements 33–50 is an application of one of the postulates F-1 *through* F-11. *Justify the statement by citing the appropriate postulate.*

Example $2(3 + 1) = 2 \cdot 3 + 2 \cdot 1$

Solution   Distributive law, F-8

33. ab is a real number   **34.** $7 + 0 = 7$
35. $(2 \cdot 3) \cdot 4 = 2 \cdot (3 \cdot 4)$ **36.** $(5 + 4) + 1 = (4 + 5) + 1$
37. $3 + (-3) = 0$ **38.** $5 + (-2) = (-2) + 5$
39. $a\left(\dfrac{1}{a}\right) = \left(\dfrac{1}{a}\right)a \quad (a \neq 0)$ **40.** $\dfrac{1}{c}(a + b) = \dfrac{1}{c} \cdot a + \dfrac{1}{c} \cdot b \quad (c \neq 0)$
41. $(a + b) + c = c + (a + b)$ **42.** $(a + b) + c = (b + a) + c$
43. $a + (b + c)d = a + bd + cd$ **44.** $a + (b + c)d = a + d(b + c)$
45. $a + (b + c)d = (b + c)d + a$ **46.** $(a + b) + [-(a + b)] = 0$
47. $a(b + c) = (b + c)a$ **48.** $a[b + (c + d)] = ab + a(c + d)$
49. $ab + a(c + d) = ab + ac + ad$ **50.** $ab + ac = ba + ac$

Is the given set closed under the stated operation?

Example   {integers}, division

Solution   Not closed. The set is not closed, because the quotient of two integers is not always an integer; for example, the quotient $-2/3 \notin J$.

51. {even natural numbers}, addition
52. {odd natural numbers}, addition
53. {odd natural numbers}, multiplication
54. {even natural numbers}, multiplication
55. {0, 1}, multiplication
56. {0, 1}, addition
57. {natural numbers}, division
58. {natural numbers}, subtraction

Express the given difference or quotient as a sum or product.

Examples   a. $4x - 3y$ b. $\dfrac{4x}{x - y}$

Solutions   a. $4x - 3y = 4x + (-3y)$ b. $\dfrac{4x}{x - y} = 4x \cdot \dfrac{1}{x - y}$

59. $2y - 5$ **60.** $3x - 9y$ **61.** $-2x - (y + z)$

62. $-y - (x - z)$ **63.** $\dfrac{2}{3}$ **64.** $\dfrac{-4}{7}$

65. $\dfrac{x}{3y + 2}$ **66.** $\dfrac{-2x}{x - 1}$

1.3   Field Properties

The field postulates together with the postulates for equality imply other properties of the real numbers. Such implications are generally stated as **theorems.** A theorem is simply an assertion of a fact that follows logically from the postulates (or axioms) and other theorems. We shall list (without proof) some of the theorems ordinarily encountered in lower-level algebra courses. We have numbered these theorems to provide an efficient way to refer to them later and have also named those that have commonly accepted names.

Addition law for equality   First, consider the following result, which reaffirms the uniqueness of the sum of two real numbers.

Theorem 1.1   *If $a, b, c \in R$ and $a = b$, then*

$$a + c = b + c \quad \text{and} \quad c + a = c + b.$$

Multiplication law for equality   A theorem closely analogous to Theorem 1.1 can be stated as follows.

Theorem 1.2   *If $a, b, c \in R$ and $a = b$, then*

$$ac = bc \quad \text{and} \quad ca = cb.$$

Additional properties   The next theorem asserts that the additive inverse of a real number and the multiplicative inverse of a nonzero real number are unique—that is, that a given real number has only one additive inverse and only one multiplicative inverse.

Theorem 1.3

I   *If $a, b \in R$ and $a + b = 0$, then*

$$b = -a \quad and \quad a = -b.$$

II   *If $a, b \in R$ and $a \cdot b = 1$, then*

$$a = \frac{1}{b} \quad and \quad b = \frac{1}{a} \quad (a, b \neq 0).$$

The proofs of the succeeding theorems follow directly from the previous results.

Theorem 1.4   *If $a, b, c \in R$ and $a + c = b + c$, then $a = b$.*

Theorem 1.5   *If $a, b, c \in R$, $c \neq 0$, and $ac = bc$, then $a = b$.*

Theorem 1.6   *For every $a \in R$, $a \cdot 0 = 0$.*

Theorem 1.7   *If $a, b \in R$ and $a \cdot b = 0$, then either $a = 0$ or $b = 0$ or both.*

Combining Theorems 1.6 and 1.7, we see that for $a, b \in R$ we have $ab = 0$ *if and only if* at least one of the factors is 0.

The following theorem concerns the familiar "laws of signs" for operating with real numbers.

Theorem 1.8   *If $a, b \in R$, then*

I $-(-a) = a$,   II $(-a) + (-b) = -(a + b)$,

III $(-a)(b) = -(ab)$,   IV $(-a)(-b) = ab$,

V $\dfrac{-a}{b} = \dfrac{a}{-b} = -\dfrac{a}{b} \quad (b \neq 0)$,   VI $\dfrac{-a}{-b} = \dfrac{a}{b} \quad (b \neq 0)$.

Theorem 1.9   *If a $b, c \in R$, then*

$$\frac{a}{b} = \frac{c}{d} \quad if\ and\ only\ if \quad ad = bc \quad (b, d \neq 0).$$

As a direct consequence of this characterization of equal quotients, we have a theorem that is sometimes referred to as the **fundamental principle of fractions.**

Theorem 1.10   *If $a, b, c \in R$, then*

$$\frac{ac}{bc} = \frac{a}{b} \quad (b, c \neq 0).$$

Finally, let us group a number of assertions about quotients into a single theorem.

Theorem 1.11   *If a, b, c, d ∈ R, then*

I $\dfrac{1}{a} \cdot \dfrac{1}{b} = \dfrac{1}{ab}$ $(a, b \neq 0)$,   II $\dfrac{a}{b} \cdot \dfrac{c}{d} = \dfrac{ac}{bd}$ $(b, d \neq 0)$,

III $\dfrac{a}{c} + \dfrac{b}{c} = \dfrac{a+b}{c}$ $(c \neq 0)$,   IV $\dfrac{a}{b} + \dfrac{c}{d} = \dfrac{ad+bc}{bd}$ $(b, d \neq 0)$,

V $\dfrac{a}{b} - \dfrac{c}{d} = \dfrac{ad-bc}{bd}$ $(b, d \neq 0)$,   VI $\dfrac{1}{\frac{a}{b}} = \dfrac{b}{a}$ $(a, b \neq 0)$,

VII $\dfrac{\frac{a}{b}}{\frac{c}{d}} = \dfrac{a}{b} \div \dfrac{c}{d} = \dfrac{ad}{bc}$ $(b, c, d \neq 0)$.

Exercise 1.3

In Exercises 1–20, each statement is justified by one part of Theorems 1.1–1.11. Cite an appropriate justification. All variables denote elements of the set R of real numbers.

Example   If $p + q + 3 = 4 + 3$,   then $p + q = 4$.

Solution   Theorem 1.4

1. If $x = 3$,   then $x + 9 = 3 + 9$.   2. $-2 - q = -(2 + q)$

3. $\dfrac{-2}{5} = -\dfrac{2}{5}$ 4. $\dfrac{4(x+y)}{6} = \dfrac{2(x+y)}{3}$

5. $\dfrac{x}{3} + \dfrac{y+z}{3} = \dfrac{x+y+z}{3}$ 6. $\dfrac{p}{3} \cdot \dfrac{q}{4} = \dfrac{pq}{12}$

7. If $4p = 0$,   then $p = 0$.   8. $(-5)(-6) = 5 \cdot 6$

9. If $3 \cdot 5 = 2y$,   then $\dfrac{3}{2} = \dfrac{y}{5}$.   10. $\dfrac{-x}{-3} = \dfrac{x}{3}$

11. $\dfrac{\frac{2}{5}}{\frac{3}{7}} = \dfrac{2 \cdot 7}{5 \cdot 3}$ 12. $\dfrac{\frac{1}{2}}{\frac{3}{2}} = \dfrac{3}{2}$

13. $(-3)(p) = -(3p)$ 14. $x - (-y) = x + y$

15. $\dfrac{x}{3} + \dfrac{y}{2} = \dfrac{2x + 3y}{3 \cdot 2}$

16. $\dfrac{x}{-3} = -\dfrac{x}{3}$

17. If $p + q = 7$, then $4(p + q) = 4 \cdot 7$.

18. If $(r + s) + 7 = 0$, then $r + s = -7$.

19. If $6(x - y) = 3z$, then $2(x - y) = z$.

20. $\dfrac{x + 1}{4} - \dfrac{y + 3}{3} = \dfrac{3(x + 1) - 4(y + 3)}{4 \cdot 3}$

1.4   *Order and Completeness in R*

There is a one-to-one correspondence between the real numbers and the points on a geometric line (to each real number there corresponds one and only one point on the line, and vice versa). To illustrate this, we imagine the line scaled in convenient units, with the positive direction (from 0 toward 1) denoted by an arrowhead. The line is then called a **number line**; the real number corresponding to a point on the line is called the **coordinate** of the point, and the point is called the **graph** of the number. For example, a number-line representation of $\{1, 3, 5\}$ is shown in Figure 1.2.

Figure 1.2

A horizontal number-line graph directed to the right can be used to illustrate the separation of the real numbers into three disjoint subsets: {negative real numbers}, {0}, {positive real numbers}. The point associated with 0 is called the **origin.** The set of numbers whose elements are associated with the points on the right-hand side of the origin belong to the set R_+ of **positive real numbers,** and the set whose elements are associated with the points on the left-hand side belong to the set R_- of **negative real numbers.**

Notice that the word "negative" has now been used in two ways. In one case, we refer to the *negative of a number*, as in Postulate F-4, whereas, in the other, we refer to a *negative number*, which is the negative of a positive number.

Order postulates   It is possible to categorize the set of positive real numbers without recourse to geometric considerations, though of course we shall continue to find it convenient to refer also to the number line. With this in mind, let us state two more postulates that apply to real numbers.

O-1   If *a* is a real number, then exactly one   *Trichotomy law.*
      of the following statements is true:
      a is positive, *a* is zero, or −*a* is positive.

O-2   If *a* and *b* are positive real numbers,   *Closure law for positive numbers.*
      then *a* + *b* is positive and *ab* is positive.

The first of these postulates asserts that every real number belongs to one of the sets R_+, {0}, or R_-, but to only one of them. The second asserts that the set R_+ of positive real numbers is closed with respect to the binary operations of addition and multiplication.

Since the set *R* of real numbers satisfies Postulates O-1 and O-2 as well as Postulates F-1 through F-11, we say that *R* is an **ordered field.** Similarly, the set *Q* of rational numbers is an ordered field.

By Postulate O-2, if *a* and *b* are positive real numbers, then *ab* is a positive real number. This fact, together with Parts III and IV of Theorem 1.8, is sufficient to establish that the product of a positive real number and a negative real number is a negative real number, while the product of two negative real numbers is a positive real number.

Less than and greater than

The addition of a positive real number *d* to a real number *a* can be visualized on a number-line graph as the process of locating the point corresponding to *a* on the line and then moving along the line *d* units to the right to arrive at the point corresponding to *a* + *d* (Figure 1.3). With this idea in mind, we define what is meant by "less than."

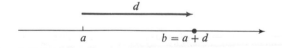

Figure 1.3

Definition 1.11   *If a, b ∈ R, then a is **less than** b if and only if there exists a positive real number d such that a + d = b.*

The fact that *d* is positive implies, for any two real numbers *a* and *b*, that if the graph of *a* lies to the left of the graph of *b*, then *a* is less than *b*. The inequality symbol < is used to denote the phrase "is less than," and *a* < *b* is read "*a* is less than *b*." The inequality symbol > means "is greater than." The statements *a* < *b* and *b* > *a* are taken as equivalent.

Definition 1.11 and the field postulates have the following implications, which we shall not prove.

Theorem 1.12   *For any a, b, c ∈ R:*

I   *If a < b and b < c, then a < c.*   II   *If a < b, then a + c < b + c.*

III   *If a < b and c > 0, then ac < bc.*   IV   *If a < b and c < 0, then ac > bc.*

As you can see, Theorem 1.12-I is comparable to the transitive law for equality. That is, we can say that "less than" is a transitive relationship.

Each of the symbols $\leq$ and $\geq$ (read "is less than or equal to" and "is greater than or equal to," respectively) is a contraction for two symbols, one of equality and one of inequality, connected by the word "or." For example, $x \leq 7$ is the statement that x is less than 7 *or* x equals 7.

We often write together two inequalities that express a transitive relationship. Thus $a < b$ and $b < c$ are written together as $a < b < c$ and read "a is less than b and b is less than c." Similarly, $0 < x$ and $x \leq 1$ are written together as $0 < x \leq 1$.

Absolute value   The graphs of the numbers a and $-a$ on a number line lie the same distance from the origin, but on opposite sides of it. If we wish to refer to the *distance* of the graph of a number from the origin, and not to the side of the origin on which it is located, then we use the term **absolute value.** Thus, the absolute value of a and the absolute value of $-a$ are the same nonnegative number. The symbol $|a|$ is used to denote the absolute value of a. We formalize the definition as follows:

Definition 1.12   *If $a \in R$, then the **absolute value** of a is given by*

$$|a| = \begin{cases} a, & \text{if } a \geq 0, \\ -a, & \text{if } a < 0. \end{cases}$$

For example, $|-3| = -(-3) = 3$, $|7| = 7$, and $|0| = 0$.

We have noted in the foregoing discussion that certain algebraic statements concerning the order of real numbers can be interpreted geometrically. We summarize some of the more common correspondences in the following table, where in each case a, b, $c \in R$.

Algebraic statement	Geometric statement
1. a is positive	1. The graph of a lies to the right of O.
2. a is negative	2. The graph of a lies to the left of O.
3. $a > b$	3. The graph of a lies to the right of the graph of b.
4. $a < b$	4. The graph of a lies to the left of the graph of b.
5. $a < c < b$	5. The graph of c is to the right of the graph of a and to the left of the graph of b.
6. $|a| < c$	6. The graph of a is less than c units from the origin.
7. $|a - b| < c$	7. The graphs of a and b are less than c units from each other.
8. $|a| < |b|$	8. The graph of a is closer to the origin than the graph of b is.

Completeness postulate   On page 14 we mentioned in passing a property of the set R of real numbers that is of fundamental importance and necessary for establishing the existence of irrational numbers. Let us formalize this property by assuming the following **completeness property.**

O-3   There is a one-to-one correspondence between the set of real numbers and the set of points on a geometric line.

Although this property might be asserted more precisely, for our purposes this formulation is quite sufficient. Because of the completeness property, the set R of real numbers is said to be a **complete ordered field.** Note, however, that while the set Q of rational numbers is an ordered field, and between any two rational numbers there are infinitely many other rational numbers, Q is not a complete ordered field—because between any two rational numbers there are also infinitely many irrational numbers; and the set of points corresponding to Q does not completely "fill" the number line.

Exercise 1.4

For $x, y \in R$, justify the given statement by citing one part of Theorem 1.12.

Example   If $2 < 3x$, then $-4 > -6x$.

Solution   Part IV.   Each member of $2 < 3x$ is multiplied by -2.

1. If $x < 3$ and $y < x$, then $y < 3$.   2. If $x + 1 < 0$, then $x < -1$.
3. If $y < 8$, then $3y < 24$.   4. If $y < 4$, then $y + 2 < 6$.
5. If $x < 7$, then $x - 2 < 5$.   6. If $x < 9$, then $-2x > -18$.
7. If $-6x < 12$, then $x > -2$.   8. If $x - 3 < 5$, then $x < 8$.

Express the given statement by means of symbols.

Examples   a.   5 is not greater than 7.   b. x is between 5 and 8.

Solutions   a. $5 \not> 7$, or $5 \le 7$ b. $5 < x < 8$

9. 7 is greater than 3.   10. 2 is less than 5.
11. -4 is less than -3.   12. -4 is greater than -7.
13. x is between -1 and 1, inclusive.   14. x is negative.
15. x is positive.   16. x is nonpositive.
17. x is nonnegative.   18. $2x$ is less than or equal to 8.

For $x, y \in R$, rewrite the given expression without using absolute-value notation.

Examples   a. $|-9|$ b. $|x - 6|$

Solutions   a. Since $-9 < 0$,   b. $x - 6$,   if $x - 6 \ge 0$ or $x \ge 6$;
                       $|-9| = -(-9) = 9$ $-(x - 6)$,   if $x - 6 < 0$ or $x < 6$.

19. $|-3|$ 20. $|-5|$ 21. $|7|$ 22. $|4|$
23. $-|-2|$ 24. $-|-5|$ 25. $|3x|$ 26. $|2y|$
27. $|x + 1|$ 28. $|x + 2|$ 29. $|y - 3|$ 30. $|y - 4|$

Replace the comma with an appropriate order symbol to form a true statement.

31. $|-2|, |-5|$ **32.** $|3|, |-4|$ **33.** $-7, |-1|$
34. $5, |-2|$ **35.** $|-3|, 0$ **36.** $-|-4|, 0$

For $x, y \in R$, write an equivalent relation without using the negation symbol, $/$.

37. $2 \not\geq 5$ **38.** $-1 \not\leq -2$ **39.** $7 \not\geq 8$
40. $|x| \not\leq 3$ **41.** $|x| \not\geq 3$ **42.** $x \not> |y|$

1.5   *Working with Inequalities*

A **solution** of an equation or inequality involving a variable is a value of that variable for which the equation or inequality is true. The **solution set** of the equation or inequality is the set of all its solutions.

Inequalities in a single variable can be transformed in ways similar to those used to transform equations. The necessary foundation for transformations of inequalities can be found in Theorem 1.12 on page 15.

Example

Solve $\dfrac{x-3}{4} < \dfrac{2}{3}$.

Solution

Multiplying each member of the inequality   12, we have
$$3(x-3) < 8, \quad \text{or} \quad 3x - 9 < 8.$$
Adding 9 to each member gives
$$3x < 17.$$
Multiplying each member by 1/3, we obtain $x < 17/3$, and the solution set is
$$S = \left\{ x \mid x < \frac{17}{3} \right\}.$$

Graphical representation of inequality solution set

The solution set in the foregoing example can be pictured on a number-line graph as shown in Figure 1.4. The heavy line indicates points with coordinates in the solution set. Note that the open dot indicates that 17/3 is not in the solution set.

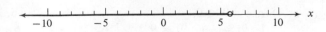

Figure 1.4

Special line graphs, called **sign graphs,** are often convenient for visualizing the solution set of inequalities, as the example in Exercise 1.5, page 20, shows.

*Equations
involving
absolute
values*

In Section 1.4, we defined the absolute value of a real number x by

$$|x| = \begin{cases} x, & \text{if } x \geq 0, \\ -x, & \text{if } x < 0. \end{cases}$$

More generally, then, for each $a \in R$, the expression $|x - a|$ satisfies

$$|x - a| = \begin{cases} x - a, & \text{if } x - a \geq 0, \text{ or, equivalently, if } x \geq a, \\ -(x - a), & \text{if } x - a < 0, \text{ or, equivalently, if } x < a, \end{cases}$$

and can be interpreted on a number line as denoting the distance the graph of x is located from the graph of a, as shown in Figure 1.5.

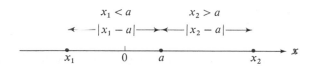

Figure 1.5

Example

Find the solution set of $|x - 3| = 5$.

Solution 1

Since $|x - 3|$ represents the distance the graph of x is located from the graph of 3, and since by the equation this distance is 5, the two solutions of the equation are $3 + 5$, or 8, and $3 - 5$, or -2. Thus, the solution set is $\{-2, 8\}$.

Solution 2

By definition, $|x - 3| = 5$ implies that $x - 3 = 5$ or $-(x - 3) = 5$.
The equation $x - 3 = 5$ is equivalent to $x = 8$, and $-(x - 3) = 5$ is equivalent to $x = -2$. Hence the solution set is given by

$$S = \{x \mid x = 8\} \cup \{x \mid x = -2\} = \{-2, 8\}.$$

*Absolute
values in
inequalities*

Inequalities involving absolute-value notation can be solved by appealing to the definition of absolute value.

Example

Find the solution set of $|x + 1| > 3$.

Solution

By definition, this inequality is equivalent to

$$x + 1 > 3 \quad \text{for} \quad x + 1 \geq 0,$$

and to

$$-(x + 1) > 3 \quad \text{for} \quad x + 1 < 0,$$

so that the solution set is given by

$$S = \{x \mid x > 2\} \cup \{x \mid x < -4\}. \tag{1}$$

The graph of the solution set is shown in the figure.

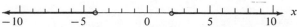

Exercise 1.5

Solve and represent the solution set on a line graph.

1. $x + 7 > 8$ **2.** $x - 5 \leq 7$ **3.** $3x - 2 > 1 + 2x$

4. $2x + 3 \leq x - 1$ **5.** $\dfrac{2x - 3}{2} \leq 5$ **6.** $\dfrac{3x + 4}{3} > 12$

Find the solution set of each inequality.

Example $\dfrac{x}{x - 2} \geq 5$

Solution   As suggested on page 18, we can approach this directly by means of a sign graph. We first rewrite the given inequality equivalently as

$$\frac{x}{x - 2} - 5 \geq 0,$$

from which we obtain

$$\frac{-4x + 10}{x - 2} \geq 0.$$

For this to be valid, $x - 2$ must not be 0, and the numerator and denominator must be of like sign (both positive or both negative) or the numerator must be 0. The sign graph shows that the quotient $(-4x + 10)/(x - 2)$ is positive or zero for x between 2 and 5/2, including 5/2 but excluding 2, a value of x for which the denominator is 0. The desired solution set is therefore $\{x \mid 2 < x \leq 5/2\}$, with graph as shown on the last line of the sign graph.

7. $\dfrac{2}{x} \leq 4$ **8.** $\dfrac{3}{x - 6} > 8$ **9.** $\dfrac{x}{x + 2} > 4$

10. $\dfrac{x + 2}{x - 2} \geq 6$ **11.** $\dfrac{2}{x - 2} \geq \dfrac{4}{x}$ **12.** $\dfrac{3}{4x + 1} > \dfrac{2}{x - 5}$

13. $x(x - 2)(x + 3) > 0$ **14.** $x^3 - 4x \leq 0$

Solve.

Example $|x + 5| = 8$

Solution   Write the equation as two first-degree equations and solve each of them.
$$x + 5 = 8 \quad \text{or} \quad -(x + 5) = 8$$
$$x = 3 \quad \text{or} \qquad\qquad x = -13$$
The solution set is {3, −13}.

15. $|x| = 6$ **16.** $|x| = 3$ **17.** $|x - 1| = 4$

18. $|x - 6| = 3$ **19.** $\left| x - \dfrac{2}{3} \right| = \dfrac{1}{3}$ **20.** $\left| x - \dfrac{3}{4} \right| = \dfrac{1}{2}$

Solve and graph each solution set on a number line.

Example $|3x - 6| < 9$

Solution   By definition, the inequality is equivalent to
$$3x - 6 < 9, \quad \text{or} \quad x < 5, \quad \text{for} \quad 3x - 6 \geq 0, \quad \text{or} \quad x \geq 2,$$
and to
$$-(3x - 6) < 9, \quad \text{or} \quad x > -1, \quad \text{for} \quad 3x - 6 < 0, \quad \text{or} \quad x < 2.$$

Hence the solution set is given by
$$\{x \mid x < 5 \quad \text{and} \quad x \geq 2\} \cup \{x \mid x > -1 \quad \text{and} \quad x < 2\},$$
or alternatively,
$$\{x \mid -1 < x < 5\}.$$

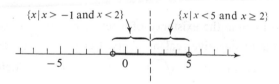

21. $|x| < 2$ **22.** $|x - 1| > 2$ **23.** $|x + 3| \geq 4$

24. $|x + 1| \leq 8$ **25.** $|2x - 5| \geq 3$ **26.** $|2x + 4| < -1$

Express inequalities with a single inequality involving an absolute-value symbol.

Example $-3 < x < 7$

Solution   Since the average of 7 and −3 is $[7 + (-3)]/2 = 4/2 = 2$, the values of x are centered about 2. Subtracting 2 from each member, we have $-5 < x - 2 < 5$, which can be expressed as
$$|x - 2| < 5.$$

27. $1 < x < 3$	**28.** $-5 \le x \le 9$	**29.** $-9 \le x \le -7$
30. $5 < x < 13$	**31.** $-7 \le 2x \le 12$	**32.** $-5 < 3x < 10$

In Exercises 33–36, consider $a, b \in R$.

33. Show that $|-a| = |a|$. *Hint:* Consider two possible cases, a nonnegative and a negative.

34. Show that $|a - b| = |b - a|$. *Hint:* Consider two possible cases, $a - b \ge 0$ and $a - b < 0$.

35. Show that $|a^2| = |a|^2 = a^2$.

36. Show that $|ab| = |a| \cdot |b|$. *Hint:* Consider four possible cases, with a nonnegative and negative, and b nonnegative and negative.

1.6   Complex Numbers

Some equations do not have solutions in the set of real numbers. For example, if $b > 0$ then $x^2 = -b$ has no real-number solution, because there is no real number whose square is negative. For this reason, the expression $\sqrt{-b}$, for $b \in R, b > 0$, is undefined in the set of real numbers. In this section, we wish to consider a set of numbers containing members whose squares are negative real numbers and also containing a subset of members that can be identified with the set of real numbers. We shall see that this new set of numbers, called the set C of **complex numbers,** provides solutions for all polynomial equations with real coefficients in one variable. It is also possible to show that the field postulates F-1 through F-11 are satisfied by the complex numbers and thus that C constitutes a field.

Pure imaginary numbers

Let us first assume the existence of a set of numbers among whose members are square roots of negative real numbers. Letting $\sqrt{-1}$ denote a number whose square is -1, so that $\sqrt{-1}\,\sqrt{-1} = -1$, we define $\sqrt{-b}$, $b > 0$, as follows:

Definition 1.13   *For* $b \in R_+$,

$$\sqrt{-b} = \sqrt{-1}\,\sqrt{b}.$$

As we shall see, it is often convenient to use the symbol i for $\sqrt{-1}$. With this convention,

$$i = \sqrt{-1} \quad \text{and} \quad i^2 = -1.$$

Next, we see from Definition 1.13 that

$$\sqrt{-b} = \sqrt{-1}\,\sqrt{b} = i\sqrt{b}.$$

Furthermore, assuming that $-\sqrt{-b} = -1 \cdot \sqrt{-b}$, and that the commutative and associative laws for multiplication hold for these numbers, we have

$$(-\sqrt{-b})(-\sqrt{-b}) = (-1 \cdot i\sqrt{b})(-1 \cdot i\sqrt{b}) = i^2 \cdot b = -b.$$

Hence, in this set of numbers, $-b$ has two square roots, $\sqrt{-b} = i\sqrt{b}$ and $-\sqrt{-b} = -i\sqrt{b}$; and, in particular, the square roots of -1 are i and $-i$.

Hence, a square root of any negative real number can be represented as the product of a real number and the number $\sqrt{-1}$, or i. For example,

$$\sqrt{-4} = \sqrt{-1} \cdot \sqrt{4} = i\sqrt{4} = 2i$$

and

$$-\sqrt{-3} = -\sqrt{-1} \cdot \sqrt{3} = -i\sqrt{3} = -\sqrt{3}\,i.$$

The numbers represented by the symbols $\sqrt{-b}$ and $-\sqrt{-b}$, where $b \in R$ and $b > 0$, are called **pure imaginary numbers.**

Complex numbers

Now, consider all possible expressions of the form $a + bi$, where $a, b \in R$ and $i = \sqrt{-1}$, that are the sums of all real numbers and all pure imaginary numbers. Such an expression names a number called a **complex number.** The set of all such numbers is denoted by C. If $b = 0$, then $a + bi = a$, and it is evident that set R of real numbers is a subset of the set C of complex numbers. If $b \neq 0$, then $a + bi$ is called an **imaginary number** (see Figure 1.6).

For example, the numbers -7, $3 + 2i$, and $-4i$ are all complex numbers. However, -7 is also a real number and $3 + 2i$ and $-4i$ are also imaginary numbers. Furthermore, $-4i$ is a pure imaginary number.

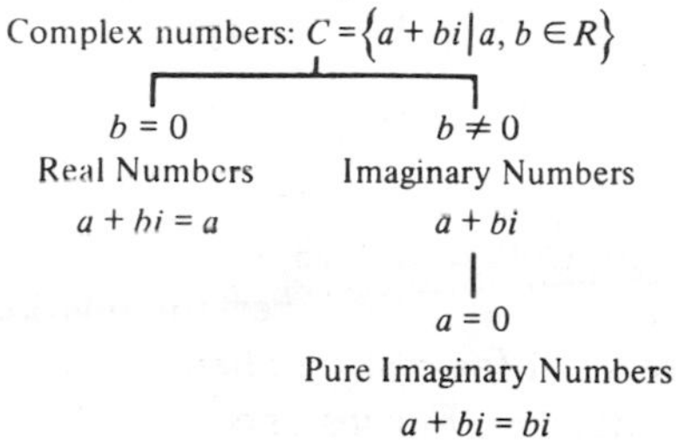

Complex numbers: $C = \{a + bi \mid a, b \in R\}$

$b = 0$	$b \neq 0$
Real Numbers	Imaginary Numbers
$a + bi = a$	$a + bi$

$a = 0$

Pure Imaginary Numbers

$a + bi = bi$

Figure 1.6

Equality

In the work that follows, we shall use z to represent an unspecified element of C. That is, $z \in C$ is a complex number. Furthermore, to avoid ambiguity in reading symbols we shall, where convenient, use the form $a + ib$ as equivalent to $a + bi$.

Definition 1.14   If $z_1 = a_1 + b_1 i$ *and* $z_2 = a_2 + b_2 i$, *then*

$$z_1 = z_2$$

if and only if

$$a_1 = a_2 \quad and \quad b_1 = b_2.$$

For example,

$$2 + 6i = (5 - 3) + (4 + 2)i$$

and

$$x + yi = 3 - 5i$$

if and only if $x = 3$ and $y = -5$.

Sums and products

Next, let us define sums and products of complex numbers. Although these definitions may appear arbitrary and unusual to start with (particularly in the case of products), we shall see that they relate to the corresponding operations in R in a direct and useful way.

Definition 1.15   If $z_1 = a_1 + b_1 i$ *and* $z_2 = a_2 + b_2 i$, *then*

I $z_1 + z_2 = (a_1 + b_1 i) + (a_2 + b_2 i) = (a_1 + a_2) + (b_1 + b_2)i$

II $z_1 \cdot z_2 = (a_1 + b_1 i) \cdot (a_2 + b_2 i) = (a_1 a_2 - b_1 b_2) + (a_1 b_2 + a_2 b_1)i$

With the operations of addition and multiplication thus defined, it can be shown that the field postulates of the set R on page 7 are valid properties for complex numbers. Hence, *the set C of complex numbers constitutes a field.* However, the complex numbers cannot be ordered in the sense that we ordered real numbers, and for this reason C is *not* an *ordered* field.

From the definitions we now have, we may rewrite expressions involving complex numbers in the "standard form" $a + bi$ in the same way that we rewrite real polynomial expressions, except that i^2 is replaced with -1.

Examples

a. $(2 + 3i) + (6 - 2i) = (2 + 6) + (3 - 2)i$

$$= 8 + i$$

b. $(2 - i)(1 + 3i) = 2 + 6i - i - 3i^2$

$$= 2 + 5i - 3(-1)$$

$$= 5 + 5i$$

The symbol $\sqrt{-b}, b > 0$

In accordance with Definition 1.13, $\sqrt{-b} = i\sqrt{b}$, $b > 0$. The symbol $\sqrt{-b}$, $b > 0$, should be used with care, since certain relationships involving the square root symbol that are valid for real numbers are not valid when the symbol does not represent a real number. For instance,

$$\sqrt{-2}\sqrt{-3} = (i\sqrt{2})(i\sqrt{3}) = i^2 \sqrt{6} = -\sqrt{6} \neq \sqrt{(-2)(-3)}.$$

To avoid difficulty with this point, you should rewrite all expressions of the form $\sqrt{-b}, b > 0$, in the form $i\sqrt{b}$ or $\sqrt{b}\,i$ before making computations involving $\sqrt{-b}$.

Examples

a. $(3 + \sqrt{-5}) + (2 - 2\sqrt{-5}) = (3 + i\sqrt{5}) + (2 - 2i\sqrt{5})$

$$= 5 - i\sqrt{5}$$

b. $(2 + \sqrt{-3})(2 - \sqrt{-3}) = (2 + i\sqrt{3})(2 - i\sqrt{3})$

$$= 4 - 3i^2 = 4 - 3(-1) = 7$$

Subtraction and division

Let us now consider two more operations, subtraction and division, for the field C. Consistent with the definition of a difference in the set of real numbers, the **difference** $z_1 - z_2$ is defined by

$$z_1 - z_2 = z_1 + (-z_2)$$

and is viewed as the result of **subtracting** z_2 from z_1.

Examples   a. $(2 + 3i) - (5 + 6i)$

$$= [2 + (-5)] + [3i + (-6i)]$$
$$= -3 - 3i$$

b. $(4 + i) - (2 - 2i) = 4 + i + (-2 + 2i)$
$$= [4 + (-2)] + (i + 2i)$$
$$= 2 + 3i$$

For $z_2 \neq 0$, the **quotient** z_1/z_2 is defined by

$$\frac{z_1}{z_2} = z_1 \cdot \frac{1}{z_2}$$

and is viewed as the result of **dividing** z_1 by z_2.

The quotient of two complex numbers can be found by using the following theorem, which is analogous to Theorem 1.10 (fundamental principle of fractions) in the set of real numbers.

Theorem 1.13   If z_1, z_2, and z_3 are elements of C, and z_2 and z_3 are not zero, then

$$\frac{z_1}{z_2} = \frac{z_1 z_3}{z_2 z_3}.$$

The notion of the *conjugate* of a complex number is also used in rewriting quotients with binomial denominators.

Definition 1.16   The **conjugate** of $z = a + bi \in C$,   denoted by $\bar{z}$ is

$$\bar{z} = a - bi.$$

Examples   a.   The conjugate of $2 + 3i$ is $2 - 3i$.
           b.   The conjugate of $-3 - i$ is $-3 + i$.

The quotient $(a + bi)/(c + di)$ can be written in standard form by using Theorem 1.13 to multiply the numerator and the denominator by $c - di$, the conjugate of the denominator.

Examples   a. $\dfrac{4 + i}{2 + 3i} = \dfrac{(4 + i)(2 - 3i)}{(2 + 3i)(2 - 3i)} = \dfrac{8 - 10i - 3i^2}{4 - 9i^2}$

$$= \frac{8 - 10i + 3}{4 + 9} = \frac{11}{13} - \frac{10}{13}i$$

b. $\dfrac{5}{-3 - i} = \dfrac{5(-3 + i)}{(-3 - i)(-3 + i)} = \dfrac{-15 + 5i}{9 - i^2}$

$$= \frac{-15 + 5i}{9 + 1} = -\frac{15}{10} + \frac{5}{10}i = -\frac{3}{2} + \frac{1}{2}i$$

Exercise 1.6

Write each expression in the form $a + bi$ *or* $a + ib$.

Examples   a. $3\sqrt{-18}$ b. $2 - 3\sqrt{-16}$

Solutions   a. $3\sqrt{-18} = 3\sqrt{-1 \cdot 9 \cdot 2}$ b. $2 - 3\sqrt{-16} = 2 - 3\sqrt{-1 \cdot 16}$

$\qquad\qquad = 3\sqrt{-1}\,\sqrt{9}\,\sqrt{2}$ $\qquad\qquad = 2 - 3\sqrt{-1}\,\sqrt{16}$

$\qquad\qquad = 3i(3)\sqrt{2}$ $\qquad\qquad = 2 - 3i(4)$

$\qquad\qquad = 9i\sqrt{2}$ $\qquad\qquad = 2 - 12i$

1. $\sqrt{-4}$ **2.** $\sqrt{-9}$ **3.** $\sqrt{-32}$

4. $\sqrt{-50}$ **5.** $3\sqrt{-8}$ **6.** $4\sqrt{-18}$

7. $4 + 2\sqrt{-1}$ **8.** $5 - 3\sqrt{-1}$ **9.** $3\sqrt{-50} + 2$

10. $5\sqrt{-12} - 1$ **11.** $\sqrt{4} + \sqrt{-4}$ **12.** $\sqrt{20} - \sqrt{-20}$

Write the given expression in the form $a + bi$.

Examples   a. $(2 + 3i) - (4 - i)$ b. $\dfrac{3}{3 + i}$

Solutions   a. $2 + 3i + (-4 + i) = [2 + (-4)] + (3 + 1)i$

$\qquad\qquad\qquad\qquad = -2 + 4i$

b. $\dfrac{3(3 - i)}{(3 + i)(3 - i)} = \dfrac{9 - 3i}{9 - i^2} = \dfrac{9 - 3i}{9 + 1}$

$\qquad\qquad = \dfrac{9}{10} - \dfrac{3}{10}\,i$

13. $(2 + 4i) + (3 + i)$ **14.** $(4 - i) - (6 - 2i)$ **15.** $(2 - i) + (3 - 2i)$

16. $(2 + i) - (4 - 2i)$ **17.** $3 - (4 + 2i)$ **18.** $(2 - 6i) - 3$

19. $\dfrac{2}{1 - i}$ **20.** $\dfrac{6}{3 + 2i}$ **21.** $\dfrac{2 + i}{1 - 3i}$

22. $\dfrac{3 - i}{2i}$ **23.** $(1 - 3i)^2$ **24.** $(2 + i)^2$

Write the given product or quotient in the form a + bi or a + ib.

Examples   a. $(1 + \sqrt{-2})(3 - \sqrt{-2})$ b. $\dfrac{1}{3 - \sqrt{-4}}$

Solutions   a. $(1 + \sqrt{-2})(3 - \sqrt{-2}) = (1 + i\sqrt{2})(3 - i\sqrt{2})$

$$= 3 + 2i\sqrt{2} - i^2 \cdot 2$$

$$= 5 + 2i\sqrt{2}$$

b. $\dfrac{1}{3 - \sqrt{-4}} = \dfrac{1(3 + 2i)}{(3 - 2i)(3 + 2i)}$

$$= \dfrac{3 + 2i}{9 - 4i^2} = \dfrac{3}{13} + \dfrac{2}{13}i$$

25. $\sqrt{-4}(1 - \sqrt{-4})$ 26. $\sqrt{-9}(3 + \sqrt{-16})$

27. $(2 + \sqrt{-9})(3 - \sqrt{-9})$ 28. $(4 - \sqrt{-2})(3 + \sqrt{-2})$

29. $(2 + \sqrt{-5})^2$ 30. $(3 - \sqrt{-2})^2$

31. $\dfrac{3}{\sqrt{-4}}$ 32. $\dfrac{-1}{\sqrt{-25}}$ 33. $\dfrac{3}{2 - \sqrt{-9}}$

34. $\dfrac{2}{3 + \sqrt{-5}}$ 35. $\dfrac{2 - \sqrt{-1}}{2 + \sqrt{-1}}$ 36. $\dfrac{1 + \sqrt{-2}}{3 - \sqrt{-3}}$

37. Simplify. (*Hint*: $i^2 = -1$ and $i^4 = 1$.)

   a. i^6 b. i^{12} c. i^{15}

38. Express each of the following with positive exponents and simplify.

   a. i^{-1} b. i^{-2} c. i^{-3} d. i^{-4} e. i^{-5}

Show that in the set C of complex numbers the following statements are true.

Example   The set is closed with respect to addition.

Solution   Let $a + bi$ and $c + di$ be complex numbers. Then, by definition,

$$(a + bi) + (c + di) = (a + c) + (b + d)i.$$

Since addition is closed in the set R of real numbers, $a + c$ and $b + d$ are real numbers. Therefore, $(a + c) + (b + d)i$ is a complex number, and the assertion is proved.

39. Addition is commutative.

40. Addition is associative.

41. There is an identity element for addition.

42. Each complex number has an additive inverse.

43. Multiplication is commutative.

44. Multiplication is associative.

45. There is an identity element for multiplication.

46. Each nonzero number has a multiplicative inverse.

Chapter Review

[1.1]   *Replace the colored comma with either $\subset$ or $\in$ to make a true statement.*

1. 6, {2, 4, 6, 8}

2. {2}, {1, 2, 3, 4}

3. $\emptyset$, {2, 4}

4. 25, {5, 10, 15, . . .}

5. List the subsets of {1, 2, 3, 4} that have {1, 3} as a subset.

6. Designate {natural-number multiples of 5} by using set-builder notation.

Let $R = \{1, 5, 9\}$, $S = \{3, 5, 7\}$, and $T = \{4, 6\}$. List the members of the following sets.

7. $(R \cup T) \cap S$

8. $(S \cap T) \cup R$

9. $S \cap (T \cup \emptyset)$

10. $T \cup (R \cap \emptyset)$

[1.2]   *Let $A = \{-8, -15/7, -\sqrt{5}, -1, 0, 3, 13/2, \sqrt{50}, 21\}$. List the members of the given set.*

11. {natural numbers in A}

12. {integers in A}

13. {rational numbers in A}

14. {irrational numbers in A}

For $a, b, c, d \in R$, justify the statement by citing an appropriate postulate.

15. $(c + d)a = (d + c)a$

16. $ca + bc = ca + cb$

17. $\dfrac{1}{a + b} \cdot (a + b) = 1 \quad (a \neq -b)$

18. $(c + d) \cdot \dfrac{1}{b} = c \cdot \dfrac{1}{b} + d \cdot \dfrac{1}{b} \quad (b \neq 0)$

19. $(4 + a) + [-(4 + a)] = 0$

20. $c[a + (b + d)] = ca + c(b + d)$

[1.3]   *For $x, y \in R$, justify the statement by citing one part of Theorems 1.1–1.11.*

21. If $x = 6$, then $x + 2 = 6 + 2$.

22. $(-5)(-x) = 5x$

23. $\dfrac{4}{5} \cdot \dfrac{x + 2}{3} = \dfrac{4(x + 2)}{5 \cdot 3}$

24. $\dfrac{4(x + 3)}{8} = \dfrac{x + 3}{2}$

25. $-\dfrac{-x}{-2} = -\dfrac{x}{2}$

26. $\dfrac{x}{3} + \dfrac{y}{5} = \dfrac{5x + 3y}{3 \cdot 5}$

27. $\dfrac{\frac{1}{3}}{\frac{3}{4}} = \dfrac{4}{3}$

28. $\dfrac{\frac{4}{7}}{\frac{2}{3}} = \dfrac{4}{7} \cdot \dfrac{3}{2}$

[1.4]   *For $x, y \in R$, justify the given statement by citing one part of Theorem 1.12.*

29.   If $x < 3$, then $-2x > -6$.

30.   If $x + 7 < 3$, then $x < -4$.

Express the given statement by means of symbols.

31. -6 is greater than -9.

32.   4 is less than or equal to y.

33. y is nonpositive.

34. $y + 2$ is nonnegative.

Rewrite the given expression without using absolute-value notation.

35. $|-7|$

36. $|x - 5|$

Write the given statement without using the negation symbol.

37. $x \not\geq 4$

38. $|x| \not< 6$

[1.5]   *Solve each inequality.*

39. $\dfrac{x - 3}{4} \leq 6$

40. $2(x - 1) > \dfrac{2}{3}x$

41.   Solve $\dfrac{2}{x - 3} < 4$.

42.   Solve $\left| x - \dfrac{1}{3} \right| = \dfrac{2}{3}$.

43.   Solve $|x - 2| > 5$, and graph the solution set on a number line.

44.   Write $-3 \leq x \leq 11$ using a single inequality involving an absolute-value symbol.

[1.6]   *Write the expression in the form $a + bi$.*

45. $(2 + 3i) + (3 - 2i)$

46. $(3 - i) - (2 + 4i)$

47. $(6 + 3i) - (-2 + 4i)$

48. $(-8 + 2i) - (-3 - i)$

49. $(5 - 2i)^2$

50. $\sqrt{-2}\sqrt{-7}$

51. $\dfrac{3 - i}{3 + i}$

52. $\dfrac{\sqrt{-4}}{1 + \sqrt{-1}}$

2

Relations and Functions

2.1 The Set R^2 of Ordered Pairs

$R \times R$, or R^2, and the geometric plane

When the order in which numbers of a number pair are to be considered is specified, the pair is called an **ordered pair**, and the pair is denoted symbolically as (3, 2), (2, 3), (−1, 5), and so forth. Each of the two numbers is called a **component** of the ordered pair, the first and second being called the **first component** and the **second component**, respectively. The set of all ordered pairs (x, y) such that $x \in A$ and $y \in B$ is called the **Cartesian product** of A and B and denoted by $A \times B$. The set of all ordered pairs of real numbers $R \times R$ is often denoted by R^2. Each member of R^2 corresponds to a point in the geometric plane, and the coordinates of each point in the geometric plane are the components of a member of R^2. The correspondence between points in the plane and ordered pairs of real numbers is usually established through a **Cartesian (or rectangular) coordinate system,** as shown in Figure 2.1. We use the correspondence to graph various sets of ordered pairs of real numbers.

Solutions of an equation in two variables

Equations in two variables, such as

$$3x + 2y = 12 \quad \text{and} \quad x^2y + 3x = y^5,$$

with $x, y \in R$, have, as solutions, ordered pairs of numbers. For example, if the components of (2, 3) are substituted for the variables x and y, in that order, in the equation

$$3x + 2y = 12, \tag{1}$$

the result is

$$3(2) + 2(3) = 12,$$

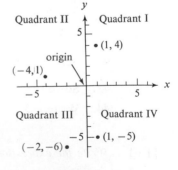

Figure 2.1

which is true. Hence, (2, 3) is a solution of Equation (1). In this book, the first component of an ordered pair is a value for the **abscissa** x, and the second component a value for the **ordinate** y. If variables other than x and y are used, their designations as first or second components will be specified. Since many equations (and inequalities) in two variables have an infinite number of solutions, we shall use the set-builder notation

$$\{(x, y) \mid \text{condition on } x \text{ and } y\}$$

to represent the set of all solutions.

Subsets of
R × R

Any condition on x and y—that is, any sentence in two variables, x and y—expresses a relationship between elements in the replacement sets of the two variables, and this relationship is precisely represented by the solution set of the sentence in two variables. For $x, y \in R$, the solution set is always a subset of $R \times R$. This leads us to the following definition.

Definition 2.1   *Any subset of R × R is called a **relation** in R.*

The relation is said to be in R because the components of the ordered pairs in the relation are elements of R. Alternatively, we frequently refer to the relationship as being *in R × R*.

The set of all first components in the ordered pairs in a relation is called the **domain** of the relation, and the set of all second components is called the **range** of the relation. Thus

$$\{(2, 5), (3, 10), (4, 15)\}$$

— elements in the domain

— elements in the range

is a relation with domain $\{2, 3, 4\}$ and range $\{5, 10, 15\}$.

If a relation is defined by an equation and the domain is not specified, we shall understand that the domain is *the set of all real numbers for which a real number exists in the range* (such relations are called real-valued relations of a real variable). For example, the domain of

$$S = \left\{ (x, y) \mid y = \frac{1}{x - 2} \right\}$$

is $\{x \mid x \in R, \ x \neq 2\}$, because for every real number x except 2, the expression $1/(x - 2)$ represents a real number.

A special kind of relation that is very important in mathematics and central to this course is called a *function*.

Definition 2.2   *A **function** is a relation in which no two ordered pairs have the same first component and different second components.*

Graphical
characteriza-
tion of a
function

A function, therefore, associates each element in its domain with one and only one element in its range. For example, $\{(1, 2), (2, 2)\}$ is a function but $\{(2, 1), (2, 2)\}$ is not. In a graphical sense, this definition implies that no two of the ordered pairs in a function graph into points on the same vertical line.

Graphs in R^2 are often continuous lines and curves. Figure 2.2 shows three such graphs. Imagine a vertical line moving across each of these from left to right. If the line at any position cuts the graph of the relation in more than one point, then the

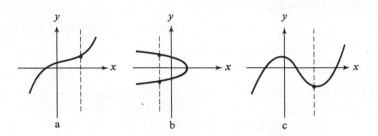

Figure 2.2

relation is not a function. Thus, although Figures 2.2-a and 2.2-c show the graphs of relations that are functions, Figure 2.2-b shows the graph of a relation that is not a function, because the vertical line shown in the figure meets the graph in two places. This means that, for the particular value of x involved, the relation associates two distinct values for y.

Algebraic characterization of a function

When a relation is defined by an equation, one way to test whether or not the relation is a function is to solve the equation explicitly for the variable y representing an element in the range. This will show whether or not more than one value of y is associated with any single value of x. This method works, of course, only if it is possible to solve for y.

Example

Since $y^2 = 1 + x^2$ implies either $y = \sqrt{1 + x^2}$ or $y = -\sqrt{1 + x^2}$, the assignment of a real value to x will result in *two* different values for y, and hence the relation defined by $y^2 = 1 + x^2$ is not a function.

Function notation

In general, functions are denoted by single symbols; for example, f, g, h, and F might designate functions.

The symbol for a function can be used in conjunction with the variable representing an element in the domain to represent the associated element in the range. Thus $f(x)$, read "f of x" or "the value of f at x," is the element in the range of f associated with the element x in the domain. Suppose, for example, that

$$f = \{(x, y) \mid y = x + 3\}.$$

The alternative notation

$$f = \{(x, f(x)) \mid f(x) = x + 3\}$$

can be used, where $f(x)$ plays the same role as y. Then for this function, $f(2)$ represents the value of $x + 3$ when x is replaced by 2:

$$f(2) = 2 + 3 = 5.$$

Similarly,

$$f(-1) = -1 + 3 = 2,$$
$$f(0) = 0 + 3 = 3,$$
$$f(a) = a + 3, \text{ etc.}$$

Exercise 2.1

Supply the missing components so that the ordered pairs

  a. (0,)   b. (1,)   c. (2,)   d. (−3,)   e. $\left(\dfrac{2}{3},\ \right)$

are solutions of the given equation.

1. $2x + y = 6$ **2.** $y = 9 - x^2$ **3.** $y = \dfrac{3x}{x^2 - 2}$

4. $y = 0$ **5.** $y = \sqrt{3x + 11}$ **6.** $y = |x - 1|$

State (a) the domain and the range of the relation and (b) whether or not the relation is a function.

Example $\{(3, 5), (4, 8), (4, 9), (5, 10)\}$

Solution

  a. Domain (the set of first components): $\{3, 4, 5\}$
     Range (the set of second components): $\{5, 8, 9, 10\}$

  b. The relation is not a function, because two ordered pairs, $(4, 8)$ and $(4, 9)$ have the same first components.

7. $\{(2, 3), (5, 7), (7, 8)\}$ **8.** $\{(-1, 6), (0, 2), (3, 3)\}$
9. $\{(2, -1), (3, 4), (3, 6)\}$ **10.** $\{(-4, 7), (-4, 8), (3, 2)\}$
11. $\{(5, 5), (6, 6), (7, 7)\}$ **12.** $\{(0, 0), (2, 4), (4, 2)\}$

Specify the maximum domain of the real numbers that would yield real numbers y for elements in the range of the relation defined by the given equation.

Examples   a. $y = \sqrt{16 - x^2}$ b. $y = \dfrac{1}{x(x + 2)}$

Solutions   a. For what values of x is   b. For what values of x is
            $16 - x^2 \geq 0$? $x(x + 2) \neq 0$?
            The domain is $\{x \mid -4 \leq x \leq 4\}$.   The domain is $\{x \mid x \neq 0, -2\}$.

13. $y = x + 7$ **14.** $y = 2x - 3$ **15.** $y = x^2$

16. $y = \dfrac{1}{x}$ **17.** $y = \dfrac{1}{x - 2}$ **18.** $y = \dfrac{1}{x^2 + 1}$

19. $y = \sqrt{x}$ **20.** $y = \sqrt{4 - x}$ **21.** $y = \sqrt{4 - x^2}$

22. $y = \sqrt{x^2 - 9}$ **23.** $y = \dfrac{4}{x(x - 1)}$ **24.** $y = \dfrac{x}{(x - 1)(x + 2)}$

State whether or not the given equation defines a function.

Examples   a. $x^2y = 3$ b. $x^2 + y^2 = 36$

Solutions   Solve explicitly for y.

a. $y = \dfrac{3}{x^2}$ b. $y = \pm\sqrt{36 - x^2}$

Yes. There is only one value   No. There are two values of y
of y associated with each   associated with values of x
value of x $(x \neq 0)$.   satisfying $|x| < 6$.

25. $x + y = 3$	**26.** $y = -x^2$	**27.** $y = \sqrt{x^2 - 5}$
28. $y = \sqrt{16 - x^2}$	**29.** $x^2 + y^2 = 16$	**30.** $y^2 - x^2 = 0$
31. $y^2 = x^3$	**32.** $y = ax^n$	

For $f(x) = x + 2$,   find the given element in the range.

Example $f(3)$

Solution   Substitute 3 for x.
$$f(3) = 3 + 2 = 5$$
The element is 5.

33. $f(0)$ **34.** $f(1)$ **35.** $f(-3)$ **36.** $f(a)$

For $g(x) = x^2 - 2x + 1$,   find the given element in the range.

37. $g(-2)$ **38.** $g(0)$ **39.** $g(a + 1)$ **40.** $g(a - 1)$

*Given that $f(x) = x + 2$ defines a function, find the element in the domain of f
associated with the given element in the range.*

Example $f(x) = 5$

Solution   Replacing $f(x)$ with $x + 2$, we have
$$x + 2 = 5,$$
$$x = 3.$$
The element is 3.

41. $f(x) = 3$ **42.** $f(x) = -2$ **43.** $f(x) = a$ **44.** $f(x) = a + 2$

For $g(x) = x^2 - 1$,   *find all elements in the domain of g associated with the given element in the range.*

45. $g(x) = 0$ **46.** $g(x) = 3$ **47.** $g(x) = 8$ **48.** $g(x) = 5$

49.   Suppose $f(x) = x + 2$ and $g(x) = x - 2$.   Find each of the following.
  a. $f(0)$ **b.** $g(2)$ **c.** $f(g(2))$ **d.** $f(g(x))$

50.   For $f(x) = x^2 - x + 1$,   find each of the following.

  a. $f(x + h) - f(x)$ **b.** $\dfrac{f(x + h) - f(x)}{h}$

Any function satisfying the condition that $f(-x) = f(x)$ for all x in the domain is called an **even function.** *Any function satisfying the condition that $f(-x) = -f(x)$ for all x in the domain is an* **odd function.** *Is the given function even or odd?*

51.   **a.** $\{(x, f(x)) \mid f(x) = x^2\}$ **b.** $\{(x, f(x)) \mid f(x) = x^3\}$
52.   **a.** $\{(x, f(x)) \mid f(x) = x^4 - x^2\}$ **b.** $\{(x, f(x)) \mid f(x) = x^3 - x\}$

2.2   *Linear Functions and Relations*

A **first-degree equation,** or **linear equation,** in x and y is an equation that can be written in the form

$$Ax + By + C = 0 \quad (A \text{ and } B \text{ not both } 0). \tag{1}$$

Graphs of first-degree equations

We shall call (1) the **standard form** for a linear equation. The graph of any such equation (technically, of its solution set) in R^2 is a straight line, although we do not prove this here. For $B \neq 0$, such an equation defines a **linear function** with domain the set of real numbers x. Since any two distinct points determine a straight line, it is evident that we need find only two solutions of such an equation to determine its graph, that is, the graph of the solution set of the equation. In practice, the two solutions easiest to find are usually those whose first and second components, respectively, are zero, that is, the solutions $(0, y)$ and $(x, 0)$. The x-coordinate of the point at which the graph crosses the x-axis is called the x-**intercept,** and the y-coordinate of the point at which the graph crosses the y-axis is called the y-**intercept.** As an example, consider the function

$$f = \{(x, y) \mid 3x + 4y = 12\}. \tag{2}$$

If $y = 0$, we have $x = 4$, and the x-intercept is 4. If $x = 0$, then $y = 3$, and the y-intercept is 3. Thus the graph of (2) appears as in Figure 2.3.

If a graph intersects both axes at or near the origin, either the intercepts do not represent two separate points, or the points are too close together to be of much use in drawing the graph. It is then necessary to plot at least one other point at a distance far enough removed from the origin to establish the line accurately.

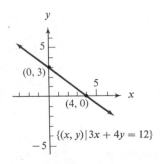

Figure 2.3

Distance between two points

Any two distinct points in the plane are the endpoints of a line segment. Two fundamental properties of a line segment are its **length** and its **inclination** with respect to the x-axis.

Figure 2.4 shows the line segment joining points $P_1(x_1, y_1)$ and $P_2(x_2, y_2)$. If a line

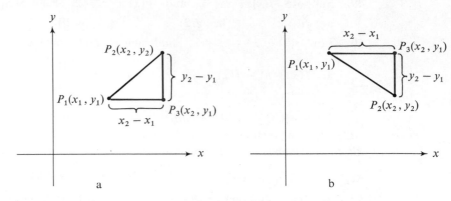

Figure 2.4

parallel to the x-axis is constructed through P_1, and a line parallel to the y-axis through P_2, then these lines will intersect at a point P_3 with coordinates (x_2, y_1). The distance from P_1 to P_3 is then $|x_2 - x_1|$, and the distance from P_2 to P_3 is $|y_2 - y_1|$. The Pythagorean theorem applied to these distances yields the length d of the line segment joining P_1 and P_2, namely,

$$d = \sqrt{|x_2 - x_1|^2 + |y_2 - y_1|^2}.$$

Since $|x_2 - x_1|^2 = (x_2 - x_1)^2$ and $|y_2 - y_1|^2 = (y_2 - y_1)^2$, we accordingly have

$$d = \sqrt{(x_2 - x_1)^2 + (y_2 - y_1)^2}.$$

This is known as the **distance formula**.

Slope of a line segment

The inclination of the line segment joining P_1 and P_2 is measured by forming the ratio of the differences $y_2 - y_1$ and $x_2 - x_1$, and is called the **slope** m of the line segment. Thus

$$m = \frac{y_2 - y_1}{x_2 - x_1} \quad (x_2 - x_1 \neq 0).$$

For the segment in Figure 2.4-a the slope is positive, while for the one in Figure 2.4-b the slope is negative. If a line segment is parallel to the x-axis, then $y_2 - y_1 = 0$ and the line segment has slope 0; but if it is parallel to the y-axis, then $x_2 - x_1 = 0$ and its slope is not defined (see Figure 2.5).

It can be shown (by similar triangles, say) that the slopes of any two line segments contained in the same line are equal, and hence we can say that the slope of a line is equal to the slope of any of its segments.

Graphs of inequalities

Sentences of the form

$$Ax + By + C \leq 0 \quad \text{and} \quad Ax + By + C < 0,$$

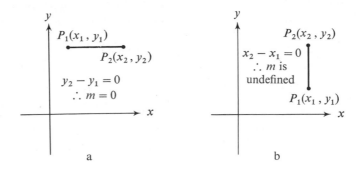

Figure 2.5

A and *B* not both 0, are inequalities of the first degree. They define the relations

$$\{(x, y) \mid Ax + By + C \le 0\} \quad \text{and} \quad \{(x, y) \mid Ax + By + C < 0\},$$

respectively. Such relations in R^2 can be represented on the plane, but the graph will be a region of the plane (a half-plane), rather than a straight line. For example, consider the relation

$$S = \{(x, y) \mid 2x + y - 3 < 0\}. \tag{3}$$

When the defining inequality is rewritten in the equivalent form

$$y < -2x + 3, \tag{4}$$

we see that for each x, solutions (x, y) are such that y is less than $-2x + 3$.

The boundary line of a graph The graph of the equation

$$y - -2x + 3 \tag{5}$$

is simply a straight line, as illustrated in Figure 2.6-a. To graph the relation S, we need only observe that any point below this line has a y-coordinate that satisfies (4), and consequently the solution set of (4), which is S, corresponds to the entire region below the line. The region is indicated on the graph with shading, as in Figure 2.6-b. That the line itself is not in the graph is indicated by means of a broken line. Were the defining inequality

$$2x + y - 3 \le 0,$$

the line would be a part of the graph and would be shown as a solid line.

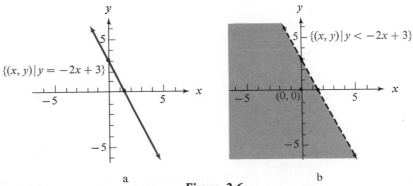

Figure 2.6

Determination of the half-plane   To determine which half-plane to shade in constructing graphs of first-degree relations, we select any point in either half-plane and find out whether or not the coordinates of the point satisfy the defining sentence. If so, then the half-plane containing the selected point is shaded; if not, the opposite half-plane is shaded. A very convenient point to use in this process is the origin. Thus, in the foregoing example, the replacement of x and y in $2x + y - 3 < 0$ by 0 gives

$$2(0) + 0 - 3 < 0, \text{ or } -3 < 0,$$

which is true, and hence the half-plane containing the origin is shaded.

Inequalities do not ordinarily define functions, according to Definition 2.2, because it usually is not true that each element of the domain is associated with a unique element in the range.

Exercise 2.2

Graph the given equation.

Example $2x - 3y = 6$

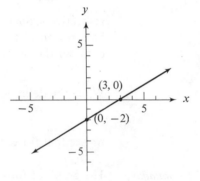

Solution   Find the intercepts of the graph.

When $x = 0$ we have $y = -2$, and when $y = 0$ we have $x = 3$. Hence the y-intercept is -2 and the x-intercept is 3.
Graph the points $(0, -2)$ and $(3, 0)$ and draw a line through the points.

1. $y = x + 5$ 2. $y = x - 7$ 3. $2x + 5y = 10$ 4. $6x - y = 12$
5. $2x - 6y = -12$ 6. $3x + y = -6$ 7. $x - 5 = 0$
8. $x + 2 = 0$ 9. $y + 4 = 0$ 10. $y - 3 = 0$

Find the distance between the given pair of points, and find the slope of the line segment joining them.

Example $(3, -5), (2, 4)$

Solution   Consider $(3, -5)$ as P_1 and $(2, 4)$ as P_2.

$$d = \sqrt{(x_2 - x_1)^2 + (y_2 - y_1)^2} \qquad\qquad m = \frac{y_2 - y_1}{x_2 - x_1}$$

$$= \sqrt{(2 - 3)^2 + [4 - (-5)]^2} = \sqrt{1 + 81} \qquad = \frac{4 - (-5)}{2 - 3} = \frac{9}{-1}$$

Distance, $\sqrt{82}$; slope, -9

11. $(1, 1), (4, 5)$ 12. $(-1, 1), (5, 9)$ 13. $(-3, 2), (2, 14)$
14. $(-4, -3), (1, 9)$ 15. $(2, 1), (1, 0)$ 16. $(-3, 2), (0, 0)$

Find the length of each side of the triangle having vertices as given.

17. (10, 1), (3, 1), (5, 9)   **18.** (0, 6), (9, −6), (−3, 0)

19. (5, 6), (11, −2), (−10, −2)   **20.** (−1, 5), (8, −7), (4, 1)

Graph the relation.

Example $\{(x, y) \mid 2x + y \geq 4\}$

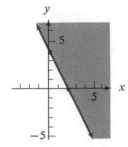

Solution   Graph the equality $2x + y = 4$.

Substitute 0 for x and y and determine that $2(0) + (0) \geq 4$ or $0 \geq 4$ is false, so the origin is not in the graph.

Shade region above the graph of $2x + y = 4$. The line is included in the graph.

21. $\{(x, y) \mid y < x\}$ **22.** $\{(x, y) \mid y > x\}$ **23.** $\{(x, y) \mid y \leq x + 2\}$

24. $\{(x, y) \mid y \geq x - 2\}$ **25.** $\{(x, y) \mid x + y < 5\}$

26. $\{(x, y) \mid 2x + y < 2\}$ **27.** $\{(x, y) \mid 3 \geq 2x - 2y\}$

28. $\{(x, y) \mid 0 \geq x + y\}$ **29.** $\{(x, y) \mid x > 0\}$

30. $\{(x, y) \mid y < 0\}$ **31.** $\{(x, y) \mid x < 0\}$ **32.** $\{(x, y) \mid x < -2\}$

33. $\{(x, y) \mid -1 < x < 5\}$ **34.** $\{(x, y) \mid 0 \leq y \leq 1\}$

35. $\{(x, y) \mid |x| < 3\}$ **36.** $\{(x, y) \mid |y| > 1\}$

37. Graph the set of points whose coordinates satisfy both

$$y \geq x - 4 \quad \text{and} \quad y \geq 0.$$

38. Graph the set of points whose coordinates satisfy both

$$2 \leq x \leq 4 \quad \text{and} \quad y \leq 2x - 6.$$

2.3   *Forms of Linear Equations*

Point-slope form   Using the slope concept, we can rewrite the defining equation for a linear function in several useful forms. Consider a line having slope m and passing through a given point (x_1, y_1), as shown in Figure 2.7. If we choose *any other* point on the line and assign to it the coordinates (x, y), it is evident that the slope of the line is given by

$$\frac{y - y_1}{x - x_1} = m,$$

from which

$$y - y_1 = m(x - x_1). \tag{1}$$

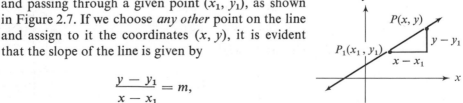

Figure 2.7

Note that (1) is satisfied also by $(x, y) = (x_1, y_1)$. Since now x and y are the co-ordinates of *any* point on the line, (1) is an equation of the line passing through (x_1, y_1) with slope m. This is called the **point-slope form** for a linear equation.

Slope-intercept form

Now consider an equation of the line with slope m that passes through a given point on the y-axis having coordinates $(0, b)$, as shown in Figure 2.8. Substituting $(0, b)$ in the point-slope form of a linear equation,

$$y - y_1 = m(x - x_1),$$

we obtain

$$y - b = m(x - 0),$$

from which

$$y = mx + b. \qquad (2)$$

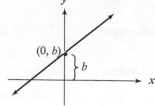

Figure 2.8

Equation (2) is called the **slope-intercept form** for a linear equation. Any linear equation in the standard form

$$Ax + By + C = 0$$

can be written equivalently in the slope-intercept form by solving for y in terms of x if $B \neq 0$.

Example

Find the slope and y-intercept of the graph of $2x + 3y - 6 = 0$.

Solution

The equation $2x + 3y - 6 = 0$ can be written equivalently as

$$y = -\frac{2}{3}x + 2$$

By comparison with $y = mx + b$, the slope of the graph of the equation, $-2/3$, and the y-intercept, 2, can now be read directly from the last form of the equation.

Intercept form

If the x- and y-intercepts of the graph of

$$y = mx + b \qquad (3)$$

are a and b $(a, b \neq 0)$, respectively, as shown in Figure 2.9, then the slope m is equal to $-b/a$. Replacing m in (3) with $-b/a$, we have

$$y = -\frac{b}{a}x + b,$$

$$ay = -bx + ab,$$

$$bx + ay = ab;$$

and multiplying each member by $1/ab$ produces

$$\frac{x}{a} + \frac{y}{b} = 1.$$

This latter form is called the **intercept form** for a linear equation.

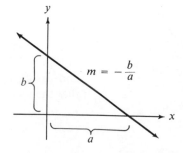

Figure 2.9

Equations of horizontal lines　　There are two special cases of linear equations worth noting. First, an equation such as

$$y - 4 = 0$$

may be considered an equation in two variables in R^2,

$$0x + y = 4.$$

For each x, this equation assigns $y = 4$. That is, any ordered pair of the form $(x, 4)$ is a solution of the equation. For instance,

$$(1, 4), (2, 4), \quad \text{and} \quad (3, 4)$$

are all solutions of the equation. If we graph these points and connect them with a straight line, we have Figure 2.10-a.

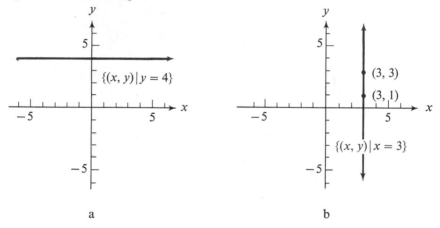

Figure 2.10

Since the equation

$$y - 4 = 0$$

assigns to each x the same value for y, the function defined by this equation is called a **constant function.**

Equations of vertical lines　　The other special case of the linear equation is of the type

$$x - 3 = 0,$$

which may be looked upon in R^2 as

$$x + 0y = 3.$$

Here, only one value is permissible for x, namely 3, whereas any value may be assigned to y. That is, any ordered pair of the form $(3, y)$ is a solution of this equation. If we choose two solutions, say $(3, 1)$ and $(3, 3)$, and complete the graph, we have Figure 2.10-b. It is clear from the graph that the equation does not define a function. This fact accounts for our restriction $B \neq 0$ on the standard form of a first-degree equation in two variables, $Ax + By + C = 0$, in order that this equation define a function.

Exercise 2.3

Find an equation, in standard form, of the line through the given point and having the given slope.

Example $(3, -5)$, $m = -2$

Solution   Substitute given values in the point-slope form of the linear equation.

$$y - y_1 = m(x - x_1)$$
$$y - (-5) = -2(x - 3)$$
$$y + 5 = -2x + 6$$
$$2x + y - 1 = 0$$

1. $(-3, -2)$, $m = \dfrac{1}{2}$ 2. $(0, 0)$, $m = 3$

3. $(-1, 0)$, $m = 1$ 4. $(2, -3)$, $m = 0$

5. $(-4, 2)$, $m = 0$ 6. $(-1, -2)$, parallel to y-axis

Write the given equation in slope-intercept form; specify the slope and y-intercept of the line.

Example $2x - 3y = 5$

Solution   Solving explicitly for y, we obtain

$$y = \frac{2}{3}x - \frac{5}{3}.$$

Comparing this equation with the general slope-intercept form,

$$y = mx + b,$$

we note that the slope is $2/3$ and the y-intercept is $-5/3$.

7. $3x + 2y = 1$ 8. $3x - y = 7$

9. $x - 3y = 2$ 10. $2x - 3y = 0$

11. $8x - 3y = 0$ 12. $-x = 2y - 5$

13. Write an equation, in standard form, of the line with the same slope as $x - 2y = 5$ and passing through the origin. Sketch the graph of this equation.

14. Write an equation, in standard form, of the line through $(0, 5)$ with the same slope as $2y - 3x = 5$.

Find the equation, in standard form, of the line with the given intercepts.

Example

$x = 3;$ $y = -1/2$

Solution

Substitute 3 and $-1/2$ for a and b, respectively, in the intercept form $x/a + y/b = 1$.

$$\frac{x}{3} + \frac{y}{-\dfrac{1}{2}} = 1$$

$$x - 6y - 3 = 0$$

15. $x = 2;$ $y = 3$ **16.** $x = 4;$ $y = -1$ **17.** $x = -2;$ $y = -5$

18. $x = -1;$ $y = 7$ **19.** $x = -\dfrac{1}{2};$ $y = \dfrac{3}{2}$ **20.** $x = \dfrac{2}{3};$ $y = -\dfrac{3}{4}$

21. Show that, for $x_2 \neq x_1$,

$$y - y_1 = \left(\frac{y_2 - y_1}{x_2 - x_1}\right)(x - x_1)$$

is an equation of the line joining the points (x_1, y_1) and (x_2, y_2). This is the **two-point form** of the linear equation. Find an equation of the lines through the given points.

a. $(2, 1)$ and $(-1, 3)$ **b.** $(3, 0)$ and $(5, 0)$

2.4   Parallel and Perpendicular Lines; Midpoint Formulas

Parallel lines

Similar triangles can be used to show that line segments (and lines) not perpendicular to the x-axis are parallel if and only if they have equal slopes. Since the proof of the following theorem depends on familiar geometric properties, it is left as an exercise.

> **Theorem 2.1**   *Two lines with slopes m_1 and m_2 are parallel if and only if $m_1 = m_2$. Two lines perpendicular to the x-axis are parallel to each other.*

Example

Find an equation of the line through $(-1, 2)$ parallel to $3x - 2y = 6$.

Solution

Make a sketch. The given equation can be written equivalently in slope-intercept form as

$$y = \frac{3}{2}x - 3,$$

and, by inspection, we see that the slope of its graph is 3/2. Then, using the point-slope form for a linear equation, with $m = 3/2$, $x_1 = -1$

(continued overleaf)

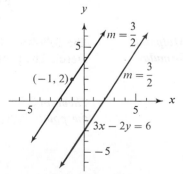

and $y_1 = 2$, we have

$$(y - 2) = \frac{3}{2}(x + 1),$$

which reduces to the standard form $3x - 2y + 7 = 0$.

Perpendicular   The slopes of two perpendicular lines are related to each other in an interesting way.
lines

> **Theorem 2.2**   *Two lines with slopes m_1 and m_2 are perpendicular if and only if*
> $m_1 \cdot m_2 = -1$. *The x- and y-axes or two lines that are parallel to the x- and y-axes,*
> *respectively, are also perpendicular to each other.*

The proof of this theorem also follows from a familiar geometric property and is left as an exercise.

Example   Find an equation of the line passing through $(3, -2)$ and perpendicular to the graph of $2x + 5y = 10$.

Solution   Make a sketch. The given equation can be written equivalently in slope-intercept form as

$$y = -\frac{2}{5}x + 2,$$

and, by inspection, the slope of its graph is $-2/5$. Hence, from Theorem 2.2, the slope of any line perpendicular to this graph must be $5/2$. Using the point-slope form for a linear equation with $m = 5/2$, $x_1 = 3$, and $y_1 = -2$, we have

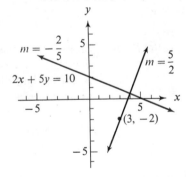

$$y - (-2) = \frac{5}{2}(x - 3),$$

which reduces to the standard form

$$5x - 2y - 19 = 0.$$

Midpoint   The following theorem concerning the midpoint of a line segment is often quite
formulas   useful. The proof is left as an exercise.

> **Theorem 2.3**   *The coordinates x and y of the midpoint of the line segment joining*
> *the points $P_1(x_1, y_1)$ and $P_2(x_2, y_2)$ are given by*

$$x = \frac{x_1 + x_2}{2} \quad and \quad y = \frac{y_1 + y_2}{2}.$$

Example   For $P_1(2, -3)$ and $P_2(4, 5)$, find the coordinates of the midpoint of the segment P_1P_2.

Solution   Using Theorem 2.3, we have $x = \dfrac{2+4}{2} = 3$, and $y = \dfrac{-3+5}{2} = 1$.

Hence the midpoint is (3, 1).

Exercise 2.4

Find an equation of the line which is (a) parallel and (b) perpendicular to the graph of the given equation, and which passes through the given point.

Example $2x + 3y = 6$; $(4, -2)$

Solution   The given equation can be written equivalently in the slope-intercept form as

$$y = -\frac{2}{3}x + 2.$$

Hence, the slope of the line is $-2/3$. An equation of the line parallel to the given line and containing the point $(4, -2)$ is given by

$$y - (-2) = -\frac{2}{3}(x - 4),$$

which is equivalent to

$$3y + 6 = -2x + 8,$$

$$2x + 3y - 2 = 0.$$

Any line perpendicular to the given line has slope

$$m = -\frac{1}{-\dfrac{2}{3}} = \frac{3}{2}.$$

An equation of the line with slope $3/2$ and containing the point $(4, -2)$ is

$$y - (-2) = \frac{3}{2}(x - 4),$$

which is equivalent to

$$2y + 4 = 3x - 12,$$

$$3x - 2y - 16 = 0.$$

1. $3x + y = 6$; $(5, 1)$ 2. $2x - y = -3$; $(-2, 1)$
3. $3x - 2y = 5$; $(0, 0)$ 4. $4x - 2y = -5$; $(0, 0)$
5. $4x + 6y = 3$; $(-2, -1)$ 6. $5x - 3y = 2$; $(-4, -1)$
7. $2x + 4y = 9$; $(-3, 4)$ 8. $3x - 9y = 4$; $(5, -7)$

Find an equation of the line that is the perpendicular bisector of the segment whose endpoints are given.

Example (7, −4) and (−5, −9)

Solution The slope of the segment is

$$m = \frac{y_2 - y_1}{x_2 - x_1} = \frac{-9 - (-4)}{-5 - 7} = \frac{-5}{-12} = \frac{5}{12}.$$

From Theorem 2.2, it follows that the slope of the desired bisector is −12/5. From Theorem 2.3, the coordinates of the midpoint of the given segment are

$$x = \frac{x_1 + x_2}{2} = \frac{7 + (-5)}{2} = \frac{2}{2} = 1,$$

$$y = \frac{y_1 + y_2}{2} = \frac{-4 + (-9)}{2} = -\frac{13}{2}.$$

Using the point-slope form for a linear equation, with $m = -12/5$ and given point $(1, -13/2)$, we have the desired equation

$$y - \left(-\frac{13}{2}\right) = -\frac{12}{5}(x - 1),$$

which simplifies to the standard form $24x + 10y + 41 = 0$.

9. (5, 3) and (9, 7)

10. (9, 1) and (−3, −5)

11. (5, −5) and (−9, 1)

12. (9, 4) and (−3, 4)

13. $(a, 0)$ and $(0, b)$

14. (a, b) and $(a + k_1, b + k_2)$

15. Use slopes to show that the triangle with vertices at $A(0, 8)$, $B(6, 2)$, and $C(-4, 4)$ is a right triangle.

16. Use slopes to show that the triangle with vertices $D(2, 5)$, $E(5, 2)$, and $F(10, 7)$ is a right triangle.

17. Use slopes to show that the quadrilateral with vertices at $P(-1, 2)$, $Q(5, 4)$, $R(8, 2)$, and $S(2, 0)$ is a parallelogram.

18. Show that the quadrilateral with vertices at $A(-5, -1)$, $B(0, 0)$, $C(1, -5)$, and $D(-4, -6)$ is a square.

19. Recall from geometry that if two parallel lines are cut by a transversal then corresponding angles are congruent. Use this fact to show, in the figure at the right, where L_1 is parallel to L_2, that $\Delta P_1Q_1P_3$ is similar to $\Delta P_2Q_2P_4$ and, hence, that the slopes, m_1 and m_2, of the lines are equal.

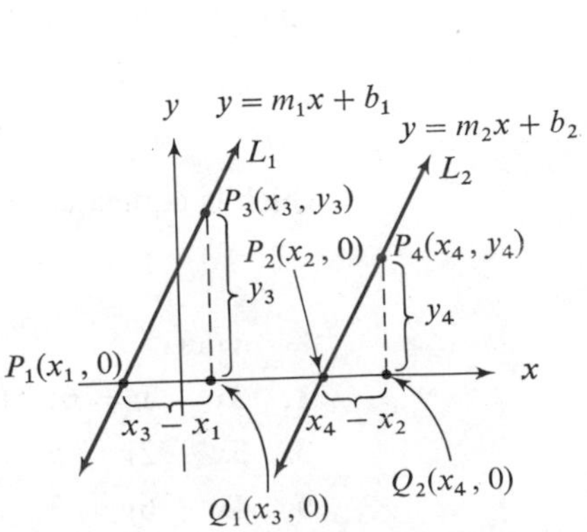

20. In the figure for Exercise 19, assume that $m_1 = m_2$. Deduce that $\Delta P_1Q_1P_3$ is similar to $\Delta P_2Q_2P_4$ and, hence, that L_1 is parallel to L_2.

21. In the adjoining figure, where L_1 is perpendicular to L_2, use the fact that $\Delta P_1 Q P_2$ is similar to $\Delta P_3 Q P_1$, so that

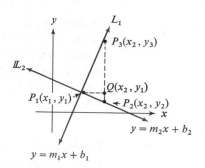

$$\frac{y_1 - y_2}{x_2 - x_1} = \frac{x_2 - x_1}{y_3 - y_1},$$

to deduce that $m_1 m_2 = -1$.

22. Use the figure in Exercise 21 to argue that, because

$$(m_1 - m_2)^2 = m_1^2 - 2m_1 m_2 + m_2^2$$

and because the converse of the Pythagorean theorem is true, the relation $m_1 m_2 = -1$ implies that L_1 is perpendicular to L_2.

23. Use the similar triangles shown in the figure to prove that

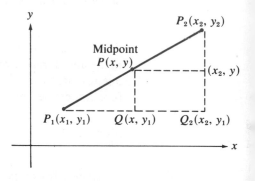

$$x = \frac{x_1 + x_2}{2} \quad \text{and} \quad y = \frac{y_1 + y_2}{2}.$$

2.5   *Inverse Relations and Functions*

Inverse relations

If the two components of each ordered pair in a relation r are interchanged, then the resulting relation is called the **inverse relation** of r and is denoted by r^{-1}. For example, each of the relations

$$r = \{(1, 2), (3, 4), (5, 5)\}$$

and

$$r^{-1} = \{(2, 1), (4, 3), (5, 5)\}$$

is the inverse of the other. (Notice that in this context r^{-1} docs not denote $1/r$.)

Inverse functions

By Definition 2.2, a function is a set of ordered pairs (x, y) such that no two have the same first components and different second components. When the components of every ordered pair in a function f are interchanged, the resulting relation may or may not be a function. For example, if

$$(1, 5), \quad (2, 5), \quad \text{and} \quad (3, 6)$$

are ordered pairs in f, then

$$(5, 1), \quad (5, 2), \quad \text{and} \quad (6, 3)$$

are members of the relation formed by interchanging the components of these ordered pairs. Clearly these latter pairs cannot be members of a function, since two of them, $(5, 1)$ and $(5, 2)$, have the same first component and different second components. If, however, a function f is *one-to-one*, that is, if no two different ordered pairs in f have the same second component (same value for y), then the relation obtained by interchanging the first and second components of every pair in the function will also be a function. This function is called the **inverse function** of f. The notation f^{-1} (read "f inverse") is frequently used to denote the inverse function of f.

Definition 2.3   *If a function f is one-to-one, then the **inverse function** f^{-1} is the set of ordered pairs obtained from f by interchanging the first and second components of each ordered pair in f.*

It is evident from this definition that the domain and range of f^{-1} are just the range and domain, respectively, of f. If $y = f(x)$ defines a function f, and if f is one-to-one, then $x = f(y)$ defines the inverse of f. For example, the inverse of the function defined by

$$y = 3x + 2 \tag{1}$$

is defined by

$$x = 3y + 2, \tag{2}$$

or, when y is expressed in terms of x, by

$$y = \frac{1}{3}x - \frac{2}{3}. \tag{2'}$$

Equations (2) and (2') are equivalent.

The graphs of inverse relations are related in an interesting way. To see this, we first observe, in Figure 2.11, that the graphs of the ordered pairs (a, b) and (b, a)

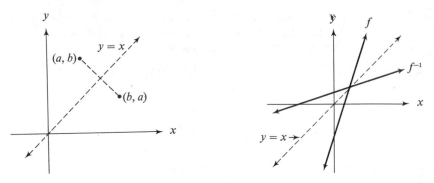

Figure 2.11   **Figure 2.12**

are always located symmetrically with respect to the graph of $y = x$. Therefore, because for every ordered pair (a, b) in a function Q the ordered pair (b, a) is in Q^{-1}, the graphs of $y = Q^{-1}(x)$ and $y = Q(x)$ are reflections of each other about the graph of $y = x$.

Figure 2.12 shows the graphs of the linear function

$$f = \{(x, y) \mid y = 4x - 3\} \tag{3}$$

and its inverse

$$f^{-1} = \{(x, y) \mid x = 4y - 3\} = \left\{(x, y) \mid y = \frac{1}{4}(x + 3)\right\},$$

together with the graph of $y = x$.

Since an element in the domain of the inverse Q^{-1} of a function Q is the range element in the corresponding ordered pair of Q, and vice versa, it follows that for every x in the domain of Q,

$$Q^{-1}(Q(x)) = x$$

(read "Q inverse of Q of x is equal to x"), and, for every x in the domain of Q^{-1},

$$Q(Q^{-1}(x)) = x$$

(read "Q of Q inverse of x is equal to x"). Using (3) above as an example, we note that if f is the linear function defined by

$$f(x) = 4x - 3,$$

then f is a one-to-one function. Now f^{-1} is defined by

$$f^{-1}(x) = \frac{1}{4}(x + 3),$$

and we see that

$$f^{-1}(f(x)) = \frac{1}{4}[(4x - 3) + 3] = x$$

and

$$f(f^{-1}(x)) = 4\left[\frac{1}{4}(x + 3)\right] - 3 = x.$$

Exercise 2.5

Graph the given function f and its inverse function f^{-1}, using the same set of axes.

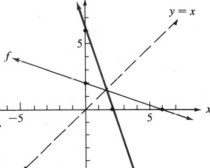

Example $f = \{(x, y) \mid x + 3y = 6\}$

Solution   Interchange the variables in the defining equation.

$$f^{-1} = \{(x, y) \mid y + 3x = 6\}$$

The graph of f and f^{-1} are shown in the figure.

1. $f = \{(-2, 3), (4, 7), (5, 9)\}$
2. $f = \{(-3, 1), (2, -1), (3, 4)\}$
3. $f = \{(-1, -1), (2, 2), (3, 3)\}$
4. $f = \{(-4, 4), (0, 0), (4, -4)\}$
5. $f = \{(x, y) \mid y = 2x + 6\}$ 6. $f = \{(x, y) \mid y = 3x - 6\}$
7. $f = \{(x, y) \mid y = 4 - 2x\}$ 8. $f = \{(x, y) \mid y = 6 + 3x\}$
9. $f = \{(x, y) \mid 3x - 4y = 12\}$ 10. $f = \{(x, y) \mid x - 6y = 6\}$

11. $f = \{(x, y) \mid 4x + y = 4\}$ **12.** $f = \{(x, y) \mid 2x - 3y = 12\}$

In Exercises 13–18, each equation defines a one-to-one function f in R × R. Find an equation defining f^{-1} and show that $f(f^{-1}(x)) = f^{-1}(f(x)) = x$. Hint: Solve explicitly for y and let $y = f(x)$.

13. $y = x$ **14.** $y = -x$ **15.** $2x + y = 4$

16. $x - 2y = 4$ **17.** $3x - 4y = 12$ **18.** $3x + 4y = 12$

2.6   *The Arithmetic of Functions*

We can define binary operations involving the functions in the set of real-valued functions of a real variable.

Definition 2.4   *If f and g are real-valued functions of a real variable, then the functions $f + g, f - g, f \cdot g$, and f/g are defined by:*

I $(f + g)(x) = f(x) + g(x),$

II $(f - g)(x) = f(x) - g(x),$

III $(f \cdot g)(x) = f(x) \cdot g(x),$

IV $\left(\dfrac{f}{g}\right)(x) = \dfrac{f(x)}{g(x)}, \quad (g(x) \neq 0).$

In each instance, the domain of the resulting function consists of the intersection of the domain of f and the domain of g (the set R, unless otherwise specified), except that for quotients we must exclude those elements for which the divisor vanishes (equals zero).

Example   Let
$$f(x) = 2x^2 - 3x - 5 \quad \text{and} \quad g(x) = 3x^2 + x - 2.$$
From Definition 2.4,

$$(f + g)(x) = f(x) + g(x) = (2x^2 - 3x - 5) + (3x^2 + x - 2)$$
$$= 5x^2 - 2x - 7,$$
$$(f - g)(x) = f(x) - g(x) = (2x^2 - 3x - 5) - (3x^2 + x - 2)$$
$$= -x^2 - 4x - 3,$$
$$(f \cdot g)(x) = f(x) \cdot g(x) = (2x^2 - 3x - 5) \cdot (3x^2 + x - 2)$$
$$= 6x^4 - 7x^3 - 22x^2 + x + 10,$$
$$\left(\frac{f}{g}\right)(x) = \frac{f(x)}{g(x)} = \frac{2x^2 - 3x - 5}{3x^2 + x - 2}$$
$$= \frac{(2x - 5)(x + 1)}{(3x - 2)(x + 1)} = \frac{2x - 5}{3x - 2}, \quad \left(x \neq \frac{2}{3}, -1\right).$$

**Composition
of functions**
Notice that the function operations defined above are simply extensions of the familiar binary operations for real numbers. There is, however, another operation that we can define on a set of functions that is uniquely a function operation. The operation is called **composition** and is denoted by the symbol ∘.

Definition 2.5   *If f and g are real-valued functions of a real variable, then*

$$(f \circ g)(x) = f(g(x)).$$

In this operation, the domain of the resulting function (called the **composite** of f and g) is the set of all real numbers x in the domain of g for which g(x) is in the domain of f.

Example
Let f and g be functions defined by $f(x) = \sqrt{4 - x}$ and $g(x) = 4 - x$, with domains the sets of real values x for which real values f(x) and g(x) are determined. Write, in simple form, a defining equation for $f \circ g$, and specify the domain of the composite function.

Solution
From Definition 2.5,

$$(f \circ g)(x) = f(g(x)) = \sqrt{4 - (4 - x)} = \sqrt{x}.$$

Since f(x) is defined only for those values of x for which $4 - x \geq 0$, or $x \leq 4$, we must restrict the domain of g to those values of x for which $g(x) \leq 4$. Thus, we require

$$4 - x \leq 4, \quad \text{or} \quad x \geq 0.$$

Thus, the domain of $f \circ g$ is $\{x \mid x \in R, x \geq 0\}$.

**An alternative
definition
of f^{-1}**
In Section 2.5, we defined the inverse function f^{-1} for each one-to-one function f. Since the domain of f^{-1} is the range of f, and the range of f^{-1} is the domain of f, and since the pairing of the elements of the two sets involves only an exchange of first and second components and otherwise does not alter the components, it follows that

$$(f^{-1} \circ f)(x) = x \tag{1}$$

for every x in the domain of f. In some treatments of functions, this relationship serves to define f^{-1}. In the same way, we have

$$(f \circ f^{-1})(y) = y$$

for every y in the range of f.

Exercise 2.6

In Exercises 1–12, find defining equations in simple form for $f + g$, $f - g$, $f \cdot g$, and f/g.

1. $f(x) = 3x - 2$; $g(x) = 2x + 1$ 2. $f(x) = x^2 - 4$; $g(x) = x - 2$
3. $f(x) = 2x + 1$; $g(x) = x^2 + 2$ 4. $f(x) = x^2 - 1$; $g(x) = x^2 + 1$
5. $f(x) = 2x^2$; $g(x) = \sqrt{x - 1}$ 6. $f(x) = x^2 - 4$; $g(x) = \sqrt{x + 4}$

7. $f(x) = 2x - 3$;

    $g(x) = \dfrac{1}{2}(x + 3)$

8. $f(x) = 4 - 5x$;

    $g(x) = \dfrac{1}{5}(4 - x)$

9. $f(x) = 3x^2 - 2x + 1$;

    $g(x) = \dfrac{1}{x}$

10. $f(x) = x^2 - 5x - 6$;

    $g(x) = 2x^2 + 5x + 3$

11. $f(x) = \dfrac{x + 1}{x - 1}$; $g(x) = \dfrac{1}{1 - x}$

12. $f(x) = x^2 - 4$; $g(x) = \dfrac{x + 2}{x - 2}$

13–24. For Exercises 1–12 above, find $f \circ g$. State the domain of the resulting function.

Given $f(x) = \dfrac{x + 1}{x - 1}$ and $g(x) = \dfrac{1}{x}$,   find each value. If the value is undefined, so state.

Example

$(f - g)(-2)$

Solution

Method 1.   First find $(f - g)(x)$.

$$(f - g)(x) = f(x) - g(x) = \frac{x + 1}{x - 1} - \frac{1}{x} = \frac{x^2 + x - x + 1}{x(x - 1)} = \frac{x^2 + 1}{x(x - 1)}.$$

Then

$$(f - g)(-2) = \frac{(-2)^2 + 1}{(-2)(-2 - 1)} = \frac{5}{6}.$$

Method 2.   First find $f(-2)$ and $g(-2)$.

$$f(-2) = \frac{-2 + 1}{-2 - 1} = \frac{1}{3} ;   \qquad g(-2) = \frac{1}{-2} = -\frac{1}{2}.$$

Then

$$(f - g)(-2) = f(-2) - g(-2) = \frac{1}{3} - \left(-\frac{1}{2}\right) = \frac{1}{3} + \frac{1}{2} = \frac{5}{6}.$$

25. $(f + g)(2)$

26. $(f \cdot g)(-5)$

27. $\left(\dfrac{f}{g}\right)(0)$

28. $\left(\dfrac{f}{g}\right)(-1)$

29. $(f \circ g)(3)$

30. $(g \circ f)(3)$

31. $(g - f)(7)$

32. $(f \cdot (f + g))(3)$

Let f be the function defined by $f(x) = 2x - 4$. In Exercises 33–38, find a defining equation for g.

33. $(f + g)(x) = 5x + 2$

34. $(f - g)(x) = x + 6$

35. $(f \cdot g)(x) = 5x^2 + 2x$

36. $\left(\dfrac{f}{g}\right)(x) = x - 2$

37. $\left(\dfrac{g}{f}\right)(x) = x + 1$

38. $(g \cdot f)(x) = 2x^2 - 10x + 12$

Using Equation 1 on page 51, show that the functions defined by the given equations are inverse functions.

39. $y = 2x + 1$ and $y = \dfrac{x - 1}{2}$

40. $y = \dfrac{3x + 2}{2}$ and $y = \dfrac{2x - 2}{3}$

Chapter Review

[2.1] *Specify the domain of the given relation.*

1. $\{(4, -1), (2, -4), (3, -5)\}$

2. $\{(1, 2), (1, 3), (1, 6)\}$

3. $\left\{(x, y) \mid y = \dfrac{1}{x + 4}\right\}$

4. $\{(x, y) \mid y = \sqrt{x - 6}\}$

Let $f(x) = x - 3$ *and* $g(x) = x^2 + 4$. *Find the value of the given expression.*

5. $g(3)$
6. $f(\ 2)$
7. $f(g(0))$
8. $g(x + h)$

[2.2] *Graph the given equation.*

9. $3x + 2y = 6$
10. $x + 4y = 8$
11. $x - 3 = 0$
12. $y - 2 = 0$

Find the distance between the given pair of points, and find the slope of the line segment joining them.

13. $(2, 0)$ and $(-3, 4)$

14. $(-6, 1)$ and $(-8, 2)$

Graph the given relation.

✳15. $\{(x, y) \mid 2x - 3y < 12\}$

16. $\{(x, y) \mid -2 < x \le 4\}$

[2.3] *Find the equation, in standard form, of the line passing through the given point and having the given slope.*

17. $(2, -7), \quad m = 4$

18. $(-6, 3), \quad m = \dfrac{1}{2}$

Write the equation in slope-intercept form; specify the slope and the y-intercept of the line.

19. $4x + y = 6$

20. $3x - 2y = 16$

Find the equation, in standard form, of the line with the given intercepts.

21. $x = -3;\quad y = 2$ | **22.** $x = \dfrac{1}{3};\quad y = -4$

[2.4]   *Find an equation of the line which passes through the given point and is* (a) *parallel to, and* (b) *perpendicular to, the graph of the given equation.*

23. $4x - 3y = 12;\quad (-2, 1)$ | **24.** $x + 6y = 0;\quad (3, -4)$

Find an equation of the line that is the perpendicular bisector of the segment whose endpoints are given.

25. $(4, -2)$ and $(-8, 6)$ | **26.** $(-1, -3)$ and $(6, -1)$

[2.5]   *Graph the function f and its inverse* f^{-1}, *using the same set of axes.*

27. $f = \{(-3, 1), (-1, 3), (2, 4)\}$ | **28.** $f = \{(x, y) \mid 2x - y = 6\}$

[2.6]   *Find defining equations in simple form for* $f + g, f - g, f \cdot g, f/g,$ *and* $f \circ g$.

29. $f(x) = 2x^2 - 2;\quad g(x) = x + 1$ | **30.** $f(x) = x^2 - x + 1;\quad g(x) = \dfrac{1}{x - 1}$

3 Conic Sections

3.1 Quadratic Functions

Graph of a quadratic function

Consider the quadratic equation in two variables

$$y = x^2 - 4. \tag{1}$$

As with linear equations in two variables, solutions of this equation must be ordered pairs (x, y). We need replacements for both x and y in order to obtain a statement we may adjudge to be true or false. As before, such ordered pairs can be found by arbitrarily assigning values to x and computing related values for y. For instance, assigning the value -3 to x in Equation (1), we obtain

$$y = (-3)^2 - 4,$$

$$y = 5,$$

and $(-3, 5)$ is a solution. Similarly, we find that

$$(-2, 0), \quad (-1, -3), \quad (0, -4), \quad (1, -3), \quad (2, 0), \quad \text{and} \quad (3, 5)$$

are also solutions of (1). Locating the corresponding points on the plane, we have the graph in Figure 3.1-a. Clearly, these points do not lie on a straight line, and we might reasonably inquire whether the graph of the solution set of (1),

$$S = \{(x, y) \,|\, y = x^2 - 4\},$$

forms any kind of meaningful pattern on the plane. By graphing additional solutions of (1)—solutions with x-components between those already found—we may be able to obtain a clearer picture. Accordingly, we find the solutions

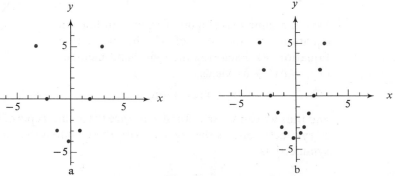

Figure 3.1

55

$$\left(\frac{-5}{2},\frac{9}{4}\right), \quad \left(\frac{-3}{2},\frac{-7}{4}\right), \quad \left(\frac{-1}{2},\frac{-15}{4}\right), \quad \left(\frac{1}{2},\frac{-15}{4}\right), \quad \left(\frac{3}{2},\frac{-7}{4}\right), \quad \left(\frac{5}{2},\frac{9}{4}\right),$$

and by graphing these points in addition to those found earlier, we have the graph in Figure 3.1-b. It now appears reasonable to connect these points in sequence, say from left to right, by a smooth curve as in Figure 3.2, and to assume that the resulting curve is a good approximation to the graph of (1). (We should realize, of course, that regardless of how many individual points are plotted, we have no absolute assurance that the smooth curve is a good approximation to the true graph; more information is needed—for example, in this case, that for $|x_2| > |x_1|$, correspondingly $y_2 > y_1$.) This curve is an example of a **parabola.**

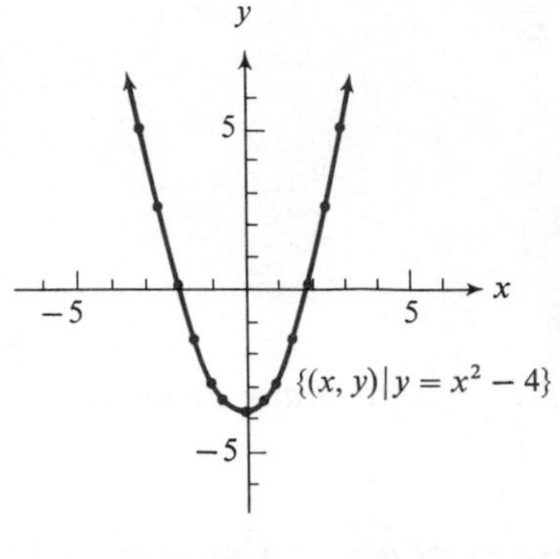

Figure 3.2

More generally, the graph of the solution set of any quadratic equation of the form

$$y = ax^2 + bx + c, \tag{2}$$

where a, b, and c are real and $a \neq 0$, is a parabola. Since for each x an equation of the form (2) will determine only one y, such an equation defines a function having as domain the entire set of real numbers and as range some subset of the reals. For example, we observe from the graph in Figure 3.2 that the range of the function defined by (1) is the set of real numbers

$$\{y \mid y \geq -4\}.$$

The parabola that is the graph of an equation of the form (2) will have a lowest (minimum) point or a highest (maximum) point, depending on whether $a > 0$ or $a < 0$, respectively. Such a point is called the **vertex** of the parabola. The line through the vertex and parallel to the y-axis is called the **axis of symmetry,** or simply the **axis,** of the parabola; it separates the parabola into two parts, each the mirror image of the other in the axis. If we observe that the graphs of

$$y = ax^2 + bx + c \tag{3}$$

and

$$y = ax^2 + bx \tag{4}$$

have the same axis (Figure 3.3), we can find an equation for the axis of (3) by inspecting Equation (4). Factoring the right-hand member of $y = ax^2 + bx$ yields

$$y = x(ax + b),$$

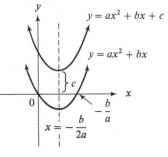

Figure 3.3

and thus we can see that 0 and $-b/a$ are the x-intercepts of the graph. Since the axis of symmetry bisects the segment with these endpoints, an equation for the axis of symmetry is

$$x = -\frac{b}{2a}.$$

Example

Find an equation for the axis of symmetry of the graph of
$$\{(x, y) \mid y = 2x^2 - 5x + 7\}.$$

Solution

By comparing the given equation to $y = ax^2 + bx + c$, we see that $a = 2$ and $b = -5$. Hence, we have

$$x = -\frac{b}{2a} = -\frac{-5}{2(2)} = \frac{5}{4},$$

and $x = 5/4$ is the desired equation for the axis.

Since the vertex of a parabola lies on its axis of symmetry, obtaining an equation for this axis will give us the *x*-coordinate of the vertex. The *y*-coordinate can then easily be obtained by substitution in the equation for the parabola.

When graphing a quadratic equation in two variables, it is desirable first to select components for the ordered pairs that ensure that the more significant parts of the graph are displayed. For a parabola, these parts include the intercepts, if they exist, and the maximum or minimum point on the curve.

Example

Graph $y = x^2 - 3x - 4$.

Solution

Setting $x = 0$, we obtain the *y*-intercept, -4. Setting $y = 0$, we have

$$0 = x^2 - 3x - 4 = (x - 4)(x + 1),$$

and the *x*-intercepts are 4 and -1. The *x*-coordinate of the minimum point is

$$x = -\frac{b}{2a} = -\frac{-3}{2(1)} = \frac{3}{2}.$$

Substituting $3/2$ for x in $y = x^2 - 3x - 4$, we obtain

$$y = \left(\frac{3}{2}\right)^2 - 3\left(\frac{3}{2}\right) - 4 = \frac{9}{4} - \frac{9}{2} - 4 = -\frac{25}{4}.$$

Graphing the intercepts and the coordinates of the minimum point and then sketching the curve produce the graph shown.

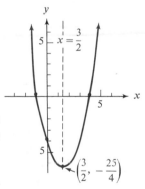

Solutions, zeros, and x-intercepts

Consider the graph of the function

$$S = \{(x, f(x)) \mid f(x) = ax^2 + bx + c\}, \tag{5}$$

for $a \neq 0$, and the solution set of the equation

$$ax^2 + bx + c = 0. \tag{6}$$

Any value of x for which $f(x) = 0$ in (5) will be a solution of (6). Since any point on the *x*-axis has *y*-coordinate zero [that is, $f(x) = 0$], the *x*-intercepts of the graph of (5) are the real solutions of (6). Values of x for which $f(x) = 0$ are called the **zeros of the function.** Thus we have three different names for a single idea:

1. The *elements in R of the solution set* of the equation $ax^2 + bx + c = 0$. These are called the *solutions* or *roots* of the equation.

2. The *zeros in R of the function* defined by $f(x) = ax^2 + bx + c$.

3. The *x-intercepts* of the graph of the equation $f(x) = ax^2 + bx + c$.

Inverse of a quadratic function

In Section 2.5 we observed that the graph of the inverse of a linear function is also a straight line, and hence the inverse is also a linear function if its graph is not vertical. Because every function is a relation, every function has an inverse, but the inverse is not always a function. For example, the graphs of

$$F = \{(x, y) \mid y = x^2\}$$

and its inverse,

$$F^{-1} = \{(x, y) \mid x = y^2\} = \{(x, y) \mid y = \pm\sqrt{x}\},$$

are shown in Figure 3.4. Since F^{-1} associates two different y's with each x for all but one value in its domain, this inverse is not a function.

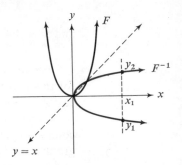

Figure 3.4

Quadratic inequalities

Relations of the form

$$\{(x, y) \mid y < ax^2 + bx + c\} \quad \text{or} \quad \{(x, y) \mid y > ax^2 + bx + c\}$$

can be graphed in the same manner in which we graphed relations defined by *linear* inequalities in two variables in Section 2.2. We first graph the relation defined by the equation having the same members as the defining inequality and then shade an appropriate region as required. For instance, to graph

$$\{(x, y) \mid y < x^2 + 2\}, \tag{7}$$

we first graph

$$\{(x, y) \mid y = x^2 + 2\}. \tag{8}$$

Then, as in the graphing of linear inequalities, the coordinates of a selected point not on the graph of the associated equation can be used to determine which of the two resulting regions of the plane is the graph of the inequality. Upon substitution of 0 for x and 0 for y in $y < x^2 + 2$, we have

$$(0) < (0)^2 + 2,$$

a true statement. Hence the region containing the origin is shaded. Since the graph of (8) is not part of the graph of (7), a broken curve is used (Figure 3.5).

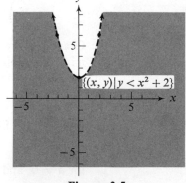

Figure 3.5

Exercise 3.1

Graph. (Obtain analytically the intercepts and the maximum or minimum point, and then sketch the curve.)

Example $\{(x, y) \mid y = x^2 - 7x + 6\}$

Solution

For $x = 0$, the y-intercept is clearly 6. Since the solutions of

$$x^2 - 7x + 6 = (x - 1)(x - 6) = 0$$

are 1 and 6, these are the x-intercepts. The axis of symmetry has equation

$$x = -\frac{b}{2a} = \frac{7}{2}.$$

Substituting $7/2$ for x in $y = x^2 - 7x + 6$, we obtain

$$y = \left(\frac{7}{2}\right)^2 - 7\left(\frac{7}{2}\right) + 6 = -\frac{25}{4}.$$

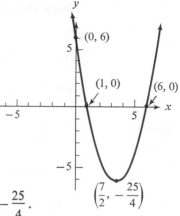

Hence the minimum point is $(7/2, -25/4)$. Using these points, we can sketch the graph as shown.

1. $\{(x, y) \mid y = x^2 - 5x + 4\}$
2. $\{(x, g(x)) \mid g(x) = x^2 - 3x + 2\}$
3. $\{(x, y) \mid y = x^2 - 6x - 7\}$
4. $\{(x, y) \mid y = x^2 + x - 12\}$
5. $\{(x, f(x)) \mid f(x) = -x^2 + 5x - 4\}$
6. $\{(x, f(x)) \mid f(x) = -x^2 - 8x + 9\}$
7. $\left\{(x, g(x)) \mid g(x) = \frac{1}{2}x^2 + 2\right\}$
8. $\left\{(x, y) \mid y = -\frac{1}{2}x^2 - 2x\right\}$

9. Find two numbers having sum 8 and product as great as possible.
10. Find the maximum possible area of a rectangle with perimeter 100 inches.
11. On a single set of axes, sketch the family of four curves that are the graphs of

$$y = x^2 + k \quad (k = -2, 0, 2, 4).$$

What effect does varying k have on the graph?

12. On a single set of axes, sketch the family of six curves that are the graphs of

$$y = kx^2 \quad \left(k = \frac{1}{2}, 1, 2, -\frac{1}{2}, -1, -2\right).$$

What effect does varying k have on the graph?

13. Graph the relation $\{(x, y) \mid x = y^2\}$. *Hint:* First assign values to y and obtain the corresponding values for x.
    a. What kind of curve is the graph?
    b. Is the given relation a function? Why or why not?
    c. Does the graph of this relation have a maximum or minimum point?

14. Graph the relation $\{(x, y) \mid x = y^2 - 2y\}$.
    a. What kind of curve is the graph?
    b. Is the given relation a function? Why or why not?
    c. Does the graph of this relation have a maximum or minimum point?

Graph the given relation. (The graph is a parabola which has an axis of symmetry parallel to the x-axis.)

15. $\{(x, y) \mid x = y^2 - 4\}$
16. $\{(x, y) \mid x = y^2 - 2y - 3\}$
17. $\{(x, y) \mid x = y^2 - 4y + 4\}$
18. $\{(x, y) \mid x = 2y^2 + 3y - 2\}$

Graph the relations r and r^{-1}, using the same set of axes.

19. $r = \{(x, y) \mid y = x^2 - 4\}$ **20.** $r = \{(x, y) \mid y = x^2 + 4\}$
21. $r = \{(x, y) \mid y = x^2 - 4x + 4\}$ **22.** $r = \{(x, y) \mid y = x^2 - 2x - 3\}$

Graph the given relation.

Example $\{(x, y) \mid y \geq x^2 + 2x\}$

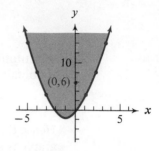

Solution   First graph $\{(x, y) \mid y = x^2 + 2x\}$.

Substitute the coordinates of any point not on the
curve for x and y in $y \geq x^2 + 2x$. We shall use $(0, 6)$.
Since

$$6 \geq (0)^2 + 2(0)$$

is true, the point $(0, 6)$ is in the graph of the inequality and the region above the
graph of $y = x^2 + 2x$ is shaded. The curve itself is included in the graph.

23. $\{(x, y) \mid y > x^2\}$ **24.** $\{(x, y) \mid y < x^2\}$
25. $\{(x, y) \mid y \geq x^2 + 3\}$ **26.** $\{(x, y) \mid y \leq x^2 + 3\}$
27. $\{(x, y) \mid y < 3x^2 + 2x\}$ **28.** $\{(x, y) \mid y > 3x^2 + 2x\}$
29. $\{(x, y) \mid y \leq x^2 + 3x + 2\}$ **30.** $\{(x, y) \mid y \geq x^2 + 3x + 2\}$

31. Graph the set of points whose coordinates satisfy both

$$y \leq 4 - x^2 \quad \text{and} \quad y \geq x^2 - 4.$$

32. Graph the set of points whose coordinates satisfy both

$$y \leq 1 - x^2 \quad \text{and} \quad y \geq -1.$$

3.2   *Central Conics*

In addition to equations typified by $y = ax^2 + bx + c$ or $x = ay^2 + by + c$
there are three other types of second-degree equations in two variables with graphs
that are of particular interest. We shall discuss each of them separately.

First, consider the relation

$$\{(x, y) \mid x^2 + y^2 = 25\}. \tag{1}$$

Solving the defining equation explicitly for y, we have

$$y = \pm\sqrt{25 - x^2}. \tag{2}$$

Assigning values to x, we find the following ordered pairs in the relation:

$$(-5, 0), \quad (-4, 3), \quad (-3, 4), \quad (0, 5), \quad (3, 4), \quad (4, 3), \quad (5, 0),$$
$$(-4, -3), \quad (-3, -4), \quad (0, -5), \quad (3, -4), \quad (4, -3).$$

Plotting these points on the plane, we have the graph shown in Figure 3.6-a. Connecting these points with a smooth curve, we obtain the graph in Figure 3.6-b, a **circle** with radius 5 and center at the origin.

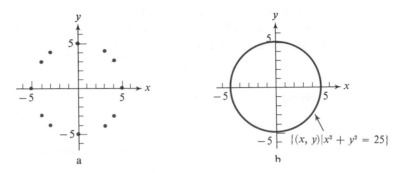

a   b

Figure 3.6

Equation of a circle

Since the number 25 in the right-hand member of the defining equation of (1) clearly determines the length of the radius of the circle, we can generalize and observe that any relation defined by an equation of the form

$$x^2 + y^2 = r^2, \quad r > 0,$$

has as its graph a circle with radius r and with center at the origin. (See Exercise 30 for a more general approach.)

Note that in the preceding example it is not necessary to assign any values to x for which $|x| > 5$, because then y^2 would have to be negative. Hence, the domain of the relation is $\{x \mid -5 \le x \le 5\}$. Since, except for -5 and $+5$, each permissible value of x is associated with two values for y—one positive and one negative—the relation (1) is not a function.

The second kind of quadratic (second-degree) relation of interest, which actually has the first as a limiting case, is that typified by

$$\{(x, y) \mid 4x^2 + 9y^2 = 36\}. \tag{3}$$

We obtain members of (3) by first solving the defining equation explicitly for y,

$$y = \pm \frac{2}{3}\sqrt{9 - x^2},$$

and then assigning values to x satisfying $-3 \le x \le 3$. (Why these values only?) We obtain, for example,

$$(-3, 0), \quad \left(-2, \frac{2}{3}\sqrt{5}\right), \quad \left(-1, \frac{4}{3}\sqrt{2}\right), \quad (0, 2), \quad \left(1, \frac{4}{3}\sqrt{2}\right), \quad \left(2, \frac{2}{3}\sqrt{5}\right), \quad (3, 0),$$

$$\left(-2, -\frac{2}{3}\sqrt{5}\right), \quad \left(-1, -\frac{4}{3}\sqrt{2}\right), \quad (0, -2), \quad \left(1, -\frac{4}{3}\sqrt{2}\right), \quad \left(2, -\frac{2}{3}\sqrt{5}\right).$$

Equation of an ellipse

Decimal approximations for each irrational number can be obtained from the table on page 283 in the Appendix. Then, locating the corresponding points on the plane and connecting them with a smooth curve, we have the graph shown in Figure 3.7. This curve is called an **ellipse.** In general, the graph of the relation

$$\{(x, y) \mid Ax^2 + By^2 = C; \ A, B, C > 0; \ A \neq B\}$$

is an ellipse with center at the origin.

The third kind of quadratic relation with which we are concerned is typified by

$$\{(x, y) \mid x^2 - y^2 = 9\}.$$

Solving the defining equation for y, we obtain

$$y = \pm\sqrt{x^2 - 9},$$

which has as a part of its solution set the ordered pairs

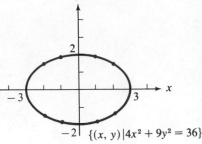

$\{(x, y) \mid 4x^2 + 9y^2 = 36\}$

Figure 3.7

$$(-5, 4), \quad (-4, \sqrt{7}), \quad (-3, 0), \quad (3, 0), \quad (4, \sqrt{7}), \quad (5, 4),$$
$$(-5, -4), \quad (-4, -\sqrt{7}), \quad (4, -\sqrt{7}), \quad (5, -4).$$

Equation of a hyperbola

Plotting the corresponding points and connecting them in a natural way, we obtain the curve shown in Figure 3.8. This curve is called a **hyperbola.** In general, the relation

$$\{(x, y) \mid Ax^2 - By^2 = C; \ A, B, C > 0\}$$

graphs into a hyperbola with center at the origin; similarly, the relation

$$\{(x, y) \mid By^2 - Ax^2 = C; \ A, B, C > 0\}$$

graphs into a hyperbola with center at the origin.

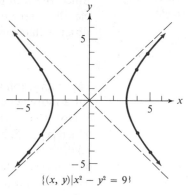

$\{(x, y) \mid x^2 - y^2 = 9\}$

Figure 3.8

Asymptotes of a hyperbola

The dashed lines shown in Figure 3.8 are called **asymptotes** of the graph. They comprise the graph of the equation $x^2 - y^2 = 0$. While we shall not discuss the notion in detail here, it is true that, in general, the graph of $Ax^2 - By^2 = C$, $A, B, C > 0$, "approaches" the two straight lines in the graph of $Ax^2 - By^2 = 0$ for large x. (See Exercises 25 and 35.)

Conic sections

The graphs of the foregoing relations, together with the parabola of the preceding section, are called **conic sections,** or **conics,** because such curves result from the intersection of a plane and a right circular cone, as shown in Figure 3.9.

The circle, ellipse, and hyperbola that we discussed above, which are the graphs of equations of the form

$$Ax^2 + By^2 = C \quad (A^2 + B^2 \neq 0),$$

are symmetric *with respect to the origin.* That is, if the point with coordinates (x_1, y_1) is on the graph, then so is the point with coordinates $(-x_1, -y_1)$. Since each graph is symmetric to a point, these conics are called **central conics.**

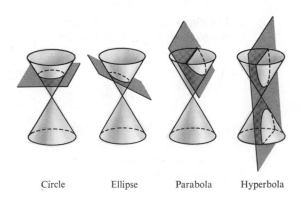

Circle   Ellipse   Parabola   Hyperbola

Figure 3.9

Use of the form of an equation in graphing

You should make use of the form of the defining equation as an aid in graphing quadratic relations. The ideas developed in this and the preceding section may be summarized as follows:

1. A quadratic equation of the form

    $$y = ax^2 + bx + c, \quad a \neq 0,$$

    has a graph that is a **parabola,** opening upward if $a > 0$ and downward if $a < 0$.

    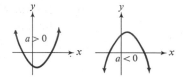

    Similarly, an equation of the form

    $$x = ay^2 + by + c, \quad a \neq 0,$$

    has a graph that is a parabola, opening to the right if $a > 0$ and to the left if $a < 0$.

    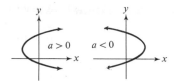

2. A quadratic equation of the form

    $$Ax^2 + By^2 = C, \quad A^2 + B^2 \neq 0, \qquad (4)$$

    has a graph that is

    (a) a **circle** if $A = B$ and A, B, and C have like signs;

    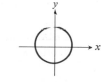

    (b) an **ellipse** if $A \neq B$ and A, B, and C have like signs;

    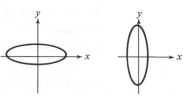

(c) a **hyperbola** if A and B are opposite in sign and $C \neq 0$;

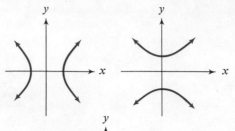

(d) **two distinct lines** through the origin if A and B are opposite in sign and $C = 0$ (see Exercise 25);

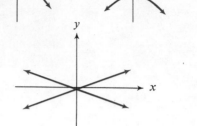

(e) **two distinct parallel lines** if one of A and $B = 0$ and the other has the same sign as C (see Exercise 26);

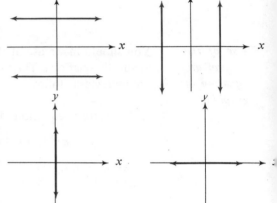

(f) **two coincident parallel lines** (one line) through the origin if one of A and $B = 0$ and also $C = 0$ (see Exercise 27);

(g) a **point** if A and B are both > 0 or both < 0 and $C = 0$ (see Exercise 28);

(h) the **null set**, $\emptyset$, if A and B are both ≥ 0 and $C < 0$, or if A and B are both ≤ 0 and $C > 0$ (see Exercise 29).

Sketch of the graph of a conic section

After you recognize the general form of the curve, the graph of a few points should suffice to sketch the complete graph. The intercepts, for instance, are always easy to locate. Consider the relation

$$\{(x, y) \mid x^2 + 4y^2 = 8\}. \tag{5}$$

By comparing the defining equation with (b) of Equation (4), we note immediately that its graph is an ellipse. If $y = 0$, then $x = \pm\sqrt{8}$; and if $x = 0$, then $y = \pm\sqrt{2}$. We can accordingly sketch the graph of (5) as in Figure 3.10.

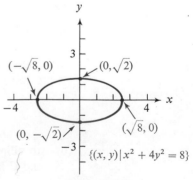

Figure 3.10

As another example, consider the relation

$$\{(x, y) \mid x^2 - y^2 = 3\}. \tag{6}$$

By comparing the defining equation with (c) of (4), we see that its graph is a hyperbola. If $y = 0$, then $x = \pm\sqrt{3}$; and if $x = 0$, then y^2 would have to be negative, an impossibility in the field of real numbers. Thus the graph does not cross the y-axis. By assigning a few other arbitrary values to one of the variables, say x, e.g., $(4,\ \ )$ and $(-4,\ \ )$, we can find additional ordered pairs

$$(4, \sqrt{13}),\ (4, -\sqrt{13}),$$
$$(-4, \sqrt{13}),\ (-4, -\sqrt{13})$$

which satisfy (6). The graph can then be sketched as shown in Figure 3.11. Observe that the asymptotes, which are the graph of $x^2 - y^2 = 0$, are given by $y = x$ and $y = -x$. They can be helpful in sketching the hyperbola.

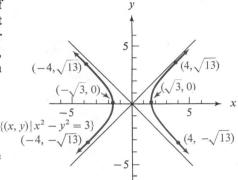

Figure 3.11

Exercise 3.2

a. *Rewrite the defining equation equivalently with y as the left-hand member.*
b. *State the domain of the relation.*

Example $\{(x, y) \mid 4x^2 + y^2 = 36\}$

Solution   **a.** $y^2 = 36 - 4x^2$
              $y = \pm2\sqrt{9 - x^2}$

              b.   The domain is $\{x \mid -3 \leq x \leq 3\}$.

1. $\{(x, y) \mid x^2 + y^2 = 4\}$
2. $\{(x, y) \mid x^2 + y^2 = 9\}$
3. $\{(x, y) \mid 9x^2 + y^2 = 36\}$
4. $\{(x, y) \mid 4x^2 + y^2 = 4\}$
5. $\{(x, y) \mid x^2 + 4y^2 = 16\}$
6. $\{(x, y) \mid x^2 + 9y^2 = 4\}$
7. $\{(x, y) \mid 2x^2 + 3y^2 = 24\}$
8. $\{(x, y) \mid 4x^2 + 3y^2 = 12\}$
9. $\{(x, y) \mid x^2 - y^2 = 1\}$
10. $\{(x, y) \mid 4x^2 - y^2 = 1\}$
11. $\{(x, y) \mid y^2 - x^2 = 9\}$
12. $\{(x, y) \mid 4y^2 - 9x^2 = 36\}$

Name and sketch the graph of the given relation. Specify the intercepts, and if the graph is a hyperbola, give equations of the asymptotes.

Example $\{(x, y) \mid 4y^2 - 36 = -9x^2\}$

Solution   Rewrite the defining equation in standard form.

$$9x^2 + 4y^2 = 36$$

By inspection, the graph is an ellipse; x-intercepts are -2 and 2; y-intercepts are -3 and 3.

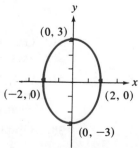

Sketch the graph.

13. $\{(x, y) \mid x^2 + y^2 = 49\}$ **14.** $\{(x, y) \mid x^2 + y^2 = 64\}$

15. $\{(x, y) \mid 3x^2 + 25y^2 = 100\}$ **16.** $\{(x, y) \mid x^2 + 2y^2 = 8\}$

17. $\{(x, y) \mid 4x^2 = 4y^2\}$ **18.** $\{(x, y) \mid x^2 - 9y^2 = 0\}$

19. $\{(x, y) \mid x^2 = 9 + y^2\}$ **20.** $\{(x, y) \mid x^2 = 2y^2 + 8\}$

21. $\{(x, y) \mid 4x^2 + 4y^2 = 1\}$ **22.** $\{(x, y) \mid 9x^2 + 9y^2 = 2\}$

23. $\{(x, y) \mid 3x^2 - 12 = -4y^2\}$ **24.** $\{(x, y) \mid 12 - 3y^2 = 4x^2\}$

25. Graph $\{(x, y) \mid 4x^2 - y^2 = 0\}$. Generalize from the result and discuss the graph of any relation of the form $\{(x, y) \mid Ax^2 - By^2 = 0, \quad A, B > 0\}$.

26. Graph $\{(x, y) \mid x^2 = 4\}$. Generalize from the result and discuss the graph of any relation of the form $\{(x, y) \mid Ax^2 = C, \quad A, C > 0\}$.

27. Graph $\{(x, y) \mid x^2 = 0\}$. Generalize from the result and discuss the graph of any relation of the form $\{(x, y) \mid Ax^2 = 0, A \neq 0\}$.

28. Graph $\{(x, y) \mid 4x^2 + y^2 = 0\}$. Generalize from the result and discuss the graph of any relation of the form $\{(x, y) \mid Ax^2 + By^2 = 0, \quad A, B > 0\}$.

29. Explain why the graph of $\{(x, y) \mid x^2 + y^2 = -1\}$ is the null set. Generalize from the result and discuss the graph of any relation of the form

$$\{(x, y) \mid Ax^2 + By^2 = C, \quad A^2 + B^2 \neq 0, \quad A, B \geq 0, \quad C < 0\}.$$

30. Use the distance formula to show that the graph of the relation given by $\{(x, y) \mid x^2 + y^2 = r^2, r \geq 0\}$ is the set of all points located a distance r from the origin.

31. Show that the graph of $\{(x, y) \mid (x - 2)^2 + (y + 3)^2 = 16\}$ is a circle with center at $(2, -3)$ and radius 4. *Hint:* Use distance formula.

32. Show that the graph of $\{(x, y) \mid (x - h)^2 + (y - k)^2 = r^2, r > 0\}$ is a circle with center at (h, k) and radius r.

33. Show that the general form of an equation for an ellipse, $Ax^2 + By^2 = C$, can be written in the intercept form $\dfrac{x^2}{a^2} + \dfrac{y^2}{b^2} = 1$, where a and b are x- and y-intercepts, respectively.

34. Show that the general form of an equation for a hyperbola, $Ax^2 - By^2 = C$, $C > 0$, can be written in the form $\dfrac{x^2}{a^2} - \dfrac{y^2}{b^2} = 1$, where a is the x-intercept.

35. By solving $Ax^2 - By^2 = C$ ($A, B, C > 0$) for y, obtain the expression

$$y = \pm \sqrt{\frac{A}{B}} \, x \left(\sqrt{1 - \frac{C}{Ax^2}} \right),$$

and argue that the graph of $Ax^2 - By^2 = C$ approaches the straight-line graphs of $y = \pm \sqrt{A/B} \, x$ as $|x|$ increases.

36. Graph $\{(x, y) \mid x^2 + y^2 = 25\}$ and $\{(x, y) \mid 4x^2 + y^2 = 36\}$ on the same set of axes. What is the significance of the coordinates of the points of intersection?

37. Graph the relation $\{(x, y) \mid x^2 + y^2 \leq 25\}$ by observing that the graph of the relation $\{(x, y) \mid x^2 + y^2 = a^2\}$ is a circle of radius a and then examining what $a \leq 5$ implies.

38. Graph the relation $\{(x, y) \mid 4x^2 + 9y^2 \leq 36\}$.

3.3   *Relations Whose Graphs Are Given*

Any set of points in the plane can be called a graph or a **locus** (plural **loci**). The words *graph* and *locus* are ordinarily used, however, to refer to the set of points meeting one or more predesignated conditions. For example, a circle can be called the set of points or the locus of points in the plane located a given distance r (the radius of the circle) from a given point (the center of the circle). A fundamental problem is that of finding equations (or inequalities) for given graphs, or loci.

Example   Find an equation of the circle having center at $P_1(3, -2)$ and radius 7.

Solution   It is best first to make a sketch showing a representative point (x, y) in the locus. By the distance formula, the distance d from P_1 to any point in the circle is

$$d = \sqrt{(x - 3)^2 + (y + 2)^2}.$$

Since this distance is given to be 7, we have

$$\sqrt{(x - 3)^2 + (y + 2)^2} = 7.$$

Squaring both members produces

$$(x - 3)^2 + (y + 2)^2 = 49,$$

$$x^2 - 6x + 9 + y^2 + 4y + 4 = 49,$$

or, finally,

$$x^2 + y^2 - 6x + 4y - 36 = 0. \tag{1}$$

Conversely, since the steps are reversible, any point whose coordinates satisfy (1) is on the circle. Hence (1) is an equation of the circle.

In a similar way, on replacing the sign $=$ with $\leq$ in the example above, you could show that the circular disc with center at $P_1(3, -2)$ and radius 7, that is, the set of points in the plane at distance not more than 7 from P_1, is the graph of the relation

$$x^2 + y^2 - 6x + 4y - 36 \leq 0.$$

Example   Find an equation of the locus of all points that are twice as far from the point $(3, -2)$ as they are from $(-2, 1)$.

Solution   Make a sketch. By the distance formula, any point $P(x, y)$ in the plane located at a distance d_1 from $(-2, 1)$ has coordinates (x, y) that satisfy

$$d_1 = \sqrt{(x + 2)^2 + (y - 1)^2},$$

and at a distance d_2 from $(3, -2)$ is given by

$$d_2 = \sqrt{(x - 3)^2 + (y + 2)^2}.$$

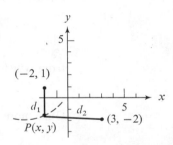

(continued overleaf)

In order for $P(x, y)$ to lie in the described locus, it is necessary and sufficient that

$$d_2 = 2d_1,$$

$$\sqrt{(x-3)^2 + (y+2)^2} = 2\sqrt{(x+2)^2 + (y-1)^2}.$$

Then, equating squares of both members, we have

$$(x-3)^2 + (y+2)^2 = 4[(x+2)^2 + (y-1)^2],$$

$$x^2 - 6x + 9 + y^2 + 4y + 4 = 4x^2 + 16x + 16 + 4y^2 - 8y + 4,$$

from which we obtain the simplified equation

$$3x^2 + 3y^2 + 22x - 12y + 7 = 0. \tag{2}$$

Again the steps are reversible, so that (2) is an equation of the given locus.

Loci as paths   Another way of considering a locus intuitively is to think of a path as having been traced out in the plane by a moving point (or particle).

Example   Find an equation of the locus of a point that moves in the plane so that it is always at the same distance from the line with equation $x = 3$ as it is from the point $(8, 0)$.

Solution   Make a sketch. The x-coordinate of P_1 is 3.
Since $\overline{P_1 P}$ is parallel to the x-axis, the distance d_1 from $P_1(3, y)$ to $P(x, y)$ is given by

$$d_1 = |x - 3|,$$

while d_2, the distance from $P_2(8, 0)$ to $P(x, y)$, is given by

$$d_2 = \sqrt{(x-8)^2 + (y-0)^2}.$$

The condition imposed on $P(x, y)$ is that

$$d_1 = d_2.$$

Hence, the condition can be expressed as

$$|x - 3| = \sqrt{(x-8)^2 + (y-0)^2}.$$

Squaring each member, we have

$$(x-3)^2 = (x-8)^2 + y^2,$$

$$x^2 - 6x + 9 = x^2 - 16x + 64 + y^2,$$

from which we obtain the simplified equation

$$y^2 - 10x + 55 = 0. \tag{3}$$

Once more, the steps are reversible, so that (3) is an equation of the given locus.

Exercise 3.3

Find an equation of tne circle with the given center and radius.

1. $(1, 0)$; $r = 1$ 2. $(4, 2)$; $r = 3$ 3. $(-3, 1)$; $r = 2$
4. $(4, -5)$; $r = 8$ 5. (h, k); $r = r$ 6. $(-h, -k)$; $r = r$

7. Find an equation of the circle which has center at $(3, 5)$ and which contains the point $(6, 9)$.
8. Find an equation of the circle which has center at $(-2, 3)$ and which contains the point $(3, -9)$.

Find an equation of the locus of a point that moves in the plane in such a way that it meets the given condition.

9. It is equidistant from $(3, 2)$ and $(5, 0)$.
10. It is equidistant from $(-3, -4)$ and $(1, 2)$.
11. Its distance from $(1, -4)$ is twice its distance from $(6, -1)$.
12. Its distance from $(5, 3)$ is three times its distance from $(5, 9)$.
13. Its distance from the line with equation $x = -2$ is equal to its distance from $(2, 0)$.
14. Its distance from the line with equation $y = 4$ is equal to its distance from the origin.
15. It is equidistant from the x-axis and the point $(-3, -2)$.
16. It is equidistant from the y-axis and the point $(4, 3)$.
17. The sum of its distances from $(2, 0)$ and $(-2, 0)$ is 8.
18. The sum of its distances from $(0, 4)$ and $(0, -4)$ is 12.
19. The difference of its distances from $(2, 0)$ and $(-2, 0)$ is 2.
20. The difference of its distances from $(0, 4)$ and $(0, -4)$ is 6.
21. Its distance from $(0, p)$ equals its distance from the x-axis.
22. Its distance from $(p, 0)$ equals its distance from the y-axis.
23. Find an equation specifying the set of points at distance greater than or equal to 3 from the point $(-4, 1)$.
24. Find a continued inequality (of the form $a < b < c$, meaning $a < b$ and $b < c$) specifying the set of points at distance from $(3, -4)$ that is greater than 5 but less than 7.

3.4   Properties of the Parabola

In Section 3.1, we examined equations of the form $y = ax^2 + bx + c$, whose graphs are conic sections. Each such graph can be categorized as a locus.

Definition 3.1   *A **parabola** is the locus of a point in the plane whose distances from a fixed line (called the **directrix**) and a fixed point (called the **focus**) are equal.*

Directrix parallel to x-axis

Since a locus in the plane is independent of the location of the coordinate axes, we can establish the axes in the way we find most convenient relative to the fixed line L and fixed point F. Figure 3.12 shows one convenient orientation, in which the origin is the midpoint of the perpendicular segment from the focus to the directrix; this midpoint is called the **vertex** of the parabola, and the line through the focus perpendicular to the directrix is called the **axis** of the parabola.

The directrix with equation $y = -p/2$ was chosen so that the focus has y-coordinate $p/2$, where $|p|$ is the distance between the focus and the directrix. This choice produces a simple form for an equation of the locus. From Definition 3.1 it follows that

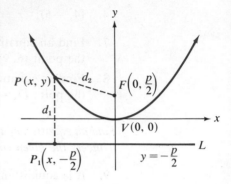

Figure 3.12

$$d_1 = d_2,$$

$$\left| y + \frac{p}{2} \right| = \sqrt{(x - 0)^2 + \left(y - \frac{p}{2} \right)^2}.$$

Squaring both members and simplifying the result produces

$$x^2 = 2py. \tag{1}$$

Because the steps in the foregoing process are reversible, (1) is an equation for the parabola. It is the standard form for the equation of a parabola with focus on the y-axis and vertex at the origin.

Example

Find the standard equation of the parabola with directrix the line with equation $y = -2$ and focus $(0, 2)$.

Solution

Make a sketch. By inspection,

$$p = 4.$$

Then Equation (1) yields

$$x^2 = 2(4)y,$$

$$x^2 = 8y.$$

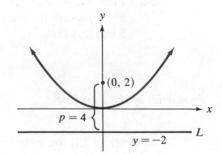

Directrix parallel to y-axis

By using Figure 3.13, we can show that for a parabola with vertex at the origin and with directrix parallel to the y-axis, an equation in standard form is

$$y^2 = 2px. \qquad (2)$$

By observing that the coefficient of the linear term in either of Equations (1) or (2) is $2p$, we can determine the coordinates of the focus and the equation for the directrix of any parabola with an equation of the given form.

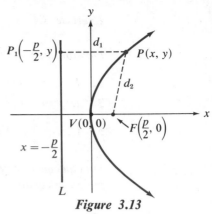

Figure 3.13

Example

Find an equation of the directrix, and the coordinates of the focus, of the parabola with equation $y^2 = 8x$. Sketch the parabola.

Solution

We first observe that the given equation is of the form (2), for a parabola with directrix parallel to the y-axis. Since $2p = 8$, we have $p = 4$ and $p/2 = 2$. Then the focus has coordinates $(2, 0)$, and an equation for the directrix is $x = -2$.

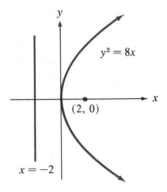

It is worthwhile observing that if $p < 0$, then the parabola with equation $x^2 = 2py$ opens downward, while the parabola with equation $y^2 = 2px$ opens to the left.

Obtaining equation from the definition

It is possible to obtain an equation for a parabola with directrix parallel to one of the coordinate axes and with vertex not at the origin, by appealing directly to Definition 3.1.

Example

Find an equation for the parabola whose directrix has the equation $x = 5$ and whose focus is the point $F(3, 0)$.

Solution

Make a sketch. By inspection,

$$d_1 = |x - 5|, \quad \text{and} \quad d_2 = \sqrt{(x - 3)^2 + (y - 0)^2}.$$

Then by definition,

$$d_1 = d_2,$$

$$|x - 5| = \sqrt{(x - 3)^2 + (y - 0)^2},$$

from which

$$y^2 = -4x + 16.$$

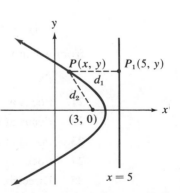

Exercise 3.4

Find an equation of the parabola with given directrix L and focus F. Sketch the curve.

1. $L: x = -4;$ $F(4, 0)$ 2. $L: x = 3;$ $F(-3, 0)$
3. $L: y = -1;$ $F(0, 1)$ 4. $L: y = 8;$ $F(0, -8)$
5. $L: x = \dfrac{3}{4};$ $F\left(-\dfrac{3}{4}, 0\right)$ 6. $L: y = -\dfrac{7}{8};$ $F\left(0, \dfrac{7}{8}\right)$

Find an equation of the directrix and the coordinates of the focus of the parabola with the given equation. Sketch the curve.

7. $y^2 = 6x$ 8. $y^2 = -10x$ 9. $x^2 = -5y$
10. $x^2 = 7y$ 11. $4y^2 = -3x$ 12. $5x^2 = 2y$

Find an equation of the parabola, with vertex at the origin, which has the given additional properties.

13. Focus $(5, 0)$ 14. Focus $\left(0, -\dfrac{2}{3}\right)$

15. Directrix $y = -\dfrac{4}{3}$ 16. Directrix $y = 6$
17. Axis the x-axis and containing $(3, -2)$
18. Axis the x-axis and containing $(-4, 2)$
19. Axis the y-axis and containing $(-1, 4)$
20. Axis the y-axis and containing $(4, -2)$

Use Definition 3.1 to find an equation of the parabola with features as given.

Example   Focus at $(0, 4)$ and directrix $y = -3$

Solution   Make a sketch. By inspection,

$$d_1 = |y - (-3)| = |y + 3|,$$

and

$$d_2 = \sqrt{(x - 0)^2 + (y - 4)^2}.$$

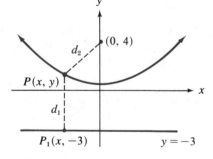

Then by definition, $d_1 = d_2$. Hence,

$$|y + 3| = \sqrt{(x - 0)^2 + (y - 4)^2},$$

from which we obtain

$$x^2 = 14y - 7.$$

21. Focus at $(3, -2)$ and directrix $x = 0$
22. Focus at $(4, 4)$ and directrix $x = 0$
23. Focus at $(-4, 2)$ and vertex at $(-2, 2)$
24. Focus at $(-3, -6)$ and vertex at $(-3, -4)$
25. Directrix $x = 5$ and vertex at $(3, 2)$
26. Directrix $y = -2$ and vertex at $(1, 2)$

3.5   Properties of the Ellipse

A defining condition that a locus be an ellipse is set forth as follows:

Definition 3.2   *An **ellipse** is the locus of a point in the plane the sum of whose distances from two fixed points (called the **foci**) is a constant.*

Foci on the x-axis

As in Section 3.4, we select the location of the axis as conveniently as possible. In this case (Figure 3.14), we choose the origin to be the midpoint of the segment whose endpoints are the fixed points, and let the x-axis contain this segment. From Definition 3.2, we wish the sum $d_1 + d_2$ to be a constant. For reasons that will be evident later, we select $2a$ to represent this constant. Then using the distance formula, we have

$$\sqrt{(x - c)^2 + (y - 0)^2} + \sqrt{(x + c)^2 + (y - 0)^2} = 2a.$$

Rewriting this equation equivalently as

$$\sqrt{(x + c)^2 + y^2} = 2a - \sqrt{(x - c)^2 + y^2},$$

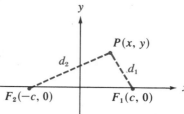

Figure 3.14

we can square each member and simplify to obtain

$$a\sqrt{(x - c)^2 + y^2} = a^2 - cx.$$

Again squaring and simplifying, we have

$$(a^2 - c^2)x^2 + a^2y^2 = a^2(a^2 - c^2). \tag{1}$$

Examining Figure 3.14 again, we note that $d_1 + d_2$ is greater than the distance between the foci, so that $2c < 2a$, or $c < a$. It follows that $a^2 - c^2 > 0$. If we now let

$$b^2 = a^2 - c^2, \tag{2}$$

with $b > 0$, Equation (1) can be written

$$b^2x^2 + a^2y^2 = a^2b^2.$$

Upon multiplication of each member by $1/a^2b^2$, this becomes

$$\frac{x^2}{a^2} + \frac{y^2}{b^2} = 1. \tag{3}$$

Because the steps are reversible, (3) is an equation of the ellipse. This equation is the standard form for the equation of an ellipse with foci on the x-axis and symmetrically located with respect to the origin. In this case, the origin is called the **center** of the ellipse.

The points at which the ellipse cuts the line containing the foci, in this case the points $(-a, 0)$ and $(a, 0)$ on the x-axis, are called the **vertices** of the ellipse (Figure 3.15). The segment with the vertices as endpoints is called the **major axis** of the ellipse, and the segment through the center, perpendicular to the major axis, and with endpoints on the

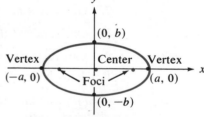

Figure 3.15

ellipse—in this case the segment with end-points $(0, -b)$ and $(0, b)$—is called the **minor axis.** (The *lines* on which the major and minor axes lie are also sometimes called axes of the ellipse.)

Foci on the y-axis

If we orient the axes so that the foci are on the y-axis, with coordinates $(0, c)$ and $(0, -c)$, the standard form of the equation for the ellipse becomes

$$\frac{x^2}{b^2} + \frac{y^2}{a^2} = 1, \tag{4}$$

with graph as shown in Figure 3.16. Notice that the ellipse with equation

$$\frac{x^2}{k_1^2} + \frac{y^2}{k_2^2} = 1$$

has vertices on the x-axis if $k_1^2 > k_2^2$, and on the y-axis if $k_1^2 < k_2^2$.

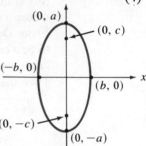

Figure 3.16

Example

Find an equation of the ellipse with center at the origin, one focus at $(-2, 0)$, and one vertex at $(-3, 0)$.

Solution

For the given focus, we have $c = 2$; and for the given vertex, $a = 3$. From Equation (2), it follows that

$$b^2 = a^2 - c^2 = 3^2 - 2^2 = 5,$$

from which $b = \sqrt{5}$. Substituting for a and b in Equation (3), we obtain

$$\frac{x^2}{9} + \frac{y^2}{5} = 1.$$

Exercise 3.5

Find the standard form of the equation of the ellipse having center at the origin and meeting the given conditions.

1. One focus at $(4, 0)$; one vertex at $(5, 0)$
2. One focus at $(0, -3)$; one vertex at $(0, -5)$
3. One focus at $(0, \sqrt{5})$; one vertex at $(0, -3)$
4. One focus at $(\sqrt{3}, 0)$; one vertex at $(-5, 0)$
5. One vertex at $(4, 0)$; length of minor axis 6
6. One vertex at $(-8, 0)$; length of minor axis 8
7. One vertex at $(5, 0)$; containing $(2, 3)$
8. One vertex at $(0, 4)$; containing $(-\sqrt{5}, 3)$
9. One focus at $(3, 0)$; containing $(2, \sqrt{2})$
10. Length of major axis 10; length of minor axis 6

Find the coordinates of the foci and vertices of the ellipse with the given equation, and sketch the curve.

Example

$25x^2 + 9y^2 = 225$

Solution   The equation can be written equivalently in standard form as

$$\frac{x^2}{9} + \frac{y^2}{25} = 1.$$

Since $9 < 25$, the foci of this ellipse are on the y-axis. Comparing this equation with Equation (4), we see that

$a^2 = 25,$ or $a = 5,$ and $b^2 = 9,$ or $b = 3.$

From (2), we obtain

$$9 = 25 - c^2, \quad \text{or} \quad c = 4.$$

Thus, the foci are the points $(0, 4)$ and $(0, -4)$, and the vertices are the points $(0, 5)$ and $(0, -5)$. The graph appears as shown.

11. $\dfrac{x^2}{9} + \dfrac{y^2}{4} = 1$ **12.** $\dfrac{x^2}{9} + \dfrac{y^2}{16} = 1$ **13.** $\dfrac{x^2}{4} + \dfrac{y^2}{36} = 1$ **14.** $\dfrac{x^2}{25} + y^2 = 1$

15. $25x^2 + 36y^2 = 900$ **16.** $16x^2 + 25y^2 = 400$

17. $10x^2 + 5y^2 = 50$ **18.** $2x^2 + 6y^2 = 24$

19. $4x^2 + y^2 = 9$ **20.** $x^2 + 16y^2 = 25$

Use the definition of an ellipse to find an equation of the ellipse which has the given foci, and for which the sum of the distances of each point on the curve to the foci is the given number.

21. $(-3, 4), (-1, 4);$ sum of distances 6

22. $(2, 3), (-4, 3);$ sum of distances 10

23. $(5, -5), (5, 3);$ sum of distances 12

24. $(1, 3), (1, 8);$ sum of distances 7

3.6   *Properties of the Hyperbola*

To characterize a hyperbola as a locus, we have the following definition:

Definition 3.3   *A **hyperbola** is the locus of a point in the plane, the difference of whose distances from two fixed points (called the **foci**) is a constant.*

Foci on the   Orienting the coordinate axes as in the case of the ellipse, we have Figure 3.17,
x-axis   on page 76. From Definition 3.3, we wish

$$|d_2 - d_1|$$

to be a constant. Since there are two possibilities,

$$d_2 > d_1 \quad \text{or} \quad d_2 < d_1,$$

we allow for this by writing

$$d_2 - d_1 = \pm 2a,$$

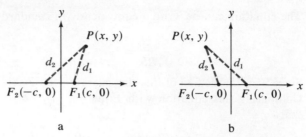

Figure 3.17

where $a > 0$. Then $2a$ applies for $x > 0$ as in Figure 3.17a, and $-2a$ applies for $x < 0$ as in Figure 3.17b. Using the distance formula, we find that

$$\sqrt{(x + c)^2 + (y - 0)^2} - \sqrt{(x - c)^2 + (y - 0)^2} = \pm 2a.$$

Rewriting this equation equivalently as

$$\sqrt{(x + c)^2 + y^2} = \pm 2a + \sqrt{(x - c)^2 + y^2},$$

we can square each member and simplify, to obtain

$$\pm a\sqrt{(x - c)^2 + y^2} = cx - a^2,$$

where the ambiguous sign $\pm$ is interpreted as above. Upon squaring each member of this latter equation and simplifying, we obtain

$$(a^2 - c^2)x^2 + a^2y^2 = a^2(a^2 - c^2). \qquad (1)$$

From Figure 3.18 with vertices at $(a, 0)$ and $(-a, 0)$, it is evident that $c > a$, so that $c^2 - a^2 > 0$. If we now set

$$c^2 - a^2 = b^2, \qquad (2)$$

with $b > 0$, Equation (1) can be written

$$-b^2x^2 + a^2y^2 = -a^2b^2,$$

which, upon multiplication by $-1/a^2b^2$, becomes

$$\frac{x^2}{a^2} - \frac{y^2}{b^2} = 1. \qquad (3)$$

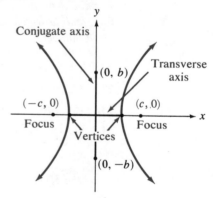

Figure 3.18

Because the steps are reversible, (3) is an equation for the hyperbola. This equation is the standard form for the equation of a hyperbola with foci on the x-axis and symmetrically located with respect to the origin. The origin is called the **center** of the hyperbola, while the points where the graph crosses the line containing the foci, in this case the points $(-a, 0)$ and $(a, 0)$ on the x-axis, are called the **vertices** of the hyperbola. The line segment with vertices as endpoints is called the **transverse axis** of the hyperbola, and, even though the points $(0, -b)$ and $(0, b)$ do not lie on the curve, the segment with these endpoints is called the **conjugate axis** of the hyperbola, as shown in Figure 3.18. (As in the case of the ellipse, the lines on which these segments lie are also sometimes called the axes of the hyperbola.)

Example   Find the standard equation of the hyperbola with foci at $(3, 0)$ and $(-3, 0)$ and with $(2, 0)$ as one vertex.

Solution   From the given foci, $c = 3$; and from the given vertex, $a = 2$. Since for a hyperbola $b^2 = c^2 - a^2$, we have

$$b^2 = 3^2 - 2^2 = 9 - 4 = 5.$$

Therefore, from Equation (3), the required equation is

$$\frac{x^2}{4} - \frac{y^2}{5} = 1.$$

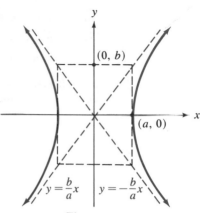

The *asymptotes* of the hyperbola are of help in sketching the curve. As suggested in Exercise 35, page 66, each hyperbola has two asymptotes, whose equations are $y = \pm(\frac{b}{a})x$. As shown in Figure 3.19, these asymptotes are extensions of the diagonals of a centrally oriented rectangle of dimensions $2a$ and $2b$. If the rectangle is drawn first, the asymptotes can be drawn as an aid in sketching the hyperbola.

Figure 3.19

Example   Sketch the graph of $\dfrac{x^2}{9} - \dfrac{y^2}{16} = 1.$

Solution   By inspection, $a = 3$ and $b = 4$. Sketching a 6-by-8 rectangle with center at the origin, we obtain the asymptotes and vertices and sketch the curve, as shown.

Foci on the   If we orient the coordinate axes so that the
y-axis   center of the hyperbola is the origin but the foci are located on the y-axis, the resulting equation takes the form

$$\frac{y^2}{a^2} - \frac{x^2}{b^2} = 1,$$

with $b^2 = c^2 - a^2$, as before.

Exercise 3.6

Find the standard form of the equation of the hyperbola with center at the origin and meeting the given conditions.

1. One focus at $(4, 0)$; one vertex at $(3, 0)$
2. One focus at $(-6, 0)$; one vertex at $(-4, 0)$
3. One focus at $(0, -2)$; one vertex at $(0, -1)$
4. One focus at $(0, 3)$; one vertex at $(0, 1)$

5. One focus at $(5, 0)$;   length of transverse axis 4
6. One focus at $(0, -8)$;   length of transverse axis 14
7. One focus at $(0, 6)$;   length of conjugate axis 10
8. One focus at $(-5, 0)$;   length of conjugate axis 8
9. One vertex at $(\sqrt{3}, 0)$;   containing $(3, 2/3)$
10. One vertex at $(0, -2)$;   containing $(3, \sqrt{13})$
11. Length of transverse axis 12;   length of conjugate axis 10
12. Length of transverse axis 8;   length of conjugate axis 10

Find: (a) the coordinates of the foci and vertices for the hyperbola whose equation is given, and (b) equations of the asymptotes. Sketch the curve.

13. $x^2 - y^2 = 1$ 14. $\dfrac{x^2}{4} - y^2 = 1$ 15. $x^2 - \dfrac{y^2}{9} = 1$

16. $\dfrac{x^2}{25} - \dfrac{y^2}{4} = 1$ 17. $y^2 - x^2 = 1$ 18. $\dfrac{y^2}{9} - x^2 = 1$

19. $\dfrac{y^2}{9} - \dfrac{x^2}{4} = 1$ 20. $y^2 - \dfrac{x^2}{36} = 1$ 21. $4x^2 - 16y^2 = 64$

22. $4x^2 - 9y^2 = 36$ 23. $36y^2 - 4x^2 = 144$ 24. $9y^2 - 25x^2 = 225$

25. $4x^2 - 4y^2 = 9$ 26. $9y^2 - 16x^2 = 25$

Use the definition of a hyperbola to find an equation for the hyperbola having the given foci and for which the difference of the distances from each point on the curve to the foci is as given.

27. Foci $(2, 2)$, $(2, -2)$;   difference of distances ± 2
28. Foci $(4, -2)$, $(-4, -2)$;   difference of distances ± 6
29. Foci $(-2, -5)$, $(1, -5)$;   difference of distances ± 2
30. Foci $(-1, 0)$, $(-1, 4)$;   difference of distances ± 1

Chapter Review

[3.1]   *Find the x-intercepts, the axis of symmetry, and the maximum or minimum point of the graph of the given function by analytic methods.*

1. $\{(x, f(x)) \mid f(x) = x^2 - x - 6\}$ 2. $\{(x, f(x)) \mid f(x) = -x^2 + 7x - 10\}$

3. Graph the function of Exercise 1.   4. Graph the function of Exercise 2.
5. Graph the relation $\{(x, y) \mid x = y^2 - 9\}$.
6. Graph the relation $\{(x, y) \mid x = y^2 - 5y + 4\}$.

Graph f and f^{-1} on the same set of axes.

7. $f = \{(x, y) \mid y = x^2 - 9\}$ 8. $f = \{(x, y) \mid y = x^2 - 4x - 5\}$

Graph the relation defined by the given inequality.

9. $\{(x, y) \mid y < x^2 - 9\}$ **10.** $\{(x, y) \mid y \geq x^2 + 6x + 5\}$

[3.2] *Name and graph the relation defined by the given equation.*

11. $x^2 - 16 = 4y^2$ **12.** $2y^2 = 8 - 2x^2$

13. $x^2 = 36 - 4y^2$ **14.** $x^2 - y - 9 = 0$

15. $x^2 - 9y^2 = 0$ **16.** $x^2 + 9y^2 = 0$

[3.3] *Find an equation of the circle with the given center and radius.*

17. $(2, 3)$; $r = 4$ **18.** $(-3, 4)$; $r = 2$

19. Find the locus of a point in the plane whose distance from $(2, -1)$ is twice its distance from $(0, 4)$.

20. Find the locus of a point in the plane, the sum of whose distances from $(-2, 6)$ and $(0, 4)$ is 8.

[3.4] *Find an equation of the parabola with given directrix L and given focus F.*

21. $L : y = 2$; $F(0, 4)$ **22.** $L: x = -3$; $F(6, 0)$

23. Find an equation of the directrix and the coordinates of the focus of the parabola with the equation $x^2 = -8y$.

24. Find an equation of the parabola which has vertex at the origin, has axis of symmetry the y-axis, and contains $(-2, 6)$.

25. Find an equation of the parabola with focus at $(2, -4)$ and directrix $y = 0$.

26. Find an equation of the parabola with directrix $y = 4$ and vertex at $(2, 5)$.

[3.5] *Find an equation of the ellipse having center at the origin and meeting the given conditions.*

27. Focus at $(\sqrt{5}, 0)$; vertex at $(3, 0)$

28. Vertex at $(0, -6)$; length of minor axis, 8

Find the coordinates of the foci and vertices of the ellipse with the given equation.

29. $\dfrac{x^2}{16} + \dfrac{y^2}{25} = 1$ **30.** $x^2 + 8y^2 = 32$

31. Find an equation of the ellipse which has foci at $(-6, 2)$ and $(-6, 10)$, and for which the sum of the distances of each point on the curve to the foci is 10.

32. Graph the equation of Exercise 31.

[3.6] *Find an equation of the hyperbola that meets the given conditions and has its center at the origin.*

33. Focus at $(-4, 0)$; vertex at $(-1, 0)$

34. Focus at $(0, 6)$; length of conjugate axis, 8

Find (a) the coordinates of the foci and vertices for the hyperbola whose equation is given, and (b) the equations of the asymptotes; then (c) sketch the curve.

35. $\dfrac{x^2}{9} - y^2 = 1$ **36.** $9y^2 - x^2 = 36$

4

Polynomials and Rational Functions

4.1 *Quotients of Polynomials*

Some results obtained by working with quotients of polynomials will be useful in our further work.

Long division You may recall that if the divisor of a quotient contains more than one term, the familiar **long-division algorithm,** involving successive subtractions, can be used to rewrite the quotient. For example, the computation

$$
\begin{array}{r}
x - 3 \\
x^2 + 2x - 1\ \overline{\smash{\big)}\ x^3 -\ \ x^2 - 7x + 3} \\
\underline{x^3 + 2x^2 -\ \ x\ \ \ \ \ } \\
-3x^2 - 6x + 3 \\
\underline{-3x^2 - 6x + 3} \\
0
\end{array}
$$

shows that, for $x^2 + 2x - 1 \neq 0$,

$$
\frac{x^3 - x^2 - 7x + 3}{x^2 + 2x - 1} = x - 3.
$$

It is most convenient to arrange the dividend and the divisor in descending powers of the variable before using the division algorithm and to leave an appropriate space for any missing terms (terms with coefficient 0) in the dividend.

When the divisor is not a factor of the dividend, the division process will produce a nonzero remainder. For example, from

$$
\begin{array}{r}
x^3 - 3x^2 + 10x - 28 \\
x + 3\ \overline{\smash{\big)}\ x^4\ \ \ \ \ \ \ \ \ + \ x^2 + \ 2x - \ 1} \\
\underline{x^4 + 3x^3\ \ \ \ \ \ \ \ \ \ \ \ \ \ \ \ \ \ \ } \\
-3x^3 + \ x^2 \\
\underline{-3x^3 - 9x^2\ \ \ \ \ \ \ } \\
10x^2 + \ 2x \\
\underline{10x^2 + 30x\ \ \ \ } \\
-28x - \ 1 \\
\underline{-28x - 84} \\
83 \quad \text{(remainder)}
\end{array}
$$

we see that

$$
\frac{x^4 + x^2 + 2x - 1}{x + 3} = x^3 - 3x^2 + 10x - 28 + \frac{83}{x + 3} \qquad (x \neq -3).
$$

Observe in the foregoing example that a space is left in the dividend for a term involving x^3, even though the dividend contains no such term.

Synthetic division

If the divisor is of the form $x + c$, the process of dividing one polynomial by another can be simplified by a process called **synthetic division.** Consider the foregoing example. If we omit writing the variables and write only the coefficients of the terms, and use zero for the coefficient of any missing power, we have

$$
\begin{array}{r}
1 - 3 + 10 - 28 \\
1 + 3 \overline{\smash{\big)}\, 1 + 0 + 1 \ + \ 2 \ - \ 1} \\
\underline{1 + 3} \\
-3 + (1) \\
\underline{-3 - \ 9} \\
10 + (2) \\
\underline{10 + 30} \\
-28 - (1) \\
\underline{-28 - 84} \\
83 \quad \text{(remainder)}.
\end{array}
$$

Now, observe that the numerals shown in color are repetitions of the numerals written immediately above and are also repetitions of the coefficients of the associated variable in the quotient; the numbers in parentheses are repetitions of the coefficients of the dividend. Therefore, the whole process can be written in compact form as

$$
\begin{array}{llrrrrl}
(1) & 3\,\rule[-1mm]{0.4pt}{4mm}\ 1 & 0 & 1 & 2 & -1 & \\
(2) & & 3 & -9 & 30 & -84 & \\
\cline{2-6}
(3) & 1 & -3 & 10 & -28 & 83 & \text{(remainder: 83)},
\end{array}
$$

where the repetitions are omitted and where 1, the coefficient of x in the divisor, has also been omitted.

The entries in line (3), which are the coefficients of the variables in the quotient and the remainder, have been obtained by *subtracting* the **detached coefficients** in line (2) from the detached coefficients of terms of the same degree in line (1). We could obtain the same result by replacing 3 with -3 in the divisor and *adding* instead of subtracting at each step, and this is what is done in the *synthetic-division* process. The final form then appears as

$$
\begin{array}{llrrrrl}
(1) & -3\,\rule[-1mm]{0.4pt}{4mm}\ 1 & 0 & 1 & 2 & -1 & \\
(2) & & -3 & 9 & 30 & 84 & \\
\cline{2-6}
(3) & 1 & -3 & 10 & -28 & 83 & \text{(remainder: 83)}.
\end{array}
$$

Comparing the results using synthetic division with those obtained by the same process using long division, we observe that the entries in line (3) are the coefficients of the polynomial $x^3 - 3x^2 + 10x - 28$ and that there is a remainder of 83.

Example

Write $\dfrac{3x^3 - 4x - 1}{x - 2}$ in the form $Q + \dfrac{r}{D}$, where r is a constant.

Solution   Using synthetic division we first write

$$2\rfloor\ 3\quad 0\quad -4\quad -1,$$

where 0 has been inserted in the position that would be occupied by the coefficient of a second-degree term if such a term were present in the dividend. The divisor, $x - 2$, is indicated by the negative of -2, or 2. Continuing the synthetic division, we write

$$
\begin{array}{llrrrr}
(1) & 2\rfloor & 3 & 0 & -4 & -1 \\
(2) & & & 6 & 12 & 16 \\
\hline
(3) & & 3 & 6 & 8 & 15 \quad \text{(remainder: 15).}
\end{array}
$$

This process employs these steps:

1. 3 is "brought down" from line (1) to line (3).
2. 6, the product of 2 and 3, is written in the next position on line (2).
3. 6, the sum of 0 and 6, is written on line (3).
4. 12, the product of 2 and 6, is written in the next position on line (2).
5. 8, the sum of -4 and 12, is written on line (3).
6. 16, the product of 2 and 8, is written in the next position on line (2).
7. 15, the sum of -1 and 16, is written on line (3).

We can use the first three entries in line (3) as coefficients to write a polynomial of degree one less than the degree of the dividend. This polynomial is the quotient lacking the remainder. The last number is the remainder. Thus, for $x - 2 \neq 0$, the quotient of $3x^3 - 4x - 1$ divided by $x - 2$ is

$$3x^2 + 6x + 8,$$

with a remainder of 15; that is,

$$\frac{3x^3 - 4x - 1}{x - 2} = 3x^2 + 6x + 8 + \frac{15}{x - 2} \qquad (x \neq 2).$$

The foregoing example illustrates a theorem that we shall state without proof.

Theorem 4.1   *If $P(x)$ is a real polynomial of degree $n \geq 1$ and c is any real number, then there exist a unique real polynomial $Q(x)$ of degree $n - 1$ and a unique real number r, such that*

$$P(x) = (x - c)Q(x) + r.$$

Although this theorem does not directly involve the quotient $\dfrac{P(x)}{x - c}$, it does assure us that, if $x \neq c$,

$$\frac{P(x)}{x - c} = Q(x) + \frac{r}{x - c}.$$

Remainder theorem   Now consider the following important result.

Theorem 4.2   *If $P(x)$ is a real polynomial, then for every real number c there exists a unique polynomial $Q(x)$ such that*

$$P(x) = (x - c)Q(x) + P(c).$$

Proof From Theorem 4.1, we know that, for every real number c, there exists a real polynomial $Q(x)$ and a real number r such that

$$P(x) = (x - c)Q(x) + r.$$

Since this is true for all $x \in R$, it must be true for $x = c$ Thus we have

$$P(c) = (c - c)Q(c) + r = 0 \cdot Q(c) + r = r,$$

and the theorem is proved.

    This theorem is called the **remainder theorem** because it asserts that the remainder, when $P(x)$ is divided by $x - c$, is the value of P at c, that is, $P(c)$. Since synthetic division offers a quick means of obtaining this remainder, we can usually find values $P(c)$ more rapidly by synthetic division than by direct substitution.

Example   Given $P(x) = x^3 - x^2 + 3$,   find $P(3)$ by the remainder theorem.

Solution   Synthetically dividing $x^3 - x^2 + 3$ by $x - 3$, we have

$$
\begin{array}{r|rrrr}
3 & 1 & -1 & 0 & 3 \\
  &   & 3 & 6 & 18 \\
\hline
  & 1 & 2 & 6 & 21
\end{array}
$$

and, by inspection, $r = P(3) = 21$.

Exercise 4.1

Write the given quotient P/D in the form $Q + r/D$, where the degree of r is less than the degree of D.

Example $\dfrac{y^3 + y^2 - 5y + 2}{y^2 - 2y}$

Solution

$$
\begin{array}{r}
y + 3 \\
y^2 - 2y \overline{\smash{\big)}\, y^3 + y^2 - 5y + 2} \\
\underline{y^3 - 2y^2} \\
3y^2 - 5y \\
\underline{3y^2 - 6y} \\
y + 2
\end{array}
\qquad
y + 3 + \frac{y + 2}{y^2 - 2y} \quad (y \neq 0, 2)
$$

1. $\dfrac{4y^3 + 12y + 5}{2y + 1}$ 2. $\dfrac{2x^4 + 13x^3 - 7}{2x - 1}$

3. $\dfrac{4y^5 - 4y^2 - 5y + 1}{2y^2 + y + 1}$ 4. $\dfrac{2x^3 - 3x^2 - 15x - 1}{x^2 + 5}$

Use synthetic division to write the given quotient $P(x)/D(x)$ in the form $Q(x) + r/D(x)$, where $Q(x)$ is a polynomial and r is a constant.

Examples

a. $\dfrac{2x^4 + x^3 - 1}{x + 2}$

b. $\dfrac{x^3 - 1}{x - 1}$

Solutions

a.

$$
\begin{array}{r|rrrrr}
-2 & 2 & 1 & 0 & 0 & -1 \\
   &   & -4 & 6 & -12 & 24 \\
\hline
   & 2 & -3 & 6 & -12 & 23
\end{array}
$$

b.

$$
\begin{array}{r|rrrr}
1 & 1 & 0 & 0 & -1 \\
  &   & 1 & 1 & 1 \\
\hline
  & 1 & 1 & 1 & 0
\end{array}
$$

$2x^3 - 3x^2 + 6x - 12 + \dfrac{23}{x + 2}$ $(x \neq -2)$ $x^2 + x + 1$ $(x \neq 1)$

5. $\dfrac{x^4 - 3x^3 + 2x^2 - 1}{x - 2}$

6. $\dfrac{x^4 + 2x^2 - 3x + 5}{x - 3}$

7. $\dfrac{2x^3 + x - 5}{x + 1}$

8. $\dfrac{3x^3 + x^2 - 7}{x + 2}$

9. $\dfrac{2x^4 - x + 6}{x - 5}$

10. $\dfrac{3x^4 - x^2 + 1}{x - 4}$

11. $\dfrac{x^3 + 4x^2 + x - 2}{x + 2}$

12. $\dfrac{x^3 - 7x^2 - x + 3}{x + 3}$

13. $\dfrac{x^6 + x^4 - x}{x - 1}$

14. $\dfrac{x^6 + 3x^3 - 2x - 1}{x - 2}$

15. $\dfrac{x^5 - 1}{x - 1}$

16. $\dfrac{x^5 + 1}{x + 1}$

Use synthetic division in Exercises 17–22.

17. $P(x) = x^3 + 2x^2 + x - 1$;   find $P(1)$, $P(2)$, and $P(3)$.
18. $P(x) = x^3 - 3x^2 - x + 3$;   find $P(1)$, $P(2)$, and $P(3)$.
19. $P(x) = 2x^4 - 3x^3 + x + 2$;   find $P(-2)$, $P(2)$, and $P(4)$.
20. $P(x) = 3x^4 + 3x^2 - x + 3$;   find $P(-2)$, $P(2)$, and $P(4)$.
21. $P(x) = 3x^5 - x^3 + 2x^2 - 1$;   find $P(-3)$, $P(2)$, and $P(3)$.
22. $P(x) = 2x^6 - x^4 + 3x^3 + 1$;   find $P(-3)$, $P(2)$, and $P(3)$.

4.2   Polynomial Functions

Graphs of polynomial functions

In Sections 2.2 and 3.1 we graphed linear and quadratic functions. These are special cases of **polynomial functions**. A polynomial function is one of the form

$$\{(x, P(x)) \mid P(x) = a_0 x^n + a_1 x^{n-1} + \cdots + a_n, \quad a_0 \neq 0\}.$$

If the coefficients of the polynomial in the defining equation are real numbers, then the polynomial is called a **polynomial over the real–number field**, or simply a

polynomial over R. If the replacement set of the variable is R, then the polynomial is said to be a **polynomial in the real variable** x. If *both* the coefficients and the variable are restricted to real value, then the polynomial is a **real polynomial.**

In the case of sketching graphs of the general polynomial function, we shall lean more heavily on the plotting of points than we did in the case of linear and quadratic functions. There is one fact about polynomials, however, that can be useful. This involves *turning points*, or local maximum and minimum values of y. Thus, for example, in Figure 4.1-a there is one local maximum as well as one local

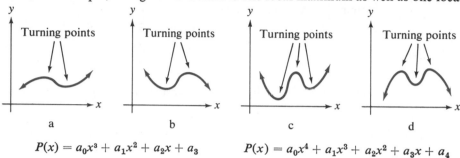

$$P(x) = a_0x^3 + a_1x^2 + a_2x + a_3 \qquad\qquad P(x) = a_0x^4 + a_1x^3 + a_2x^2 + a_3x + a_4$$

Figure 4.1

minimum, or a total of two turning points; in Figure 4.1-c are one local maximum and two local minima, for a total of three turning points. In general, we have the following theorem.

Theorem 4.3   *If $P(x) = a_0x^n + a_1x^{n-1} + \cdots + a_n$ is a real polynomial equation of degree n, then the graph of*

$$\{(x, P(x)) \mid P(x) = a_0x^n + a_1x^{n-1} + \cdots + a_n, \quad a_0 \neq 0\}$$

is a smooth curve that has at most $n - 1$ turning points.

The proof of this theorem is omitted because it involves ideas we shall not discuss herein. In accordance with this theorem, the graphs of third- and fourth-degree polynomial functions might appear as in a, b, and c, d, respectively, of Figure 4.1.

General form of a poly-nomial graph   To ascertain whether or not the graph of a polynomial function ultimately goes up to the right, taking the general form a or c, we can examine the leading coefficient, a_0, of the right-hand member of the defining equation; if $a_0 > 0$, then we can look for a form similar to a or c. But if $a_0 < 0$, we can expect something similar to b or d, where the ultimate direction is down. In other words, the graph ultimately goes up or down to the right according as $a_0 > 0$ or $a_0 < 0$.

If $a_0 > 0$, then the graph ultimately goes down to the left as in a and d, or up to the left as in b and c, according as n is odd or even. If $a_0 < 0$, the situation is reversed.

For the actual graphing process, we can obtain ordered pairs $(x, f(x))$ for any polynomial function either by direct substitution or by using the remainder theorem. Such a procedure generally gives us a good approximation to the graph of the function. However, note (as in the example below) that the ordered pairs selected do not always correspond to the significant features of the graph, such as the points where the curve intersects the x-axis and the high and low points.

Example   Graph $\{(x, P(x)) \mid P(x) = 2x^3 + 13x^2 + 6x\}$.

Solution   Since P is defined by a cubic polynomial with positive leading coefficient, we expect
a graph having a form similar to Figure 4.1-a. Now, to find points $(x, P(x))$ lying on
the graph, we shall use the process of synthetic division and find $P(x)$ by the remainder theorem. Since we know nothing about where to look for turning points,
let us start with $x = 0$. By inspection, $P(0) = 0$, so that the graph includes the
origin. For $x = 1$, we have

$$\underline{1\rfloor} \quad \begin{array}{cccc} 2 & 13 & 6 & 0 \\ & 2 & 15 & 21 \\ \hline 2 & 15 & 21 & 21 \end{array}$$

and $P(1) = 21$, so that $(1, 21)$ is on the graph. For $x = 2$, we have

$$\underline{2\rfloor} \quad \begin{array}{cccc} 2 & 13 & 6 & 0 \\ & 4 & 34 & 80 \\ \hline 2 & 17 & 40 & 80 \end{array}$$

and $(2, 80)$ is on the graph. Since the signs involved at each step in the last row of
the division process here are positive, it is evident that for values $x > 2$, $P(x)$ is
positive and will grow increasingly large; consequently, let us turn our attention to
negative values of x. For $x = -1$, we have

$$\underline{-1\rfloor} \quad \begin{array}{cccc} 2 & 13 & 6 & 0 \\ & -2 & -11 & 5 \\ \hline 2 & 11 & -5 & \end{array}$$

and $(-1, 5)$ is on the graph. Similarly we find that the following points lie on the
graph of P:

$$(-2, 24), \quad (-3, 45), \quad (-4, 56), \quad (-5, 45), \quad \text{and} \quad (-6, 0).$$

The graphs of the ordered pairs are shown in the figure on the left below. These
points make the general appearance of the graph clear, and there remains only the
question of whether or not there is a zero for the function between -1 and 0. If to x
we assign values $-3/4$, $-1/4$, and $-1/2$, we obtain the additional pairs
$(-3/4, 63/32)$, $(-1/4, -23/32)$, and $(-1/2, 0)$. For $-1/2 < x < 0$, we have
$P(x) < 0$, and the graph can be sketched as shown in the right-hand figure.

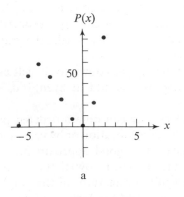

a

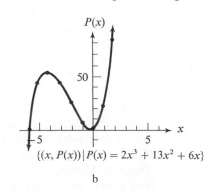

$\{(x, P(x)) \mid P(x) = 2x^3 + 13x^2 + 6x\}$

b

Exercise 4.2

Graph. Include approximations to all turning points.

1. $\{(x, P(x)) \mid P(x) = x^2 - 4x\}$ 2. $\{(x, P(x)) \mid P(x) = x^2 - 4x + 2\}$
3. $\{(x, P(x)) \mid P(x) = -x^2 - 2x\}$ 4. $\{(x, P(x)) \mid P(x) = -x^2 - 2x - 3\}$
5. $\{(x, P(x)) \mid P(x) = x^3 - 4\}$ 6. $\{(x, P(x)) \mid P(x) = -x^3 + 4x\}$
7. $\{(x, P(x)) \mid P(x) = -x^3 + 4x^2\}$ 8. $\{(x, P(x)) \mid P(x) = x^3 - 4x^2 + 3x\}$
9. $\{(x, P(x)) \mid P(x) = x^3 - 2x^2 + 1\}$
10. $\{(x, P(x)) \mid P(x) = 2x^3 + 9x^2 + 7x - 6\}$
11. $\{(x, P(x)) \mid P(x) = x^4 - 4x^2\}$
12. $\{(x, P(x)) \mid P(x) = x^4 - x^3 - 4x^2 + 4x\}$

4.3   Real Zeros of Polynomial Functions

Continuity

In Section 4.2, we graphed polynomial functions over R by plotting a number of points and using the degree of the polynomial involved to help us deduce a basic pattern for the curve. In the process, we also implicitly assumed the theorem on continuity that follows.

The proofs of this and the succeeding theorems in this section are beyond the scope of this book and are omitted.

Theorem 4.4   *If $P(x) = a_0x^n + a_1x^{n-1} + \cdots + a_n$, $a_i \in R$, and k is a real number between $P(x_1)$ and $P(x_2)$, then there exists at least one $c \in R$ between x_1 and x_2 such that $P(c) = k$.*

A geometric interpretation of this theorem is shown in Figure 4.2. Essentially, $P(x)$ must assume all values between any two of its values, because of the fact that the graph of P must be a continuous curve. This continuity property of the values of a polynomial, together with the theorems given below, enables us to deduce some facts about real zeros of a polynomial function without having to graph the function.

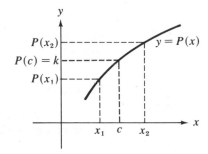

Figure 4.2

Number of zeros

To begin with, a polynomial function of degree $n \geq 1$ over R can have at most n real zeros, and may have none. In Section 4.5, you will find that, with an agreement on how to count zeros, every such function has exactly n complex zeros (zeros that are complex numbers, real or imaginary) and that when imaginary zeros (zeros that are imaginary complex numbers) exist for a polynomial over R, there is always an even number of them.

Turning now to real zeros only, let us agree that a **variation in sign** occurs in a polynomial with real coefficients if, as the polynomial is viewed from left to right, successive coefficients are opposite in sign. For example, in the polynomial

$$P(x) = 3x^5 - 2x^4 - 2x^2 + x - 1,$$

there are three variations in sign, and in

$$P(-x) = -3x^5 - 2x^4 - 2x^2 - x - 1,$$

there are no variations in sign. We have the following result, called **Descartes' rule of signs.**

Theorem 4.5   *If $P(x)$ is a polynomial over the field R of real numbers, then the number of positive real zeros of $P(x)$ either is equal to the number of variations in sign occurring in the coefficients of $P(x)$ or else is less than this number by an even natural number. Moreover, the number of negative real zeros of $P(x)$ either is equal to the number of variations in sign occurring in $P(-x)$ or else is less than this number by an even natural number.*

Example   Find an upper bound on the number of real positive zeros and real negative zeros of

$$\{(x, P(x)) \mid P(x) = 3x^4 + 3x^3 - 2x^2 + x + 1\}.$$

Solution   Since $P(x) = 3x^4 + 3x^3 - 2x^2 + x + 1$ has but two variations in sign, $P(x)$ can have no more than two positive real zeros. Since

$$P(-x) = 3(-x)^4 + 3(-x)^3 - 2(-x)^2 + (-x) + 1$$
$$= 3x^4 - 3x^3 - 2x^2 - x + 1$$

has two variations in sign, $P(x)$ can have at most two negative real zeros.

Bounds on sets of zeros   The following theorem is sometimes helpful in isolating real zeros of a polynomial function with real coefficients.

Theorem 4.6   *Let $P(x)$ be a polynomial over the field R of real numbers.*

I   *If $c_1 \geq 0$ and the coefficients of the terms in $Q(x)$ and the term $P(c_1)$ are all of the same sign in the right-hand member of*

$$P(x) = (x - c_1)Q(x) + P(c_1),$$

then $P(x)$ can have no zero greater than c_1.

II   *If $c_2 \leq 0$ and the coefficients of the terms in $Q(x)$ and the term $P(c_2)$ alternate in sign (zero suitably denoted by $+0$ or -0) in the right-hand member of*

$$P(x) = (x - c_2)Q(x) + P(c_2),$$

then $P(x)$ can have no zero less than c_2.

This theorem permits us to place upper and lower bounds on the set of zeros of the polynomial function

$$P = \{(x, P(x)) \mid P(x) = a_0 x^n + \cdots + a_n\},$$

that is, on the members of the solution set of $P(x) = 0$.

Example

Show that 2 and -2 are upper and lower bounds, respectively, for the set of zeros of

$$\{(x, P(x)) \mid P(x) = 18x^3 - 12x^2 - 11x + 10\}.$$

Solution

To find $Q(x)$, we use synthetic division to divide $P(x)$ by $(x - 2)$:

$$
\begin{array}{r|rrrr}
2 & 18 & -12 & -11 & 10 \\
  &   & 36 & 48 & 74 \\
\hline
  & 18 & 24 & 37 & 84
\end{array}
$$

Since

$$Q(x) = 18x^2 + 24x + 37$$

and

$$P(2) = 84 > 0,$$

from Theorem 4.6-I it follows that 2 is an upper bound for the set of zeros of P. Next dividing $P(x)$ by $(x + 2)$, we have

$$
\begin{array}{r|rrrr}
-2 & 18 & -12 & 11 & 10 \\
   &   & -36 & 96 & -170 \\
\hline
   & 18 & -48 & 85 & -160
\end{array}
$$

Since $Q(x) = 18x^2 - 48x + 85$ and $P(-2) = -160$, it follows from Theorem 4.6-II that -2 is a lower bound for the set of zeros of P.

Example

Find the least nonnegative integer and the greatest nonpositive integer that are, by Theorem 4.6, upper and lower bounds, respectively, for the set of zeros of

$$\{(x, P(x)) \mid P(x) = x^4 - x^3 - 10x^2 - 2x + 12\}.$$

Solution

We shall first seek an upper bound by dividing $P(x)$ successively by $(x - 1)$, $(x - 2)$, and so on. Each row after the first in the following array is the bottom row in the synthetic division involved.

$$
\begin{array}{c|rrrrr}
  & 1 & -1 & -10 & -2 & 12 \\
\hline
1 & 1 & 0 & -10 & -12 & 0 \\
\hline
2 & 1 & 1 & -8 & -18 & -24 \\
\hline
3 & 1 & 2 & -4 & -14 & -30 \\
\hline
4 & 1 & 3 & 2 & 6 & 36
\end{array}
$$

Since the numbers in the last row are all positive, 4 is an upper bound. Next, we divide by $(x + 1)$, $(x + 2)$, and so on, in search of a lower bound.

(continued overleaf)

	1	−1	−10	−2	12
−1	1	−2	−8	6	6
−2	1	−3	−4	6	0
−3	1	−4	2	−8	36

Since the signs alternate in the last row, −3 is a lower bound. Had the numbers in the row been 1, 0, 2, −8, 36, then the sign "−" could arbitrarily have been assigned to 0 to give the desired pattern of alternating signs.

To narrow the search for real zeros of a polynomial function still further, we have the following **location theorem,** which follows directly from Theorem 4.4.

Theorem 4.7   *Let $P(x)$ be a polynomial over the field R of real numbers. If x_1, $x_2 \in R$, with $x_1 < x_2$, and $P(x_1)$ and $P(x_2)$ are opposite in sign, then there exists at least one $c \in R$, $x_1 < c < x_2$, such that $P(c) = 0$.*

This theorem expresses the fact that if the graphs of $(x_1, P(x_1))$ and $(x_2, P(x_2))$ are on opposite sides of the x-axis, as shown in Figure 4.3, then the graph of $y = P(x)$ must cross the x-axis at some (at least one) point with abscissa c between x_1 and x_2.

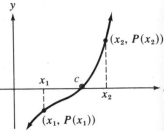

Example   Verify that $\{(x, P(x)) \mid P(x) = 2x^3 - 3x^2 + 4x - 6\}$ has a zero between 1 and 2.

Figure 4.3

Solution   We can apply synthetic division to find $P(1)$ and $P(2)$.

	2	−3	4	−6
1	2	−1	3	−3
2	2	1	6	6

Since $P(1) = -3$ and $P(2) = 6$, Theorem 4.7 assures us that P has a zero between 1 and 2.

Exercise 4.3

Use Theorem 4.5 to find bounds on the number of positive zeros and the number of negative zeros of the function defined by the given equation.

1. $P(x) = x^4 - 2x^3 + 2x + 1$ 2. $R(x) = 3x^4 + 3x^3 + 2x^2 - x + 1$
3. $Q(x) = 2x^5 + 3x^3 + 2x + 1$ 4. $P(x) = 4x^5 - 2x^3 - 3x - 2$
5. $S(x) = 3x^4 + 1$ 6. $Q(x) = 2x^5 - 1$

Find the least nonnegative integer and the greatest nonpositive integer that are, by Theorem 4.6, an upper bound and a lower bound, respectively, for the set of real zeros of the function defined by the given equation.

7. $P(x) = x^3 + 2x^2 - 7x - 8$
8. $Q(x) = x^3 - 8x + 5$
9. $Q(x) = x^4 - 2x^3 - 7x^2 + 10x + 10$
10. $R(x) = x^3 - 4x^2 - 4x + 12$
11. $S(x) = x^5 - 3x^3 + 24$
12. $R(x) = x^5 - 3x^4 - 1$
13. $P(x) = 2x^5 + x^4 - 2x - 1$
14. $G(x) = 2x^5 - 2x^2 + x - 2$

Use Theorem 4.7 to verify the given statement.

15. $\{(x, f(x)) \mid f(x) = x^3 - 3x + 1\}$ has a zero between 0 and 1.
16. $\{(x, f(x)) \mid f(x) = 2x^3 + 7x^2 + 2x - 6\}$ has a zero between -2 and -1.
17. $\{(x, g(x)) \mid g(x) = x^4 - 2x^2 + 12x - 17\}$ has a zero between -3 and -2.
18. $\{(x, g(x)) \mid g(x) = 2x^4 + 3x^3 - 14x^2 - 15x + 9\}$ has a zero between -2 and -1.
19. $\{(x, P(x)) \mid P(x) = 2x^2 + 4x - 4\}$ has one zero between -3 and -2, and one between 0 and 1.
20. $\{(x, P(x)) \mid P(x) = x^3 - x^2 - 2x + 1\}$ has one zero between -2 and -1, one between 0 and 1, and one between 1 and 2.

Using Theorems 4.3 to 4.7, sketch the graph of the function in the specified exercise.

21. Exercise 15 above
22. Exercise 16
23. Exercise 17
24. Exercise 18
25. Exercise 19
26. Exercise 20

4.4   *Rational Zeros of Polynomial Functions*

Prime and composite numbers

The results of the present section depend on the notion of a **prime number**. If a is an element of the set N of natural numbers, and $a \neq 1$, then a is a prime number if and only if a has no prime factor in N other than itself and 1; otherwise, a is a **composite number**. For example, 2 and 3 are prime numbers, and $6 = 2 \cdot 3$ is composite; but 1 is considered to be neither prime nor composite.

The **unique-factorization theorem**, or **fundamental theorem of arithmetic** (which we shall not prove), follows.

Theorem 4.8   *If a is a composite number, then a is the product of only one set of prime factors.*

Two integers a and b are said to be **relatively prime** if and only if they have no prime factors in common. For example, -6 and 35 are relatively prime, since $-6 = -2 \cdot 3$ and $35 = 5 \cdot 7$; but 6 and 8 are not, since they have the prime factor 2 in common. The fraction a/b is said to express a rational number in **lowest terms** if and only if a and b are relatively prime.

Identifying
possible
rational zeros

If all the coefficients of the defining equation

$$P(x) = a_0 x^n + a_1 x^{n-1} + \cdots + a_n$$

of a polynomial function P are integers, then we can identify all possible rational zeros of P by means of the following theorem.

Theorem 4.9   *If the rational number p/q, in lowest terms, is a solution of*

$$P(x) = a_0 x^n + a_1 x^{n-1} + \cdots + a_n = 0,$$

where $a_j \in J$, then p is an integral factor of a_n and q is an integral factor of a_0.

Proof Since p/q is a solution of $P(x) = 0$, we have

$$a_0 \left(\frac{p}{q}\right)^n + a_1 \left(\frac{p}{q}\right)^{n-1} + \cdots + a_n = 0,$$

and we can multiply each member here by q^n to obtain

$$a_0 p^n + a_1 p^{n-1} q + \cdots + a_{n-1} p q^{n-1} + a_n q^n = 0.$$

Adding $-a_n q^n$ to each member and factoring p from each term in the left-hand member of the resulting equation, we have

$$p(a_0 p^{n-1} + a_1 p^{n-2} q + \cdots + a_{n-1} q^{n-1}) = -a_n q^n.$$

Since J is closed with respect to addition and multiplication, the expression in parentheses in the left-hand member here represents an integer, say r, so that we have

$$pr = -a_n q^n,$$

where pr is an integer having p as a factor. Hence, p is a factor of $-a_n q^n$. But, by Theorem 4.8, p and q^n have no factor in common, because, by hypothesis, p/q is in lowest terms; hence p must be a factor of a_n. In a similar manner, by writing the equation

$$a_0 p^n + a_1 p^{n-1} q + \cdots + a_n q^n = 0$$

in the form

$$-a_0 p^n = a_1 p^{n-1} q + \cdots + a_n q^n,$$

we can factor q from each term in the right-hand member and show that q must be a factor of a_0.

Example

List all possible rational zeros of $\{(x, P(x)) \mid P(x) = 2x^3 - 4x^2 + 3x + 9\}$.

Solution

Rational zeros, p/q, must, by Theorem 4.8, be such that p is an integral factor of 9 and q is an integral factor of 2. Hence

$$p \in \{-9, -3, -1, 1, 3, 9\}, \quad q \in \{-2, -1, 1, 2\},$$

and the set of possible rational zeros of P is

$$\left\{-9, \ -\frac{9}{2}, \ -3, \ -\frac{3}{2}, -1, \ -\frac{1}{2}, \frac{1}{2}, \ 1, \frac{3}{2}, 3, \frac{9}{2}, 9\right\}.$$

Test for
rational zeros

It is important to observe that Theorem 4.9 does not assure us that a polynomial function with integral coefficients indeed has a rational zero; it simply enables us to identify possibilities for rational zeros. These can then be checked by synthetic division. Identification of the zeros of P in the previous example is left as an exercise.

As a special case of Theorem 4.9, it is evident that if a function P is defined by

$$P(x) = x^n + a_1 x^{n-1} + \cdots + a_n,$$

in which $a_i \in J$ and $a_0 = 1$, then any rational zero of P must be an integer, and, moreover, must be an integral factor of a_n.

Example

Find all rational zeros of $\quad \{(x, P(x)) \mid P(x) = x^3 - 4x^2 + x + 6\}.$

Solution

The only possible rational zeros of P are -6, -3, -2, -1, 1, 2, 3, and 6. Using synthetic division, we set up the following array:

$$
\begin{array}{r|rrrr}
 & 1 & -4 & 1 & 6 \\
\hline
1 & 1 & -3 & -2 & 4 \\
\hline
-1 & 1 & -5 & 6 & 0 \\
\end{array}
$$

We can cease our trials with -1, since by the remainder theorem, -1 is a zero of P, and the remaining zeros can be obtained from the equation $x^2 - 5x + 6 = 0$, whose coefficients are taken from the last line in the array. We can write the equation as $(x - 2)(x - 3) = 0$ and observe that 2 and 3 are also zeros. Note that it is often useful to start with possible roots of *least* absolute value, since the synthetic-division testing might reveal an upper or lower bound and thus eliminate the necessity of testing some of the numbers.

Exercise 4.4

Find all integral zeros of each function.

1. $\{(x, f(x)) \mid f(x) = 3x^3 - 13x^2 + 6x - 8\}$
2. $\{(x, f(x)) \mid f(x) = x^4 - x^2 - 4x + 4\}$
3. $\{(x, f(x)) \mid f(x) = x^4 + x^3 + 2x - 4\}$
4. $\{(x, f(x)) \mid f(x) = 5x^3 + 11x^2 - 2x - 8\}$
5. $\{(x, P(x)) \mid P(x) = 2x^4 - 3x^3 - 8x^2 - 5x - 3\}$
6. $\{(x, P(x)) \mid P(x) = 3x^4 - 40x^3 + 130x^2 - 120x + 27\}$

Find all rational zeros of each function.

7. $\{(x, f(x)) \mid f(x) = 2x^3 + 3x^2 - 14x - 21\}$
8. $\{(x, f(x)) \mid f(x) = 3x^4 - 11x^3 + 9x^2 + 13x - 10\}$
9. $\{(x, P(x)) \mid P(x) = 4x^4 - 13x^3 - 7x^2 + 41x - 14\}$
10. $\{(x, P(x)) \mid P(x) = 2x^3 - 4x^2 + 3x + 9\}$
11. $\{(x, Q(x)) \mid Q(x) = 2x^3 - 7x^2 + 10x - 6\}$
12. $\{(x, Q(x)) \mid Q(x) = x^3 + 3x^2 - 4x - 12\}$
13. Show that $\sqrt{3}$ is irrational. *Hint:* Consider the equation $x^2 - 3 = 0$.
14. Show that $\sqrt{2}$ is irrational.

4.5   *Complex Zeros of Polynomial Functions*

Zeros of a
polynomial
over C

In Section 4.3, we observed that every polynomial function of degree $n \geq 1$ over the field of complex numbers (i.e., a polynomial with complex coefficients) has exactly n zeros. This is a consequence of the **fundamental theorem of algebra,** which follows.

Theorem 4.10   *Every polynomial function of degree $n \geq 1$ in the complex variable x over the field C of complex numbers has at least one complex zero.*

The proofs of this and the following theorems involve concepts beyond those available to us and are omitted. As a consequence of Theorem 4.10, we have the following result.

Theorem 4.11   *Every polynomial, of degree $n \geq 1$, in the complex variable x over the field C of complex numbers can be expressed as a product of a constant and n linear factors.*

Thus if $P(x) = a_0x^n + a_1x^{n-1} + \cdots + a_n$, where $a_i \in C$, $a_0 \neq 0$, $x \in C$, and $n \in N$, $n \geq 1$, then

$$P(x) = a_0(x - x_1)(x - x_2) \cdots (x - x_n).$$

If a factor $(x - x_i)$ occurs k times in such a linear factorization, then x_i is said to be a zero of $P(x)$ of **multiplicity** k. With this agreement, Theorem 4.11 shows that every polynomial function defined by a polynomial $P(x)$ of degree n with complex coefficients has exactly n complex zeros.

Examples

a $P(x) = 2x^5 + x^3 - x^2 + 3x + 1$ has exactly five zeros.

b. $P(x) = (x + 3)(x + 3)(x - i)(x + i)$ has exactly four zeros; -3 is a zero of multiplicity 2.

Zeros and
roots

Note that any theorem stated in terms of zeros of polynomial functions applies to roots (solutions) of polynomial equations, and vice versa; a *zero* of

$$\{(x, P(x)) \mid P(x) = a_0x^n + a_1x^{n-1} + \cdots + a_n\}$$

is a *solution*, or *root*, of $P(x) = 0$.

Imaginary
zeros of real
polynomials

Another important property of a polynomial with real coefficients establishes the fact that the imaginary zeros of such a polynomial occur in conjugate pairs.

Theorem 4.12   *If $P(x)$ is a polynomial in the complex variable x over the field R of real numbers, and $P(x) = 0$ for some $x \in C$, then $P(\bar{x}) = 0$, where $\bar{x}$ is the complex conjugate of x.*

Example

Given that $-1 + i$ is a zero of $P(x) = x^2 + 2x + 2$,   then, by Theorem 4.12, $-1 - i$ is also a zero.

We can use Theorems 4.11 and 4.12 and the theorems in previous sections applicable to real zeros to help us study zeros of polynomial functions with real coefficients.

Example   Find the zeros of $\{(x, P(x)) \mid P(x) = x^3 - 2x^2 + 3x - 6\}$.

Solution   We first find all rational zeros. Since the constant term of the polynomial is -6, the only possible rational zeros of $P(x)$ are 1, -1, 2, -2, 3, -3, 6, and -6. Using synthetic division, we obtain:

		1	-2	3	-6
	1	1	-1	2	-4
	-1	1	-3	6	-12
	2	1	0	3	0

From the entries in the last row, we note that $x - 2$ is a factor of $P(x)$, and so is $x^2 + 3$. The solutions of $x^2 + 3 = 0$, which is equivalent to $x^2 = -3$, are $\pm\sqrt{-3} = \pm 3i$. Hence we can conclude that the zeros of P are 2, $\sqrt{3}i$, and $-\sqrt{3}i$.

In the foregoing example, note that the imaginary zeros of $P(x)$, a polynomial with real coefficients, occur as a conjugate pair, as required by Theorem 4.12.

Exercise 4.5

Find all complex zeros.

Example $\{(x, P(x)) \mid P(x) = x^2 - 2x + 4\}$

Solution   Since $P(x)$ is of second degree, we can use the quadratic formula

$$x = \frac{-b \pm \sqrt{b^2 - 4ac}}{2a}$$

on the defining equation for $P(x) = 0$ to obtain the zeros. Here $a = 1$, the coefficient of x^2; $b = -2$, the coefficient of x; and $c = 4$, the constant term.

$$x = \frac{-(-2) \pm \sqrt{(-2)^2 - 4(1)(4)}}{2(1)} = \frac{2 \pm 2i\sqrt{3}}{2} = 1 \pm i\sqrt{3}.$$

Thus, zeros of the function are $1 + i\sqrt{3}$ and $1 - i\sqrt{3}$.

1. $\{(x, P(x)) \mid P(x) = x^2 + 1\}$
2. $\{(x, P(x)) \mid P(x) = x^2 + x + 1\}$
3. $\{(x, P(x)) \mid P(x) = 2x^2 - 3x + 1\}$
4. $\{(x, P(x)) \mid P(x) = 3x^2 - 2x + 4\}$
5. $\{(x, P(x)) \mid P(x) = 3x^3 - 5x^2 - 14x - 4\}$
6. $\{(x, P(x)) \mid P(x) = x^3 - 4x^2 - 5x + 14\}$
7. $\{(x, P(x)) \mid P(x) = 2x^4 + 3x^3 + 2x^2 - 1\}$
8. $\{(x, P(x)) \mid P(x) = 8x^4 - 22x^3 + 29x^2 - 66x + 15\}$

One or more zeros are given for the specified polynomial function; find the other zeros.

Example $\{(x, P(x)) \mid P(x) = x^4 + 3x^3 + 4x^2 + 12x\};$ $2i$ is one zero.

Solution   From Theorem 4.12, since $2i$ is a zero, $-2i$ is also a zero. Hence $x + 2i$, $x - 2i$, and their product $x^2 + 4$ are factors of

$$P(x) = x^4 + 3x^3 + 4x^2 + 12x.$$

Dividing $P(x)$ by $x^2 + 4$, we obtain $x^2 + 3x$. Hence,

$$P(x) = (x^2 + 4)(x^2 + 3x) = (x + 2i)(x - 2i)x(x + 3).$$

Hence, the three other zeros in addition to $2i$ are $-2i$, 0, and -3.

9. $\{(x, P(x)) \mid P(x) = x^2 + 4\};$ $2i$ is one zero.
10. $\{(x, P(x)) \mid P(x) = 3x^2 + 27\};$ $-3i$ is one zero.
11. $\{(x, P(x)) \mid P(x) = x^3 - 3x^2 + x - 3\};$ 3 and i are zeros.
12. $\{(x, Q(x)) \mid Q(x) = x^3 - 5x^2 + 7x + 13\};$ -1 and $3 - 2i$ are zeros.
13. $\{(x, Q(x)) \mid Q(x) = x^4 + 5x^2 + 4\};$ $-i$ and $2i$ are zeros.
14. $\{(x, P(x)) \mid P(x) = x^4 + 11x^2 + 18\};$ $3i$ and $\sqrt{2}\,i$ are zeros.
15. $\{(x, Q(x)) \mid Q(x) = x^4 + 3x^3 + 4x^2 + 27x - 45\};$ $-3i$ is a zero.
16. $\{(x, Q(x)) \mid Q(x) = x^5 - 2x^4 + 8x^3 - 16x^2 + 16x - 32\};$ $2i$ (multiplicity 2) and 2 are zeros.
17. A cubic equation with real coefficients has roots -2 and $1 + i$. What is the third root? Write the equation in the form $P(x) = 0$, given that the **leading coefficient** (the coefficient of the highest power of x) is 1.
18. A cubic equation with real coefficients has roots 4 and $2 - i$. What is the third root? Write the equation in the form $P(x) = 0$, given that the leading coefficient is 1.
19. One zero of $P(x) = 2x^3 - 11x^2 + 28x - 24$ is $2 - 2i$. Find the remaining zeros.
20. One zero of $Q(x) = 3x^3 - 10x^2 + 7x + 10$ is $2 + i$. Find the remaining zeros.

4.6   *Rational Functions*

A function defined by an equation of the form

$$y = \frac{P(x)}{Q(x)}, \tag{1}$$

where $P(x)$ and $Q(x)$ are polynomials in x, and $Q(x)$ is not the zero polynomial, is called a **rational function.** Rational functions with real coefficients (that is, rational functions over R) are of importance in the calculus and provide some interesting problems with respect to their graphs. We shall consider them only briefly here.

Graph near vertical asymptotes

Since $P(x)/Q(x)$ is not defined for values of x for which $Q(x) = 0$, it is evident that we shall not be able to find points in $R \times R$ having such x-coordinates. We can, however, consider the graph for values of x as close as we please to a value, say x_0, for which $Q(x_0) = 0$, but still with $x \neq x_0$. This consideration is usually described by saying that x "approaches" x_0, "grows close" to x_0, etc., and correspondingly that $Q(x)$ approaches 0. Thus, the closer $Q(x)$ approaches 0, if $P(x)$ does not approach 0 at the same time, then the larger $|y|$ in (1) becomes. For example, Figure 4.4-a shows the behavior of

$$\left\{ (x, y) \mid y = \frac{2}{x - 2} \right\} \tag{2}$$

as x approaches 2 from the right. In situations such as this, the vertical line that the curve approaches is called a **vertical asymptote**.

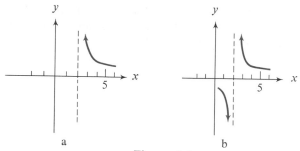

Figure 4.4

We are also interested in the behavior of the graph as x approaches 2 from the left. Figure 4.4-b illustrates this. As long as $x > 2$, we have $x - 2 > 0$ and $2/(x - 2) > 0$; but if $x < 2$, then we have $x - 2 < 0$ and $2/(x - 2) < 0$.

Theorem 4.13   *The graph of the rational function over R defined by $y = P(x)/Q(x)$ has a vertical asymptote at $x = a$ for each value a at which $Q(x)$ vanishes and $P(x)$ does not vanish.*

Horizontal asymptotes

The graphs of some rational functions have **horizontal asymptotes** which can, in general, be identified by using the following theorem.

Theorem 4.14   *The graph of the rational function over R defined by*

$$y = \frac{a_0 x^n + a_1 x^{n-1} + \cdots + a_n}{b_0 x^m + b_1 x^{m-1} + \cdots + b_m},$$

where $a_0, b_0 \neq 0$ and n, m are nonnegative integers, has

    I   *a horizontal asymptote at $y = 0$ if $n < m$,*

    II   *a horizontal asymptote at $y = a_0/b_0$ if $n = m$,*

    III   *no horizontal asymptotes if $n > m$.*

Though we shall not give a rigorous proof of this theorem, we can certainly make the results plausible. If $n < m$, we can divide the numerator and denominator of the right-hand member of

$$y = \frac{a_0 x^n + a_1 x^{n-1} + \cdots + a_n}{b_0 x^m + b_1 x^{m-1} + \cdots + b_m}$$

by x^m to obtain, for $x \neq 0$,

$$y = \frac{\dfrac{a_0}{x^{m-n}} + \dfrac{a_1}{x^{m-n+1}} + \cdots + \dfrac{a_n}{x^m}}{b_0 + \dfrac{b_1}{x} + \cdots + \dfrac{b_m}{x^m}}.$$

Now, as $|x|$ grows larger and larger, each term containing an x in its denominator grows closer and closer to 0, and we find the expression on the right approaching $0/b_0$, so that y approaches 0. But if, as y grows close to 0, $|x|$ is increasing without bound, then y is approaching the line $y = 0$ asymptotically, and actually must approach 0 from one of the four directions shown in Figure 4.5-a. A similar argument shows that if $n = m$, then, as $|x|$ increases without bound, y approaches the line $y = a_0/b_0$ from one of the directions shown in Figure 4.5-b. If $n > m$, then as $|x|$ becomes larger and larger, so does $|y|$.

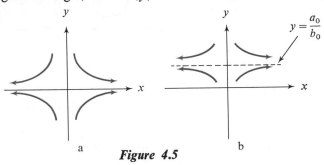

Figure 4.5

Oblique asymptotes

In particular, if $n = m + 1$, that is, if the numerator is of degree one greater than the denominator, we can argue that though the graph has no horizontal asymptote, it does have an **oblique asymptote**. We shall illustrate a special case only, but the technique involved is quite general.

Example

Find all asymptotes for the graph of $\left\{(x, y) \mid y = \dfrac{x^2 - 4}{x - 1}\right\}$.

Solution

First we note that, by Theorem 4.13, there is a vertical asymptote at $x = 1$ and that, by Theorem 4.14, there are no horizontal asymptotes. If, however, we rewrite the equation $y = (x^2 - 4)/(x - 1)$ by dividing $x^2 - 4$ by $x - 1$, we obtain

$$y = x + 1 - \frac{3}{x - 1}.$$

Now as $|x|$ grows larger and larger, $|3/(x - 1)|$ approaches zero and the graph of $y = (x^2 - 4)/(x - 1)$ approaches the graph of $y = x + 1$. Hence, the graph of $y = x + 1$, which is an oblique line, is an asymptote to the curve.

Helpful items for graphing

Identifying asymptotes is one aid to the graphing of a rational function over R. Other helpful items are the following:

1. The zeros of the function, because these give us the x-intercepts.

2. The domain and range, because these let us know where we can expect to find parts of the graph and where we cannot.

3. Some specific points on the graph, because these give us guidelines in sketching. The y-intercept is easily found by setting $x = 0$.

Example   Graph $\left\{(x, y) \mid y = \dfrac{x - 1}{x - 2}\right\}$.

Solution   We can begin by observing that the numerator of the right-hand member will be equal to 0 when x is equal to 1. Therefore, when $x = 1$, we have $y = 0$, and 1 is an x-intercept. Also, when $x = 0$, we have $y = 1/2$, so that $1/2$ is a y-intercept. Thus we can begin our graph as shown in Figure a. By inspection, $y = (x - 1)/(x - 2)$ is defined for all real x except $x = 2$, so that $\{x \mid x \neq 2\}$ is the domain. Similarly, if we solve the defining equation for x in terms of y, we have $x = (2y - 1)/(y - 1)$, which is defined for all values of y except 1. Hence $\{y \mid y \neq 1\}$ is the range of the function. From the defining equation, Theorem 4.13, and Theorem 4.14 we see that there is a vertical asymptote at $x = 2$ and a horizontal asymptote at $y = 1$. We can then add this information to our graph, as indicated in Figure b.

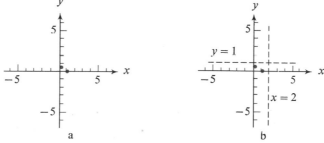

a   b

Next we call our powers of observation into play. That the vertical asymptote, for example, is approached downward instead of upward from the left can be confirmed by observing that if x is just less than 2, say $2 - p$, $0 < p < 1/10$, then the denominator $x - 2$ in the expression for y is $2 - p - 2 = -p$, which is barely negative, whereas the numerator $x - 1 = 2 - p - 1$ is definitely positive; hence y is negative and $|y|$ large. The graph appears in Figure c. To find the curve when $x > 2$, that is, to the right of the vertical asymptote, we simply find one or two points associated with such values of x. For example, we might choose 3 and 4. If $x = 3$, then

$$y = (3 - 1)/(3 - 2) = 2,$$

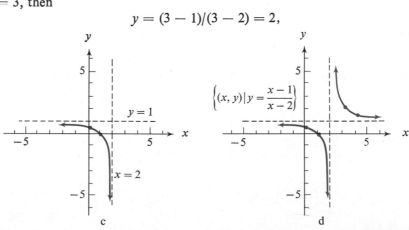

c   d

(continued overleaf)

and (3, 2) is on the curve. If $x = 4$, then

$$y = (4 - 1)/(4 - 2) = 3/2,$$

and (4, 3/2) is on the curve. Again, the knowledge that $x = 2$ and $y = 1$ are asymptotes, together with the location of the points (3, 2) and (4, 3/2), leads us to the complete graph of $\{(x, y) \mid y = (x - 1)/(x - 2)\}$, as shown in Figure d.

Example   Graph $\left\{(x, y) \mid y = \dfrac{x^2 - 4}{x - 1}\right\}$.

Solution   This is the same function we investigated earlier in the example on page 98, where we found the vertical asymptote $x = 1$ and the oblique asymptote $y = x + 1$.

By inspection, if $x = 0$, then $y = 4$, and if $y = 0$, then $x = 2$ or $x = -2$, so that the y-intercept is 4, and the x-intercepts are 2 and -2. We therefore have the situation shown in Figure a. Without any further information, we have strong reason to suspect that the graph will appear as shown in Figure b. The plotting of a few check points, say $(-1, 3/2)$, $(1/2, 15/2)$, $(3/2, -7/2)$, and $(3, 5/2)$, would tend to confirm our conjecture.

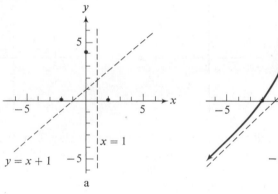

a

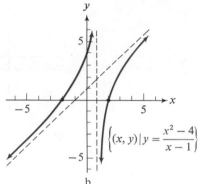

b

Exercise 4.6

Determine the vertical asymptotes of the graph of the specified function.

Example $\{(x, y) \mid x^2y - 4y = 1\}$

Solution   Express y explicitly in terms of x by means of an equivalent equation.

$$y(x^2 - 4) = 1 \qquad y = \frac{1}{x^2 - 4}$$

The denominator $x^2 - 4$ vanishes for $x = 2, -2$. Therefore, by Theorem 4.3, there are vertical asymptotes at $x = 2$ and $x = -2$.

1. $\left\{(x, y) \mid y = \dfrac{1}{x - 3}\right\}$ 2. $\left\{(x, y) \mid y = \dfrac{1}{x + 4}\right\}$

3. $\left\{(x, y) \mid y = \dfrac{4}{(x + 2)(x - 3)}\right\}$

4. $\left\{(x, y) \mid y = \dfrac{8}{(x - 1)(x + 3)}\right\}$

5. $\left\{(x, y) \mid y = \dfrac{2x - 1}{x^2 + 5x + 4}\right\}$

6. $\left\{(x, y) \mid y = \dfrac{x + 3}{2x^2 - 5x - 3}\right\}$

7. $\{(x, y) \mid xy + y = 4\}$

8. $\{(x, y) \mid x^2 y + xy = 3\}$

Graph.

9. $\left\{(x, y) \mid y = \dfrac{1}{x}\right\}$

10. $\left\{(x, y) \mid y = \dfrac{1}{x + 4}\right\}$

11. $\left\{(x, y) \mid y = \dfrac{1}{x - 3}\right\}$

12. $\left\{(x, y) \mid y = \dfrac{1}{x - 6}\right\}$

13. $\left\{(x, y) \mid y = \dfrac{4}{(x + 2)(x - 3)}\right\}$

14. $\left\{(x, y) \mid y = \dfrac{8}{(x - 1)(x + 3)}\right\}$

15. $\left\{(x, y) \mid y = \dfrac{2}{(x - 3)^2}\right\}$

16. $\left\{(x, y) \mid y = \dfrac{1}{(x + 4)^2}\right\}$

Determine any vertical, horizontal, or oblique asymptotes of the graphs of the specified function.

Example $\left\{(x, y) \mid y = \dfrac{6x^2 + 1}{2x^2 + 5x - 3}\right\}$

Solution   The defining equation can be written equivalently as $y = \dfrac{6x^2 + 1}{(2x - 1)(x + 3)}$.

By Theorem 4.13, there are vertical asymptotes at $x = 1/2$ and $x = -3$. By Theorem 4.14-II, there is a horizontal asymptote at $y = 6/2$, or 3. There are no oblique asymptotes.

17. $\left\{(x, y) \mid y = \dfrac{x}{x^2 - 4}\right\}$

18. $\left\{(x, y) \mid y = \dfrac{3x + 6}{x^2 + 3x + 2}\right\}$

19. $\left\{(x, y) \mid y = \dfrac{x^2 - 9}{x - 4}\right\}$

20. $\left\{(x, y) \mid y = \dfrac{x^3 - 27}{x^2 - 1}\right\}$

21. $\left\{(x, y) \mid y = \dfrac{x^2 - 3x + 2}{x^2 - 3x - 4}\right\}$

22. $\left\{(x, y) \mid y = \dfrac{x^2}{x^2 - x - 6}\right\}$

Graph. Use information concerning the zeros of the specified function and any vertical, horizontal, or oblique asymptotes.

23. $\left\{(x, y) \mid y = \dfrac{x}{x - 2}\right\}$

24. $\left\{(x, y) \mid y = \dfrac{x - 1}{x + 3}\right\}$

25. $\left\{(x, y) \mid y = \dfrac{2x - 4}{x^2 - 9}\right\}$

26. $\left\{(x, y) \mid y = \dfrac{3x}{x^2 - 5x + 4}\right\}$

27. $\left\{(x, y) \mid y = \dfrac{x^2 - 4}{x^3}\right\}$

28. $\left\{(x, y) \mid y = \dfrac{x - 2}{x^2}\right\}$

29. $\left\{(x, y) \mid y = \dfrac{x^2 - 4x + 4}{x - 1}\right\}$

30. $\left\{(x, y) \mid y = \dfrac{x^2 + 4}{x - 2}\right\}$

31. $\left\{ (x, y) \mid y = \dfrac{x + 1}{x(x^2 - 4)} \right\}$ **32.** $\left\{ (x, y) \mid y = \dfrac{x^2 + x - 2}{x(x^2 - 9)} \right\}$

Chapter Review

[4.1]   *Write the given quotient as a polynomial.*

1. $\dfrac{2n^2 - 5n - 3}{n - 3}$ **2.** $\dfrac{2r^2 + 3r + 1}{2r + 1}$

Use synthetic division to write the given quotient $P(x)/D(x)$ either in the form $Q(x)$ or in the form $Q(x) + \dfrac{r}{D(x)}$, where $Q(x)$ is a polynomial and r is a constant.

3. $\dfrac{2x^3 - 11x^2 + 13x - 4}{x - 4}$ **4.** $\dfrac{3y^3 + 7y^2 - 3y + 10}{y + 3}$

[4.2]   **5.** Use synthetic division to find $P(2)$, where $P(x) = x^3 - 3x^2 + 2x + 1$.
   6. Use synthetic division to find $P(-3)$, where $P(x) = x^4 - 3x^2 - 1$.
   7. Graph the function $\{(x, P(x)) \mid P(x) = x^3 + 4x^2 + x - 6\}$.
   8. Graph the function $\{(x, P(x)) \mid P(x) = x^4 + 3x^3 - 4x^2 - 12x\}$.

[4.3]   *Use Theorem 4.5 to find bounds on the number of positive zeros and the number of negative zeros of the function defined by each equation.*

9. $P(x) = 4x^3 - 3x^2 + 2x - 7$ **10.** $Q(x) = x^4 - 5x^3 - 6x - 9$

Find the least nonnegative integer and greatest nonpositive integer that Theorem 4.6 shows to be upper and lower bounds for the set of real zeros of the function defined by each equation.

11. $R(x) = x^3 - 4x^2 - 4x + 12$ **12.** $S(x) = x^4 + 5x^3 - 5x - 6$

Use Theorem 4.7 to verify each statement.

13. $\{(x, f(x)) \mid f(x) = x^3 - 2x^2 + x - 3\}$ has a zero between 2 and 3.
14. $\{(x, g(x)) \mid g(x) = x^3 - 4x + 1\}$ has a zero between 0 and 1.

[4.4]   **15.** Find all integral zeros of $\{(x, Q(x)) \mid Q(x) = x^4 - 3x^3 - x^2 - 11x - 4\}$.
   16. Find all rational zeros of $\{(x, P(x)) \mid P(x) = 2x^3 - 11x^2 + 12x + 9\}$.

[4.5]   *Find the other zeros of the given polynomial function if one zero is as given.*

17. $\{(x, P(x)) \mid P(x) = x^3 - 2x^2 + 4x - 8\}$; $2i$ is one zero.
18. $\{(x, Q(x)) \mid Q(x) = 2x^3 - 11x^2 + 28x - 24\}$; $2 + 2i$ is one zero.

[4.6]   *Determine any asymptotes and graph the specified function.*

19. $\left\{ (x, y) \mid y = \dfrac{3}{x + 2} \right\}$ **20.** $\left\{ (x, y) \mid y = \dfrac{2x}{(x + 3)(x - 4)} \right\}$

21. $\left\{ (x, y) \mid y = \dfrac{x + 2}{x - 3} \right\}$ **22.** $\left\{ (x, y) \mid y = \dfrac{2x}{x^2 - 3x + 2} \right\}$

5
Exponential and Logarithmic Functions

5.1 Properties of Exponents

In this section we shall be concerned with powers having rational exponents. First, however, we should recall how powers of real numbers are defined for *natural-number* exponents.

Definition 5.1   For $a \in R$, $a \neq 0$, and $m \in N$,

$$a^m = a \cdot a \cdot a \cdots a \quad (m \text{ factors}).$$

We should also recall some simple properties of products and quotients that follow from this definition.

Theorem 5.1   For $a, b \in R$, and $m, n \in N$,

$$\text{I} \quad a^m \cdot a^n = a^{m+n},$$

$$\text{II} \quad \frac{a^m}{a^n} = a^{m-n} \quad (a \neq 0, m > n),$$

$$\text{III} \quad (a^m)^n = a^{mn},$$

$$\text{IV} \quad (ab)^m = a^m b^m,$$

$$\text{V} \quad \left(\frac{a}{b}\right)^m = \frac{a^m}{b^m} \quad (b \neq 0).$$

Examples

a. $x^3 \cdot x^2 = x^{3+2} = x^5$ b. $\dfrac{x^6}{x^2} = x^{6-2} = x^4$ c. $(x^2)^3 = x^{2 \cdot 3} = x^6$

d. $(x^2 y^3)^2 = (x^2)^2 (y^3)^2$ e. $\left(\dfrac{x^3}{y^4}\right)^2 = \dfrac{(x^3)^2}{(y^4)^2}$

$\qquad\qquad = x^4 y^6$ $\qquad\qquad\qquad = \dfrac{x^6}{y^8}$

Reason for defining a^0 to be 1

Powers having integral and rational exponents can be defined in a manner consistent with the properties stated in Theorem 5.1. If Theorem 5.1-II is to hold for $m = n$, then we must have

$$\frac{a^n}{a^n} = a^{n-n} = a^0.$$

Since $a^n/a^n = 1$, we make the following definition.

Definition 5.2   *If $a \in R$, $a \neq 0$, then*

$$a^0 = 1.$$

In the same way, if Theorem 5.1-II is also to extend to $m = 0$, then we must have

$$\frac{a^0}{a^n} = a^{0-n} = a^{-n}.$$

Reason for defining a^{-n} to be $1/a^n$

Since $a^0 = 1$, we therefore make the following definition.

Definition 5.3   *If $a \in R$, $a \neq 0$, and $n \in N$, then*

$$a^{-n} = \frac{1}{a^n}.$$

Next, by Theorem 5.1-III, for $a \in R$, and $m, n \in N$, we have

$$(a^m)^n = a^{mn}. \tag{1}$$

If (1) is to hold for $m = 1/n$, and $a^{1/n}$ is a real number, then we must have

$$(a^{1/n})^n = a^{(1/n)(n)} = a^{n/n} = a^1 = a,$$

so that the nth power of $a^{1/n}$ must be a. A number having a as its nth power is called an **nth root** of a. In particular, for $n = 2$ or 3, an nth root is called a **square root** or a **cube root,** respectively.

Number of roots

For n odd, each $a \in R$ has just one real nth root. Thus

$$(-2)^3 = -8 \quad \text{and} \quad 2^3 = 8,$$

so that -2 is the cube root of -8, and 2 is the cube root of 8.
    For n even and $a > 0$, a has two real nth roots. Thus,

$$(-2)^4 = 16 \quad \text{and} \quad 2^4 = 16,$$

so that -2 and 2 are both fourth roots of 16.
    For n even and $a < 0$, a has no real nth root. Thus -1 has no real square root, since the square of each real number is nonnegative.
    If $a = 0$, then a has exactly one real nth root, namely 0.
    We therefore make the following definition.

Definition 5.4   *If $a \in R$, $n \in N$, then $a^{1/n}$ is the real number if one exists, and is the positive real number if two exist, such that*

$$(a^{1/n})^n = a.$$

Examples   a. $25^{1/2} = 5$ b. $-25^{1/2} = -5$ c. $(-25)^{1/2}$ is not a real
                                                                                number.

                    d. $27^{1/3} = 3$ e. $-27^{1/3} = -3$ f. $(-27)^{1/3} = -3$

Notice in parts b and e that $-25^{1/2}$ and $-27^{1/3}$ denote $-(25^{1/2})$ and $-(27^{1/3})$, respectively.

To generalize from rational exponents of the form $1/n$, for $n \in N$, to rational exponents of the form m/n, for $m \in J$, $n \in N$, we need the results expressed in the following theorem, which we state without proof.

Theorem 5.2   *If $a^{1/n} \in R$, $m \in J$, and $n \in N$, with $a \neq 0$ for $m \leq 0$, then*

$$(a^{1/n})^m = (a^m)^{1/n}.$$

Examples   a. $(16^{1/2})^3 = 4^3 = 64$ b. $(16^3)^{1/2} = 4096^{1/2} = 64$
                    c. $(16^{1/4})^6 = 2^6 = 64$ d. $(16^6)^{1/4} = 16{,}777{,}216^{1/4} = 64$

We now make the following definition.

Definition 5.5   *If $a^{1/n} \in R$, $m \in J$, and $n \in N$, then*

$$a^{m/n} = (a^{1/n})^m.$$

Observe that requiring $n \in N$ does not alter the fact that m/n can represent every rational number, since all that is done is to restrict the denominator of the fraction representing the rational number to be positive, which is always possible in light of Theorems 1.8-V and 1.8-VI. From Theorem 5.2 and Definition 5.5, we see that for $a^{1/n} \in R$,

$$a^{m/n} = (a^{1/n})^m = (a^m)^{1/n}.$$

Examples   a. $8^{2/3} = (8^{1/3})^2$ b. $16^{-3/4} = (16^{1/4})^{-3}$
                         $= 2^2 = 4$ $= 2^{-3} = \dfrac{1}{2^3} = \dfrac{1}{8}$

Because we define $a^{1/n}$ to be the positive nth root of a for a positive and n an even natural number, and since a^m is positive for a negative and m an even natural number, it follows that, for m and n *even* natural numbers and a *any* real number,

$$(a^m)^{1/n} = |a|^{m/n}.$$

For the special case $m = n$ (m and n even),

$$(a^n)^{1/n} = |a|.$$

Properties of   Powers with rational exponents have the same fundamental properties as powers
rational powers with integral exponents, as long as the powers are real numbers. Theorem 5.1, appropriately reworded for rational-number exponents, can be invoked to rewrite exponential expressions involving rational exponents.

Examples   a. $\dfrac{x^{2/3}}{x^{1/3}} = x^{2/3 - 1/3}$ b. $\left(\dfrac{a^3 b^6}{c^{12}}\right)^{2/3} = \dfrac{(a^3)^{2/3}(b^6)^{2/3}}{(c^{12})^{2/3}}$ c. $(a^6)^{1/2} = |a|^{6/2}$

         $= x^{1/3} \quad (x \neq 0)$ $= \dfrac{a^2 b^4}{c^8} \quad (c \neq 0)$ $= |a|^3$

Observe that in the last example it was necessary to use absolute-value notation because the expression has been defined to be positive, since 6 and 2 are both even.

Exercise 5.1

Write the given expression as a power with exponent 1.

Examples   a. $64^{1/2}$ b. $\left(\dfrac{8}{27}\right)^{-2/3}$

Solutions   a.   8   b. $\left[\left(\dfrac{8}{27}\right)^{1/3}\right]^{-2} = \left(\dfrac{2}{3}\right)^{-2} = \dfrac{9}{4}$

1. $(32)^{1/5}$ 2. $(-27)^{1/3}$ 3. $(81)^{-3/4}$ 4. $(81)^{-1/2}$

5. $\left(\dfrac{1}{8}\right)^{-5/3}$ 6. $\left(\dfrac{1}{8}\right)^{5/3}$ 7. $\left(\dfrac{4}{9}\right)^{3/2}$ 8. $\left(\dfrac{4}{9}\right)^{-3/2}$

Write the given expression as a product or quotient of powers in which each variable occurs but once and in which all exponents are positive. Assume all variable bases are positive and all variable exponents are natural numbers.

Examples   a. $\dfrac{(x^{1/2} y^2)^2}{(x^{2/3} y)^3}$ b. $(y^{2n} y^{n/2})^4$

Solutions   a. $\dfrac{x y^4}{x^2 y^3} = \dfrac{y}{x}$ b. $y^{8n} y^{2n} = y^{10n}$

9. $x^{1/3} x^{5/3}$ 10. $x^{4/3} x^{1/2}$ 11. $a^{2/3} a^{3/4}$ 12. $x^{1/2} x^{5/6}$

13. $\dfrac{x^{5/6}}{x^{1/2}}$ 14. $\dfrac{x^{1/2}}{x^{1/3}}$ 15. $\dfrac{x^{-2/5}}{x^{2/3}}$ 16. $\left(\dfrac{a^6}{c^3}\right)^{-2/3}$

17. $\left(\dfrac{y^4}{x^2}\right)^{1/2}$ 18. $\left(\dfrac{16}{ab^2}\right)^{1/4}$ 19. $\left(\dfrac{x^5 y^8}{y^{13}}\right)^{1/4}$ 20. $\left(\dfrac{125 x^3 y^4}{27 x^{-6} y}\right)^{1/3}$

21. $(x^2)^{n/2} (y^{2n})^{2/n}$ 22. $(x^{n/2})^2 (y^n)^{5/n}$ 23. $\dfrac{x^{2n}}{x^{n/2}}$

24. $\left(\dfrac{a^n}{b}\right)^{1/2} \cdot \left(\dfrac{b}{a^{2n}}\right)^{3/2}$ 25. $\dfrac{x^{3n} y^{2m-1}}{(x^n y^m)^{1/2}}$ 26. $\left(\dfrac{x^{2n^2}}{y^{4n}}\right)^{1/n}$

Apply the distributive law to write the given product as a sum.

Examples   a. $x^{1/2}(x^{1/2} - x)$ b. $(x^{1/3} - x)(x^{1/3} + x)$

Solutions   a. $x^{1/2}x^{1/2} - x^{1/2}x$ b. $x^{1/3}x^{1/3} - x^2$

                      $x - x^{3/2}$ $x^{2/3} - x^2$

27. $x^{1/3}(x^{2/3} - x^{1/3})$ 28. $y^{2/3}(y^{2/3} + y^{1/3})$

29. $(x^{1/2} - y^{-1/2})(x^{1/2} - y^{-1/2})$ 30. $(x^{1/2} + y^{1/2})(x^{1/2} - y^{1/2})$

31. $(x + y)^{1/2}[(x + y)^{1/2} - (x + y)]$

32. $(a - b)^{2/3}[(a - b)^{-1/3} + (a - b)]$

33. $(x^{1/3} + y^{1/3})(x^{2/3} - x^{1/3}y^{1/3} + y^{2/3})$

34. $(a^{1/3} - b^{1/3})(a^{2/3} + a^{1/3}b^{1/3} + b^{2/3})$

In the previous problems, the variables were restricted to represent positive numbers. In Exercises 35–40, consider variable bases to denote any elements of the set of real numbers and simplify.

Examples   a. $[(-3)^2]^{1/2}$ b. $[u^2(u + 5)]^{1/2}$

Solutions   a. $|-3| = 3$ b. $|u| (u + 5)^{1/2}$

35. $[(-5)^2]^{1/2}$ 36. $[(-3)^{12}]^{1/4}$ 37. $[4x^2]^{1/2}$

38. $[x^2(x - 1)]^{1/2}$ 39. $\dfrac{2}{[x^2(x + 1)]^{1/2}}$ 40. $\left[\dfrac{9}{x^6(x^2 + 1)}\right]^{1/2}$

5.2   Exponential Functions

Powers of the form b^x, where $b \in R$, $b > 0$, and x denotes a *rational number*, were discussed in Section 5.1. These can be used to define functions. Notice that the base b is restricted here to positive real numbers to ensure that b^x be real for all rational numbers x.

Since we now want to define a function over R using b^x, we must be able to interpret powers with *irrational exponents*, such as

$$b^{\sqrt{2}}, \quad b^{-\sqrt{3}}, \quad \text{and} \quad b^{\pi},$$

to be real numbers. The following theorem, which is presented without proof, will be useful in doing this.

Theorem 5.3   Let $x, y \in Q$, and let $x > y$. Then

$$b^x > b^y \text{ if } b > 1, \quad b^x = b^y \text{ if } b = 1, \quad \text{and} \quad b^x < b^y \text{ if } 0 < b < 1.$$

Powers with
real-number
exponents

You should recall that irrational numbers can be approximated by rational numbers to as great a degree of accuracy as desired. For example, $\sqrt{2} \approx 1.4$, or $\sqrt{2} \approx 1.414$, etc. Since b^x ($b > 0$) is defined for rational x, by Theorem 5.3 we can write the sequence of inequalities

$$2^1 < 2^2,$$
$$2^{1.4} < 2^{1.5},$$
$$2^{1.41} < 2^{1.42},$$
$$2^{1.414} < 2^{1.415},$$

$$\cdot$$
$$\cdot$$
$$\cdot$$

This process can be continued indefinitely, and it seems plausible and is actually true, though we shall not prove it, that the difference between the number on the left and that on the right can be made as small as we please. This being the case, the completeness property O-3 of the real numbers guarantees that there is just one number, which we denote by $2^{\sqrt{2}}$, that lies between the number on the left and the number on the right, no matter how long this process of approximation is continued. Since we can produce the same type of argument for any irrational exponent x, we shall assume that b^x ($b > 0$) is defined in this way for all real values of x and that the laws of exponents given in Theorem 5.1, appropriately reworded for real-number exponents, are valid for such powers.

Since for any given $b > 0$ and for each $x \in R$, there is only one value for b^x, the equation

$$f(x) = b^x \quad (b > 0) \tag{1}$$

defines a function. Because $1^x = 1$ for all $x \in R$, (1) defines a constant function if $b = 1$. If $b \neq 1$, we say that (1) defines an **exponential function.** Exponential functions can perhaps be visualized most clearly by considering their graphs. We illustrate two typical examples, in which $0 < b < 1$ and $b > 1$, respectively. Assigning values to x in the equations

$$f(x) = \left(\frac{1}{2}\right)^x \quad \text{and} \quad f(x) = 2^x,$$

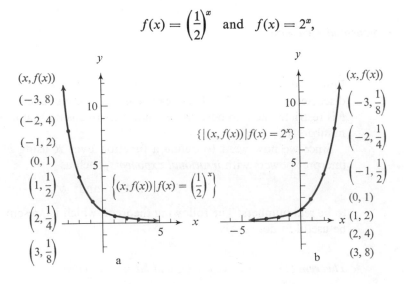

Figure 5.1

we find some ordered pairs in each solution set and sketch the graphs of

$$\{(x, f(x)) \mid f(x) = (1/2)^x\} \quad \text{and} \quad \{(x, f(x)) \mid f(x) = 2^x\},$$

shown in Figures 5.1-a and 5.1-b, respectively.

Increasing and decreasing functions   Notice that, in accordance with Theorem 5.3, the graph of the function determined by $f(x) = (1/2)^x$ goes *down* to the right, and the graph of the function determined by $f(x) = 2^x$ goes *up* to the right. For this reason, we say that the former function is a *decreasing* function and the latter is an *increasing* function. In either case, the domain is the set of real numbers, and the range is the set of positive real numbers.

Exercise 5.2

Find the second component of each of the ordered pairs that makes the pair a solution of the corresponding equation.

1. $y = 3^x$; $(0, \ ), (1, \ ), (2, \ )$
2. $y = -2^x$; $(0, \ ), (1, \ ), (2, \ )$
3. $y = -5^x$; $(0, \ ), (1, \ ), (2, \ )$
4. $y = 4^x$; $(0, \ ), (1, \ ), (2, \ )$
5. $y = \left(\frac{1}{2}\right)^x$; $(-3, \ ), (0, \ ), (3, \ )$
6. $y = \left(\frac{1}{3}\right)^x$; $(-3, \ ), (0, \ ), (3, \ )$
7. $y = 10^x$; $(-2, \ ), (1, \ ), (0, \ )$
8. $y = 10^{-x}$; $(0, \ ), (1, \ ), (2, \ )$

Graph the specified function.

9. $\{(x, y) \mid y = 4^x\}$
10. $\{(x, y) \mid y = 5^x\}$
11. $\{(x, y) \mid y = 10^x\}$
12. $\{(x, y) \mid y = 10^{-x}\}$
13. $\{(x, y) \mid y = 2^{-x}\}$
14. $\{(x, y) \mid y = 3^{-x}\}$
15. $\left\{(x, y) \mid y = \left(\frac{1}{3}\right)^x\right\}$
16. $\left\{(x, y) \mid y = \left(\frac{1}{4}\right)^x\right\}$
17. $\left\{(x, y) \mid y = \left(\frac{1}{2}\right)^{-x}\right\}$
18. $\left\{(x, y) \mid y = \left(\frac{1}{3}\right)^{-x}\right\}$

19. Graph $\{(x, f(x)) \mid f(x) = 1^x\}$. Is this an exponential function? Name the function.

20. Graph $F = \{(x, y) \mid y = 10^x, x > 0\}$ and $F^{-1} = \{(x, y) \mid x = 10^y, x > 0\}$ on the same set of axes.

21. Solve for x by inspection.

   a. $10^x = \dfrac{1}{100}$ b. $\left(\frac{1}{2}\right)^x = 16$ c. $16^x = 8$

22. Determine an integer n such that $n < x < n + 1$.

   a. $3^x = 16.2$ b. $4^x = 87.1$ c. $10^x = 0.016$

5.3   *Logarithmic Functions*

*Inverse of an
exponential
function*

In the exponential function

$$\{(x, y) \mid y = b^x \quad (b > 0, b \neq 1)\}, \tag{1}$$

illustrated in Figure 5.1 for $b = 1/2$ and $b = 2$, there is only one x associated with each y. Thus by Definition 2.3 we have the inverse function

$$\{(x, y) \mid x = b^y \quad (b > 0, b \neq 1)\}. \tag{2}$$

Observe that the relation $x > 0$ is implied by Equation (2), because there is no real number y for which b^y is not positive.

The graphs of functions of this form can be illustrated by the example

$$\{(x, y) \mid x = 10^y\}.$$

We consider x the variable denoting an element in the domain and, in the defining equation, we assign arbitrary values for x, say,

$$(0.01, \quad), (0.1, \quad), (1, \quad), (10, \quad), (100, \quad),$$

to obtain the ordered pairs

$$(0.01, -2), (0.1, -1), (1, 0), (10, 1), (100, 2).$$

These can be graphed and connected with a smooth curve as shown in Figure 5.2.

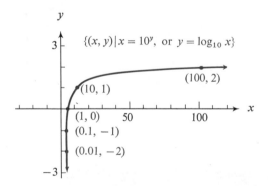

Figure 5.2

It is always useful to be able to express the variable denoting an element in the range explicitly in terms of the variable denoting an element in the domain. To do this in an equation such as that defining the function (2), we use the notation

$$y = \log_b x \quad (x > 0, b > 0, b \neq 1). \tag{3}$$

Here, $\log_b x$ is read "the logarithm to the base b of x," or "the logarithm of x to the base b." Functions defined by such equations are called **logarithmic functions**.

Properties of logarithmic functions

From the graph in Figure 5.2, we generalize from $\log_{10} x$ to $\log_b x$, and observe that for $b > 1$, a logarithmic function has the following properties:

1. The domain is the set of positive real numbers, and the range is the set of all real numbers.
2. $\log_b x < 0$ for $0 < x < 1$, $\log_b x = 0$ for $x = 1$, and $\log_b x > 0$ for $x > 1$.

Logarithmic and exponential statements

It should be recognized Equations (2) and (3) are different equations determining the same function, in the same way that $x = y + 4$ and $y = x - 4$ determine the same function, and we may use whichever equation suits our purpose. Thus, exponential statements may be written in logarithmic form, and logarithmic statements may be written in exponential form.

Examples

a. $5^2 = 25$ can be written as $\log_5 25 = 2$.

b. $8^{1/3} = 2$ can be written as $\log_8 2 = \dfrac{1}{3}$.

c. $3^{-3} = \dfrac{1}{27}$ can be written as $\log_3 \dfrac{1}{27} = -3$.

Examples

a. $\log_{10} 100 = 2$ can be written as $10^2 - 100$.
b. $\log_3 81 = 4$ can be written as $3^4 = 81$.

c. $\log_2 \dfrac{1}{2} = -1$ can be written as $2^{-1} = \dfrac{1}{2}$.

The logarithmic function associates with each positive number x the exponent y such that the power b^y is equal to x. In other words, we can think of $\log_b x$ as an exponent on b. Thus

$$b^{\log_b x} = x.$$

Laws of logarithms

Since a logarithm is an exponent, the following theorem follows directly from the properties of powers with real-number exponents.

Theorem 5.4   *If $x_1, x_2 \in R$, $x_1 > 0$, $x_2 > 0$, $b > 0$, $b \neq 1$, $m \in R$, then*

$$\text{I} \quad \log_b (x_1 x_2) = \log_b x_1 + \log_b x_2,$$

$$\text{II} \quad \log_b \frac{x_2}{x_1} = \log_b x_2 - \log_b x_1,$$

$$\text{III} \quad \log_b (x_1)^m = m \log_b x_1.$$

The validity of I can be shown as follows: Since

$$x_1 = b^{\log_b x_1} \quad \text{and} \quad x_2 = b^{\log_b x_2},$$

it follows that

$$x_1 x_2 = b^{\log_b x_1} \cdot b^{\log_b x_2} = b^{\log_b x_1 + \log_b x_2},$$

and, by the definition of a logarithm,

$$\log_b (x_1 x_2) = \log_b x_1 + \log_b x_2.$$

The validity of II and III can be established in a similar way.

Exercise 5.3

Express in logarithmic notation.

1. $4^2 = 16$ 　　　　　2. $5^3 = 125$ 　　　　　3. $3^3 = 27$

4. $8^2 = 64$ 　　　　　5. $\left(\dfrac{1}{2}\right)^2 = \dfrac{1}{4}$ 　　　　　6. $\left(\dfrac{1}{3}\right)^2 = \dfrac{1}{9}$

7. $8^{-1/3} = \dfrac{1}{2}$ 　　　　8. $64^{-1/6} = \dfrac{1}{2}$ 　　　　9. $10^2 = 100$

10. $10^0 = 1$ 　　　　　11. $10^{-1} = 0.1$ 　　　　12. $10^{-2} = 0.01$

Express in exponential notation.

13. $\log_2 64 = 6$ 　　　　14. $\log_5 25 = 2$ 　　　　15. $\log_3 9 = 2$

16. $\log_{16} 256 = 2$ 　　　17. $\log_{1/3} 9 = -2$ 　　　18. $\log_{1/2} 8 = -3$

19. $\log_{10} 1000 = 3$ 　　20. $\log_{10} 1 = 0$

Find the value of each expression.

21. $\log_7 49$ 　　　　　22. $\log_2 32$ 　　　　　23. $\log_4 64$

24. $\log_3 \dfrac{1}{3}$ 　　　　25. $\log_5 \dfrac{1}{5}$ 　　　　26. $\log_3 3$

27. $\log_2 2$ 　　　　　28. $\log_{10} 10$ 　　　　29. $\log_{10} 100$

30. $\log_{10} 1$ 　　　　　31. $\log_{10} 0.1$ 　　　　32. $\log_{10} 0.01$

Solve for x, y, or b.

Examples　　a. $\log_2 x = 3$ 　　　　　　　　b. $\log_b 2 = \dfrac{1}{2}$

Solutions　　Determine the solution by inspection or by first writing the equation equivalently in exponential form.

a. $2^3 = x$ 　　　　　　　　　　　b. $\quad b^{1/2} = 2$
　　$x = 8$ 　　　　　　　　　　　　　$(b^{1/2})^2 = (2)^2$
　　　　　　　　　　　　　　　　　　　　　$b = 4$

33. $\log_3 9 = y$ **34.** $\log_5 125 = y$ **35.** $\log_b 8 = 3$

36. $\log_b 625 = 4$ **37.** $\log_4 x = 3$ **38.** $\log_{1/2} x = -5$

39. $\log_2 \dfrac{1}{8} = y$ **40.** $\log_5 5 = y$ **41.** $\log_b 10 = \dfrac{1}{2}$

42. $\log_b 0.1 = -1$ **43.** $\log_2 x = 2$ **44.** $\log_{10} x = -3$

45. Show that $\log_b 1 = 0$ for $b > 0$. *Hint:* Express in exponential form.

46. Show that $\log_b b = 1$ for $b > 0$.

Express as the sum or difference of simpler logarithmic quantities.

Example $\log_b \left(\dfrac{xy}{z}\right)^{1/2}$

Solution   By Theorem 5.4-III,

$$\log_b \left(\frac{xy}{z}\right)^{1/2} = \frac{1}{2} \log_b \left(\frac{xy}{z}\right).$$

By Theorems 5.4-I and 5.4-II,

$$\frac{1}{2} \log_b \left(\frac{xy}{z}\right) = \frac{1}{2} [\log_b x + \log_b y - \log_b z].$$

47. $\log_b (xy)$ **48.** $\log_b (xyz)$ **49.** $\log_b \left(\dfrac{x}{y}\right)$

50. $\log_b \left(\dfrac{xy}{z}\right)$ **51.** $\log_b x^5$ **52.** $\log_b x^{1/2}$

53. $\log_b \sqrt[3]{x}$ **54.** $\log_b \sqrt[3]{x^2}$ **55.** $\log_b \sqrt{\dfrac{x}{z}}$

56. $\log_b \sqrt{xy}$ **57.** $\log_{10} \sqrt[3]{\dfrac{xy^2}{z}}$ **58.** $\log_{10} \sqrt[5]{\dfrac{x^2 y}{z^3}}$

Express as a single logarithm with coefficient 1.

Example $\dfrac{1}{2}(\log_b x - \log_b y)$

Solution   By Theorems 5.4-II and 5.4-III,

$$\frac{1}{2}(\log_b x - \log_b y) = \frac{1}{2} \log_b \left(\frac{x}{y}\right) = \log_b \left(\frac{x}{y}\right)^{1/2}$$

59. $\log_b 2x + 3 \log_b y$ **60.** $3 \log_b x - \log_b 2y$

61. $\dfrac{1}{2} \log_b x + \dfrac{2}{3} \log_b y$ **62.** $\dfrac{1}{4} \log_b x - \dfrac{3}{4} \log_b y$

63. $3 \log_b x + \log_b y - 2 \log_b z$ **64.** $\dfrac{1}{3}(\log_b x + \log_b y - 2 \log_b z)$

65. $\log_{10}(x - 2) + \log_{10} x - 2 \log_{10} z$ **66.** $\dfrac{1}{2}(\log_{10} x - 3 \log_{10} y - 5 \log_{10} z)$

5.4   *Logarithms to the Base 10*

There are two logarithmic functions of special interest; one is defined by

$$y = \log_{10} x, \tag{1}$$

and the other by

$$y = \log_e x, \tag{2}$$

where e is an irrational number with decimal approximation 2.7182818 to eight digits. Because these functions possess similar properties, and because we are more familiar with the number 10, we shall, for the present, confine our attention to (1).

Determination of $\log_{10} x$ Values for $\log_{10} x$ are called **logarithms to the base 10,** or **common logarithms.** From the definition of $\log_{10} x$,

$$10^{\log_{10} x} = x \quad (x > 0); \tag{3}$$

that is, $\log_{10} x$ is the exponent that must be placed on 10 so that the resulting power is x. Our concern in this section is that of finding $\log_{10} x$ for each positive x. First, $\log_{10} x$ can easily be determined for all values of x that are *integral* powers of 10 by inspection, directly from the definition of a logarithm:

$$\log_{10} 10 \quad = \log_{10} 10^1 = 1,$$
$$\log_{10} 100 \quad = \log_{10} 10^2 = 2,$$

and so forth; similarly,

$$\log_{10} 1 \quad = \log_{10} 10^0 \ = 0,$$
$$\log_{10} 0.1 \quad = \log_{10} 10^{-1} = -1,$$
$$\log_{10} 0.01 \quad = \log_{10} 10^{-2} = -2,$$
$$\log_{10} 0.001 = \log_{10} 10^{-3} = -3.$$

A table of logarithms is used to find $\log_{10} x$ for $1 \leq x \leq 10$ (see page 284). Consider the excerpt from this table shown in Figure 5.3. Each number in the column headed by x represents the first two significant digits of x, while each number in the row opposite x contains the third significant digit of x. The digits located at the intersection of a row and a column form the logarithm of x. For example, to find $\log_{10} 4.25$, we look at the intersection of the row opposite 4.2 under x and the column headed by 5. Thus

$$\log_{10} 4.25 = 0.6284.$$

The equality sign is used here in an inexact sense; $\log_{10} 4.25 \approx 0.6284$ is more proper, because $\log_{10} 4.25$ is irrational and cannot be precisely represented by a rational number. We shall follow customary usage, however, writing $=$ instead of $\approx$, and leave the intent to the context.

x	0	1	2	3	4	5	6	7	8	9
3.8	.5798	.5809	.5821	.5832	.5843	.5855	.5866	.5877	.5888	.5899
3.9	.5911	.5922	.5933	.5944	.5955	.5966	.5977	.5988	.5999	.6010
4.0	.6021	.6031	.6042	.6053	.6064	.6075	.6085	.6096	.6107	.6117
4.1	.6128	.6138	.6149	.6160	.6170	.6180	.6191	.6201	.6212	.6222
4.2	.6232	.6243	.6253	.6263	.6274	.6284	.6294	.6304	.6314	.6325
4.3	.6335	.6345	.6355	.6365	.6375	.6385	.6395	.6405	.6415	.6425
4.4	.6435	.6444	.6454	.6464	.6474	.6484	.6493	.6503	.6513	.6522
4.5	.6532	.6542	.6551	.6561	.6571	.6580	.6590	.6599	.6609	.6618
4.6	.6628	.6637	.6646	.6656	.6665	.6675	.6684	.6693	.6702	.6712

Figure 5.3

Characteristic and mantissa

Now suppose we wish to find $\log_{10} x$ for values of x outside the range of the table—that is, for $0 < x < 1$ or $x > 10$ as shown in color in Figure 5.4. This can be

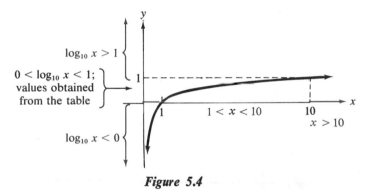

Figure 5.4

done quite readily by first representing the number in scientific notation—that is, as the product of a number between 1 and 10 and a power of 10—and then applying Theorem 5.4-I. For example,

$$\log_{10} 42.5 = \log_{10} (4.25 \times 10^1) = \log_{10} 4.25 + \log_{10} 10^1$$
$$= 0.6284 + 1 = 1.6284,$$

$$\log_{10} 425 = \log_{10} (4.25 \times 10^2) = \log_{10} 4.25 + \log_{10} 10^2$$
$$= 0.6284 + 2 = 2.6284.$$

Observe that the decimal portion of the logarithm is always 0.6284, and the integral portion is just the exponent on 10 when the number is written in scientific notation.

This process can be reduced to a mechanical one by considering $\log_{10} x$ to consist of two parts, an integral part (called the **characteristic**) and a nonnegative decimal fraction part (called the **mantissa**). Thus the table of values for $\log_{10} x$ for $1 < x < 10$ can be looked upon as a table of mantissas for $\log_{10} x$ for all $x > 0$.

For example, to find $\log_{10} 43700$, we first write

$$\log_{10} 43700 = \log_{10} (4.37 \times 10^4).$$

Upon examining the table of logarithms, we find that $\log_{10} 4.37 = 0.6405$, so that

$$\log_{10} 43700 = 4.6405,$$

where we have prefixed the characteristic 4.

Now consider an example of the form $\log_{10} x$ for $0 < x < 1$. To find $\log_{10} 0.00402$, we first write

$$\log_{10} 0.00402 = \log_{10} (4.02 \times 10^{-3}).$$

Examining the table, we find that $\log_{10} 4.02$ is 0.6042. Upon adding 0.6042 to the characteristic -3, we obtain

$$\log_{10} 0.00402 = -2.3958,$$

where the decimal portion of the logarithm is no longer 0.6042 as it is in the case of all numbers $x > 1$ for which the first three significant digits of x are 402. To circumvent this situation, it is customary to write the logarithm in a form in which the decimal part is positive. In this example, we write

$$\begin{aligned}
\log_{10} 0.00402 &= 0.6042 - 3 \\
&= 0.6042 + (7 - 10) \\
&= 7.6042 - 10,
\end{aligned}$$

and the decimal part is positive. The logarithm

$$6.6042 - 9$$

is an equally valid representation, but $7.6042 - 10$, in which a multiple of 10 is subtracted, is customary in most cases.

Determination of antilog$_{10}$ *x*   It is possible to reverse the process described in this section. In other words, being given $\log_{10} x$, an element in the range of the function, we can find x, an element in the domain. In this event, x is referred to as the **antilogarithm** (antilog$_{10}$) of $\log_{10} x$. For example, antilog$_{10}$ 1.6395 can be obtained by locating the mantissa, 0.6395, in the body of the $\log_{10}$ tables and observing that the associated antilog$_{10}$ is 4.36. Thus

$$\text{antilog}_{10}\, 1.6395 = 4.36 \times 10^1 = 43.6.$$

Linear interpolation   If we seek the common logarithm of a number that is not an entry in the table (for example, $\log_{10} 3712$), or if we seek x when $\log_{10} x$ is not an entry in the table, it is customary to use a procedure called **linear interpolation.** Observe that a table of common logarithms is a set of ordered pairs. For each number x there is an associated number $\log_{10} x$, and we have $(x, \log_{10} x)$ displayed in convenient tabular form. Because of space limitations, only three digits for the number x and four for the number $\log_{10} x$ appear in Table II (pages 284 and 285). By means of linear interpolation, however, the table can be used to find approximations to logarithms for four-digit numbers.

Let us examine geometrically the concepts involved. A portion of the graph of

$$y = \log_{10} x$$

is shown in Figure 5.5a. The curvature is exaggerated to illustrate the principle involved. We propose to use the line segment joining the points P_1 and P_2 as an approximation to the curve passing through the points. If a large graph of $y = \log_{10} x$ were available, the value of, say, $\log_{10} 4.257$ could be found by using the

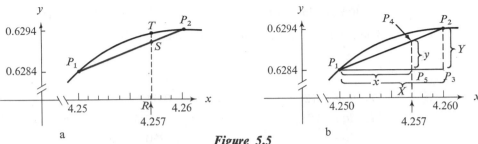

Figure 5.5

value of the ordinate (RT) to the curve for $x = 4.257$. Since there is no way to accomplish this with a table of values only, we shall instead use the value of the ordinate (RS) to the line segment as an approximation to the ordinate of the curve.

This can be accomplished directly from the set of numbers available in the table of logarithms. Consider Figure 5.5-b, where $P_2 P_3$ and $P_4 P_5$ are perpendicular to $P_1 P_3$. From geometry, we see that $\triangle P_1 P_4 P_5$ is similar to $\triangle P_1 P_2 P_3$, where the lengths of the corresponding sides are proportional, and hence

$$\frac{x}{X} = \frac{y}{Y}. \tag{4}$$

If we know any three of these numbers, the fourth can be determined. For the purpose of interpolation, we assume all of our numbers now have four-digit numerals; that is, we consider 4.250 instead of 4.25, and 4.260 instead of 4.26. We note that the point corresponding to 4.257 is located just 7/10 of the distance between the points corresponding to 4.250 and 4.260, respectively, and the value Y (0.0010) is just the difference between the logarithms 0.6284 and 0.6294. It follows from (4) that

$$\frac{7}{10} = \frac{y}{0.0010},$$

from which

$$y = \frac{7}{10}(0.0010) = 0.0007.$$

We now add 0.0007 to 0.6284 and thus obtain a good approximation to the required logarithm. That is,

$$\log_{10} 4.257 = 0.6291.$$

An example in Exercise 5.4 shows a convenient arrangement for the calculations involved in the example presented here. The antilogarithm of a number can be found by a similar procedure. With practice, however, it is possible to interpolate mentally in both cases.

Exercise 5.4

Write the characteristic of the given expression.

Examples   a. $\log_{10} 248$ b. $\log_{10} 0.0057$

Solution overleaf

Solutions   Represent the number in scientific notation.

a. $\log_{10} (2.48 \times 10^2)$ b. $\log_{10} (5.7 \times 10^{-3})$

The exponent on the base 10 is the characteristic.

2 -3, or $7 - 10$

1. $\log_{10} 312$ **2.** $\log_{10} 0.02$ **3.** $\log_{10} 0.00851$
4. $\log_{10} 8.012$ **5.** $\log_{10} 0.00031$ **6.** $\log_{10} 0.0004$
7. $\log_{10} (15 \times 10^3)$ **8.** $\log_{10} (820 \times 10^4)$

Find the specified logarithm.

Examples   a. $\log_{10} 16.8$ b. $\log_{10} 0.043$

Solutions   Represent the number in scientific notation.

a. $\log_{10} (1.68 \times 10^1)$ b. $\log_{10} (4.3 \times 10^{-2})$

Determine the mantissa from the table.

0.2253   0.6335

Add the characteristic as determined by the exponent on the base 10.

1.2253 $8.6335 - 10$

9. $\log_{10} 6.73$ **10.** $\log_{10} 891$
11. $\log_{10} 0.813$ **12.** $\log_{10} 0.00214$
13. $\log_{10} 0.08$ **14.** $\log_{10} 0.000413$
15. $\log_{10} (2.48 \times 10^2)$ **16.** $\log_{10} (5.39 \times 10^{-3})$

Find the antilogarithm.

Example $\text{antilog}_{10} 2.7364$

Solution   Locate the mantissa in the body of the table of mantissas and determine the associated antilog_{10} (a number between 1 and 10); write the characteristic 2 as an exponent on the base 10.

$$5.45 \times 10^2 = 545$$

17. $\text{antilog}_{10} 0.6128$ **18.** $\text{antilog}_{10} 0.2504$
19. $\text{antilog}_{10} 0.5647$ **20.** $\text{antilog}_{10} 3.9258$
21. $\text{antilog}_{10} (8.8075 - 10)$ **22.** $\text{antilog}_{10} (3.9722 - 5)$
23. $\text{antilog}_{10} 3.7388$ **24.** $\text{antilog}_{10} 2.0086$
25. $\text{antilog}_{10} (6.8561 - 10)$ **26.** $\text{antilog}_{10} (1.8156 - 4)$

Find the logarithm.

Example $\log_{10} 4.257$

Solution   Interpolate mentally or use the following procedure.

$$
\begin{array}{cc}
x & \log_{10} x \\
\end{array}
$$

$$
10\left\{ 7\left\{\begin{array}{c|c} 4.250 & 0.6284 \\ 4.257 & ? \end{array}\right\}y \atop 4.260 \quad 0.6294 \right\}0.0010
$$

Set up a proportion and solve for y.

$$
\frac{7}{10} = \frac{y}{0.0010}
$$

$$
y = 0.0007
$$

Add 0.0007, the value of y, to 0.6284.

$$
\log_{10} 4.257 = 0.6284 + 0.0007 = 0.6291
$$

27. $\log_{10} 4.213$ **28.** $\log_{10} 8.184$ **29.** $\log_{10} 1522$

30. $\log_{10} 203.4$ **31.** $\log_{10} 37110$ **32.** $\log_{10} 72.36$

33. $\log_{10} 0.5123$ **34.** $\log_{10} 0.008351$

Find the antilogarithm.

Example $\text{antilog}_{10} 0.6446$

Solution   Interpolate mentally or use the following procedure.

$$
\begin{array}{cc}
x & \text{antilog}_{10} x \\
\end{array}
$$

$$
0.0010\left\{ 0.0002\left\{\begin{array}{c|c} 0.6444 & 4.410 \\ 0.6446 & ? \end{array}\right\}y \atop 0.6454 \quad 4.420 \right\}0.010
$$

Set up a proportion and solve for y.

$$
\frac{0.0002}{0.0010} = \frac{y}{0.010}
$$

$$
y = 0.002
$$

Add 0.002, the value of y, to 4.410.

$$
\text{antilog}_{10} 0.6446 = 4.410 + 0.002 = 4.412
$$

35. $\text{antilog}_{10} 0.5085$ **36.** $\text{antilog}_{10} 0.8087$

37. $\text{antilog}_{10} 1.0220$ **38.** $\text{antilog}_{10} 3.0759$

39. antilog$_{10}$ (8.7055 — 10)   **40.** antilog$_{10}$ (9.8742 — 10)
41. antilog$_{10}$ (2.8748 — 3)   **42.** antilog$_{10}$ (7.7397 — 10)

43. If we use linear interpolation to find log$_{10}$ 3.751 and log$_{10}$ 3.755, which of the resulting approximations should we expect to be more nearly correct? Why?

44. If we use linear interpolation to find log$_{10}$ 1.025 and log$_{10}$ 9.025, which of the resulting approximations should we expect to be more nearly correct? Why?

Find the value of the given power by means of the table of logarithms to the base 10.

Example $10^{0.6263}$

Solution   The logarithmic function is the inverse of the exponential function. Hence, the element $10^{0.6263}$ in the range of the exponential function is antilog$_{10}$ 0.6263 in the domain of the logarithmic function. From the table of logarithms, we find that

$$10^{0.6263} = \text{antilog}_{10}\ 0.6263 = 4.23.$$

45. $10^{0.9590}$ **46.** $10^{0.8241}$ **47.** $10^{3.6990}$
48. $10^{2.3874}$ **49.** $10^{2.0531}$ **50.** $10^{1.7396}$

51. Write the following logarithms to the base 10 in a form in which both the integral and fractional parts are negative.
  a. 8.7321 — 10   **b.** 6.4187 — 10

52. Write the following logarithms to the base 10 in a form in which the fractional part is positive.
  a. −2.7113   **b.** −4.6621

5.5   Applications of Logarithms

Computations   The use of the slide rule and the advent of high-speed computing devices have almost removed the need to perform routine numerical computations with pencil and paper by logarithms. Nevertheless, we introduce the techniques involved in making such computations because the writing of the logarithmic equations involved sheds light on the properties of the logarithmic function and on the usefulness of Theorem 5.4, which we reproduce here using the base 10.

If x_1 and x_2 are positive real numbers and $m \in R$, then

$$\text{I}\quad \log_{10}(x_1 x_2) = \log_{10} x_1 + \log_{10} x_2,$$

$$\text{II}\quad \log_{10} \frac{x_1}{x_2} = \log_{10} x_1 - \log_{10} x_2,$$

$$\text{III}\quad \log_{10}(x_1)^m = m \log_{10} x_1.$$

Before illustrating the use of these laws, let us make two observations:

L-1   If $M = N$ ($M, N > 0$),   then $\log_b M = \log_b N$.
L-2   If $\log_b M = \log_b N$,   then $M = N$.

These statements follow from the fact that the variables in a logarithmic function are in one-to-one correspondence.

Example

Find the product of 3.825 and 0.00729, using logarithms and linear interpolation.

Solution

Let $N = (3.825)(0.00729)$. By Property L-1,

$$\log_{10} N = \log_{10} [(3.825)(0.00729)].$$

Now, by Theorem 5.4-I,

$$\log_{10} N = \log_{10} 3.825 + \log_{10} 0.00729,$$

and from the table of logarithms we obtain

$$\log_{10} 3.825 = 0.5826 \quad \text{and} \quad \log_{10} 0.00729 = 7.8627 - 10,$$

so that

$$\log_{10} N = (0.5826) + (7.8627 - 10)$$
$$= 8.4453 - 10.$$

The computation is completed by referring to the table for

$$N = \text{antilog}_{10} (8.4453 - 10) = 2.788 \times 10^{-2}$$
$$= 0.02788.$$

Thus we have

$$N = (3.825)(0.00729) = 0.02788.$$

Actual computation shows the product to be 0.02788425, so that the result obtained by use of logarithms is correct to four significant digits. Some error should be expected when we employ a table of logarithms, because we are using approximations and linear interpolations.

Example

Compute $\dfrac{(8.21)^{1/2}(2.17)^{2/3}}{(3.14)^3}$.

Solution

Setting

$$N = \frac{(8.21)^{1/2}(2.17)^{2/3}}{(3.14)^3},$$

we obtain

$$\log_{10} N = \log_{10} \frac{(8.21)^{1/2}(2.17)^{2/3}}{(3.14)^3}$$
$$= \log_{10} (8.21)^{1/2} + \log_{10} (2.17)^{2/3} - \log_{10} (3.14)^3$$
$$= \frac{1}{2} \log_{10} 8.21 + \frac{2}{3} \log_{10} 2.17 - 3 \log_{10} 3.14.$$

Table II provides values for the logarithms involved here, and the remainder of the computation is routine. In order to avoid confusion in computations of this sort, a systematic approach of some kind is desirable (see the example in the exercise set).

As observed on page 116, in computations involving numbers less than 1, it is sometimes convenient to use expressions other than such differences as $7.\_\_\_ - 10$ and $8.\_\_\_ - 10$ for negative characteristics.

Example   Compute $\sqrt[3]{0.0235}$.

Solution   Setting
$$N = \sqrt[3]{0.0235} = (0.0235)^{1/3},$$
we have

$$\log_{10} N = \log_{10} (0.0235)^{1/3} = \frac{1}{3} \log_{10} (0.0235).$$

From Table II, we see that the mantissa of this logarithm is 0.3711. Because we wish to multiply by 1/3, we select $7.3711 - 9$ instead of $8.3711 - 10$ for the logarithm. Thus

$$\log_{10} N = \frac{1}{3} (7.3711 - 9) \approx 2.4570 - 3,$$

where we have avoided obtaining a nonintegral negative part of the logarithm. From the table,
$$N = \text{antilog}_{10} (2.4570 - 3) = 0.2864.$$
Thus,

$$N = \sqrt[3]{0.0235} \approx 0.2864.$$

Solving exponential equations   An equation in one variable in which the variable occurs in an exponent is called an **exponential equation.** Solution sets of some such equations can be found by means of logarithms.

Example   Find the solution set of $5^x = 7$.

Solution   Since $5^x > 0$ for all x, we can apply Property L-1 and write

$$\log_{10} 5^x = \log_{10} 7;$$
and from Theorem 5.4-III,
$$x \log_{10} 5 = \log_{10} 7.$$

Dividing each member by $\log_{10} 5$, we obtain

$$x = \frac{\log_{10} 7}{\log_{10} 5} = \frac{0.8451}{0.6990} \approx 1.209.$$
The solution set is

$$\left\{ \frac{\log_{10} 7}{\log_{10} 5} \right\},$$

and a decimal approximation for the single solution is 1.209. Note that in seeking this approximation, the logarithms are *divided*, not subtracted.

Exercise 5.5

Compute by means of logarithms.

Example $\sqrt{\dfrac{(23.4)(0.\ 81)}{4.13}}$

Solution   Let $P = \sqrt{\dfrac{(23.4)(0.681)}{4.13}}$.

Then $\log_{10} P = \dfrac{1}{2}\,(\log_{10} 23.4 + \log_{10} 0.681 - \log_{10} 4.13)$.

$$\left.\begin{array}{l}\log_{10} 23.4 = \quad 1.3692 \\ \log_{10} 0.681 = \quad 9.8331 - 10\end{array}\right\}\text{add}$$

$$\left.\begin{array}{l}\log_{10} (23.4)(0.681) = 11.2023 - 10 \\ \quad\quad\log_{10} 4.13 = \underline{\ \ 0.6160\ \ } \end{array}\right\}\text{subtract}$$
$$\phantom{\log_{10} (23.4)(0.681) = 1}10.5863 - 10$$

$$\log_{10}\dfrac{(23.4)(0.681)}{4.13} = 10.5863 - 10 = 0.5863$$

$$\dfrac{1}{2}\log_{10}\dfrac{(23.4)(0.681)}{4.13} = \dfrac{1}{2}(0.5863) = 0.2931.$$

Hence, $\log_{10} P = 0.2931$, from which

$$P = \text{antilog}_{10}\ 0.2931 = 1.964.$$

1. $(2.32)(1.73)$ 2. $(83.2)(6.12)$ 3. $\dfrac{3.15}{1.37}$ 4. $\dfrac{1.38}{2.52}$

5. $\dfrac{0.0149}{32.3}$ 6. $\dfrac{0.00214}{3.17}$ 7. $(2.3)^5$ 8. $(4.62)^3$

9. $\sqrt[3]{8.12}$ 10. $\sqrt[5]{75}$ 11. $(0.0128)^5$ 12. $(0.0021)^6$

13. $\sqrt{0.0021}$ 14. $\sqrt[5]{0.0471}$ 15. $\sqrt[3]{0.0214}$ 16. $\sqrt[4]{0.0018}$

17. $\dfrac{(8.12)(8.74)}{7.19}$ 18. $\dfrac{(0.412)^2(84.3)}{\sqrt{21.7}}$ 19. $\dfrac{(6.49)^2\sqrt[3]{8.21}}{17.9}$ 20. $\dfrac{(2.61)^2(4.32)}{\sqrt{7.83}}$

Solve. Leave the solution in logarithmic form using the base 10.

Example $3^{x-2} = 16$

Solution   By Property L-1,
$$\log_{10} 3^{x-2} = \log_{10} 16.$$

By Theorem 5.4-III,
$$(x - 2)\log_{10} 3 = \log_{10} 16,$$

(Continued overleaf)

from which

$$x - 2 = \frac{\log_{10} 16}{\log_{10} 3},$$

$$x = \frac{\log_{10} 16}{\log_{10} 3} + 2.$$

The solution set is $\left\{ \dfrac{\log_{10} 16}{\log_{10} 3} + 2 \right\}$.

21. $2^x = 7$ **22.** $3^x = 4$ **23.** $3^{x+1} = 8$

24. $2^{x-1} = 9$ **25.** $7^{2x-1} = 3$ **26.** $3^{x+2} = 10$

Solve. Leave the result in the form of an equation equivalent to the given equation.

27. $y = x^n$, for n **28.** $y = Cx^{-n}$, for n

29. $y = e^{kt}$, for t **30.** $y = Ce^{-kt}$, for t

P dollars invested at an interest rate r compounded yearly yields an amount A after n years, given by $A = P(1 + r)^n$. If the interest is compounded t times yearly, the amount is given by

$$A = P\left(1 + \frac{r}{t}\right)^{tn}$$

Solve for the variable n (nearest year), r (nearest $\frac{1}{2}\%$), or A (accuracy obtainable from four-place table of mantissas).

Example $(1 + r)^{12} = 1.127$

Solution   Equate $\log_{10}$ of each member and apply Theorem 5.4-III.

$$12 \log_{10} (1 + r) = \log_{10} 1.127 = 0.0519$$

Multiply each member by 1/12.

$$\log_{10} (1 + r) = \frac{1}{12} (0.0519) = 0.0043$$

Determine antilog$_{10}$ 0.0043 and solve for r.

$$1 + r = \text{antilog}_{10} \ 0.0043 = 1.01^{-}$$

$$r = 0.01, \quad \text{or } 1\%$$

31. $(1 + 0.03)^{10} = A$ **32.** $(1 + 0.04)^8 = A$

33. $(1 + r)^6 = 1.34$ **34.** $(1 + r)^{10} = 1.48$

Example $40(1 + 0.02)^n = 51.74$

Solution   Divide each member by 40; equate $\log_{10}$ of each member, and apply Theorem 5.4, Parts II and III.

$$n \log_{10} (1.02) = \log_{10} 51.74 - \log_{10} 40$$

$$n(0.0086) = 1.7138 - 1.6021$$

$$n = \frac{0.1117}{0.0086}$$

$$n = 13 \text{ years}$$

35. $(1 + 0.04)^n = 2.19$ **36.** $(1 + 0.04)^n = 1.60$

37. $150(1 + 0.01)^{4n} = 240$ **38.** $60(1 + 0.02)^{2n} = 116$

39. Find the amount earned if \$5000 is invested at 4% for 10 years when compounded annually. When compounded semi-annually.

40. Two men, A and B, each invested \$10,000 at 4% for 20 years with a bank that computed interest quarterly. A withdrew his interest at the end of each three-month period but B let his investment be compounded. How much more did B earn than A from his investment over the period of 20 years?

Chemists define pH (hydrogen potential) of a solution by

$$p\mathrm{H} = \log_{10} \frac{1}{[\mathrm{H^+}]} = \log_{10} [\mathrm{H^+}]^{-1}.$$

Therefore,

$$p\mathrm{H} = -\log_{10} [\mathrm{H^+}],$$

where $[\mathrm{H^+}]$ is a numerical value for the concentration of hydrogen ions in aqueous solution in moles per liter.

Calculate the pH of a solution with given hydrogen-ion concentration.

41. $[\mathrm{H^+}] = 10^{-7}$ **42.** $[\mathrm{H^+}] = 4.0 \times 10^{-5}$ **43.** $[\mathrm{H^+}] = 2.0 \times 10^{-8}$

44. $[\mathrm{H^+}] = 8.5 \times 10^{-3}$ **45.** $[\mathrm{H^+}] = 6.3 \times 10^{-7}$ **46.** $[\mathrm{H^+}] = 5.7 \times 10^{-7}$

Calculate the hydrogen-ion concentration $[\mathrm{H^+}]$ of a solution with the given pH.

Example $p\mathrm{H} = 7.4$

Solution   Substitute 7.4 for pH in the relationship $p\mathrm{H} = \log_{10} \frac{1}{[\mathrm{H^+}]}$.

$$\log_{10} \frac{1}{[\mathrm{H^+}]} = 7.4$$

$$\frac{1}{[\mathrm{H^+}]} = \text{antilog}_{10}\, 7.4 = 2.5 \times 10^7$$

$$[\mathrm{H^+}] = \frac{1}{2.5 \times 10^7} = 0.4 \times 10^{-7} = 4 \times 10^{-8}$$

47. $pH = 3.0$ **48.** $pH = 4.2$ **49.** $pH = 5.6$
50. $pH = 8.3$ **51.** $pH = 7.2$ **52.** $pH = 6.9$

53. The period T of a simple pendulum is given by the formula $T = 2\pi\sqrt{L/g}$, where T is in seconds, L is the length of the pendulum in feet, and $g \approx 32$ ft/sec^2. Find the period of a pendulum 1 foot long.

54. The area $\mathscr{A}$ of a triangle in terms of the sides is given by the formula

$$\mathscr{A} = \sqrt{s(s - a)(s - b)(s - c)},$$

where a, b, and c are the lengths of the sides of the triangle and s equals one-half of the perimeter. Find the area of a triangle in which the lengths of the three sides are 2.314 inches, 4.217 inches, and 5.618 inches.

5.6   *Logarithms to the Base e*

The number e ($e \approx 2.7182818$) mentioned in Section 5.4 is of great mathematical interest and importance, and logarithms to the base e are frequently encountered in practical situations. Logarithms to the base e are called **natural logarithms.** It is possible to determine $\log_e x$ provided we have a table of $\log_{10} x$. Indeed, given a table of $\log_b x$, we can always find $\log_a x$ for any $a > 0$, $a \neq 1$. For example, if we wanted to express $\log_8 9$ in terms of $\log_{10}$, we could proceed as follows. First we let

$$\log_8 9 = N.$$

Then from our definition of a logarithm, we have

$$8^N = 9.$$

Using L-1 and Theorem 5.4-III with logarithms to the base 10, we obtain

$$N \log_{10} 8 = \log_{10} 9,$$

from which

$$N = \frac{\log_{10} 9}{\log_{10} 8}.$$

Hence,

$$\log_8 9 = \frac{\log_{10} 9}{\log_{10} 8}.$$

Conversion to another base   We can do this for the general case as follows. Since $x > 0$, $a > 0$, and $a \neq 1$, we have

$$x = a^{\log_a x}. \tag{1}$$

We can equate the logarithms to the base b, $b > 0$, and $b \neq 1$, of each member of (1) to obtain

$$\log_b x = \log_b a^{\log_a x}.$$

By Theorem 5.4-III, this yields

$$\log_b x = \log_a x \cdot \log_b a,$$

from which we obtain

$$\log_a x = \frac{\log_b x}{\log_b a}.$$ (2)

Conversion to
base e

Equation (2) gives us a means of finding $\log_a x$ when we have a table of logarithms to the base b. In particular, if $a = e$ and $b = 10$, we have

$$\log_e x = \frac{\log_{10} x}{\log_{10} e}.$$ (3)

Since $\log_{10} e \approx 0.4343$, Equation (3) can be written

$$\log_e x = \frac{\log_{10} x}{0.4343}$$

$$= \frac{1}{0.4343} \log_{10} x,$$ (4)

or

$$\log_e x = 2.3026 \log_{10} x,$$ (5)

which gives us a direct means of approximating $\log_e x$ when we have a table of logarithms to the base 10.

A table of $\log_e x$ will be found on page 287. We cannot, however, do as we did with a table of $\log_{10} x$—that is, simply list mantissas for logarithms over a certain interval and then manipulate characteristics to take care of all other intervals. Our numeration system is based on 10 and not on e. With the help of logarithms, though, we can use Formula (5), above, and the table of $\log_{10}$ on page 284, to find $\log_e x$.

Example

Find $\log_e 278$.

Solution

We shall do this in two ways.

1. Since
$$\log_e x = 2.3026 \log_{10} x,$$

using the $\log_{10}$ table, we have

$$\log_e 278 = 2.3026 \log_{10} 278$$
$$= 2.3026(2.4440)$$
$$= 5.63.$$

2. Alternatively, we observe that $278 = 2.78 \times 10^2$. Using Table IV ($\log_e N$), we have

$$\log_e 278 = \log_e(2.78 \times 10^2) = \log_e 2.78 + 2 \log_e 10$$
$$= 1.0223 + 2(2.3026)$$
$$= 1.0223 + 4.6052$$
$$= 5.6275 \approx 5.63.$$

Example   Find $\log_e 0.278$.

Solution   This time, let us simply observe that $0.278 = 2.78 \times 10^{-1}$. Hence

$$\log_e 0.278 = \log_e (2.78 \times 10^{-1}) = \log_e 2.78 + \log_e 10^{-1}$$
$$= \log_e 2.78 - \log_e 10$$
$$= 1.0224 - 2.3026$$
$$= -1.2802.$$

As one would expect, the logarithm is negative.

The process used in the preceding examples can be reversed to find antilog$_e$ x. If, however, a table for the exponential function $\{(x, y) \mid y = e^x\}$ is available (see Table III on page 286), the antilog$_e$ x can be read directly from this table, since antilog$_e$ $x = e^x$.

Examples   a. Find antilog$_e$ 3.2.   b. Find antilog$_e$ (-3.2).

Solutions   The values can be read directly from Table III.

a. antilog$_e$ $3.2 = e^{3.2} = 24.533$ b. antilog$_e$ $(-3.2) = e^{-3.2} = 0.0408$.

Exercise 5.6

Find the logarithm.

Example $\log_3 7$

Solution   Use logarithms to base 10 to evaluate. From Equation (2) on page 127,

$$\log_3 7 = \frac{\log_{10} 7}{\log_{10} 3} = \frac{0.8451}{0.4771} \approx 1.77.$$

1. $\log_2 10$	**2.** $\log_2 5$	**3.** $\log_5 240$	**4.** $\log_3 18$
5. $\log_7 8.1$	**6.** $\log_5 60$	**7.** $\log_{100} 38$	**8.** $\log_{100} 240$

Find the logarithm directly from Table IV on page 287.

9. $\log_e 3$	**10.** $\log_e 8$	**11.** $\log_e 17$	**12.** $\log_e 98$
13. $\log_e 327$	**14.** $\log_e 107$	**15.** $\log_e 450$	**16.** $\log_e 605$

Find the antilogarithm directly from Table III on page 286.

17. antilog$_e$ 0.50	**18.** antilog$_e$ 1.5	**19.** antilog$_e$ 3.4
20. antilog$_e$ 4.5	**21.** antilog$_e$ 0.231	**22.** antilog$_e$ 1.43
23. antilog$_e$ (-0.15)	**24.** antilog$_e$ (-0.95)	**25.** antilog$_e$ (-2.5)
26. antilog$_e$ (-4.2)	**27.** antilog$_e$ (-0.255)	**28.** antilog$_e$ (-3.65)

29. The amount of a radioactive element remaining at any time t is given by $y = y_0 e^{-0.4t}$, where t is in seconds and y_0 is the amount present initially. How much of the element would remain after three seconds if 40 grams were present initially?

30. The number of bacteria present in a culture is related to time by the formula $N = N_0 e^{0.04t}$, where N_0 is the amount of bacteria present at time $t = 0$, and t is time in hours. If 10,000 bacteria are present 10 hours after the beginning of an experiment, how many were present when $t = 0$?

31. The atmospheric pressure p, in inches of mercury, is given approximately by $p = 30.0(10)^{-0.09a}$, where a is the altitude in miles above sea level. What is the atmospheric pressure at sea level? At 3 miles above sea level?

32. The intensity I (in lumens) of a light beam after passing through a thickness t (in centimeters) of a medium having an absorption coefficient of 0.1 is given by $I = 1000e^{-0.1t}$. How many centimeters of the material would reduce the illumination to 800 lumens?

Chapter Review

[5.1] *Write the given expression as a product or quotient in which each variable occurs at most once in the expression and all exponents are positive. Assume that all variable bases are positive and all variable exponents are natural numbers.*

1. $(x^2 y^4)^3$

2. $\dfrac{x^2 y^{-3}}{x^{-1} y}$

3. $\dfrac{x^0 y^{-2}}{(xy)^{-2}}$

4. $\left(\dfrac{x^{-2} y^0}{y^{-3}}\right)^{-1}$

5. $x^{5/4} x^{3/2}$

6. $\left(\dfrac{x^2}{y^3}\right)^{-1/6}$

7. $\dfrac{x^{2n} x^n}{x^{n+1}}$

8. $\left(\dfrac{x^n y^{n+1}}{x y^{n-1}}\right)^2$

9. $(x^n y^{2n})^{-1/2}$

10. $\left(\dfrac{y^n}{x^n y^{2n}}\right)^{-1/n}$

Apply the distributive law to write the product as a sum.

11. $y^{3/2}(y^{3/2} - y^{1/2})$

12. $(y - y^{1/2})(y + y^{1/2})$

[5.2] 13. Graph the function $\{(x, y) \mid y = 3^x\}$.

14. Solve by inspection: $10^x = \dfrac{1}{1000}$.

[5.3] *Express in logarithmic notation.*

15. $16^{-1/2} = \dfrac{1}{4}$

16. $7^3 = 343$

Express in exponential notation.

17. $\log_2 8 = 3$

18. $\log_{10} 0.0001 = -4$

Solve.

19. $\log_2 16 = y$

20. $\log_{10} x = 3$

Express as the sum or difference of simpler logarithmic quantities.

21. $\log_{10} \sqrt[3]{xy^2}$

22. $\log_{10} \dfrac{2R^3}{\sqrt{PQ}}$

Express as a single logarithm with coefficient 1.

23. $2 \log_b x - \dfrac{1}{3} \log_b y$

24. $\dfrac{1}{3}(2 \log_{10} x + \log_{10} y) - 3 \log_{10} z$

[5.4]   *Find the logarithm. Interpolate as required.*

25. $\log_{10} 42$

26. $\log_{10} 0.00314$

27. $\log_{10} 682.4$

28. $\log_{10} 0.04142$

Find the antilogarithm. Interpolate as required.

29. $\text{antilog}_{10} \, 1.8287$

30. $\text{antilog}_{10} \, (8.6684 - 10)$

31. $\text{antilog}_{10} \, 0.4240$

32. $\text{antilog}_{10} \, (9.8224 - 10)$

33. Find the value of $10^{0.4624}$.

34. Find the value of $10^{2.9518}$.

[5.5]   *Compute by means of logarithms.*

35. $(8.73)(41.6)$

36. $\dfrac{85.9}{1.73}$

37. $\dfrac{(4.78)(62.3)^2}{12.8}$

38. $\sqrt[3]{4.3} \, \sqrt[5]{0.024}$

Solve. Leave the solution in logarithmic form, using the base 10.

39. $3^x = 2$

40. $3^{x+1} = 80$

[5.6]   *Find the value of the given expression, using Table II.*

41. $\log_3 11$

42. $\log_2 120$

43. Find $\log_e 130$, using Table IV.

44. Find $\text{antilog}_e \, 4.8$, using Table III.

45. Find $\text{antilog}_e \, (-0.24)$, using Table III.

46. Given $y = y_0 e^{-0.4t}$, find y if $y_0 = 30$ and $t = 5$.

6 Circular Functions

6.1 The Circular Functions Cosine and Sine

Notion of periodicity

The graphs of some functions display interesting cyclical characteristics. For example, the x-axis shown in Figure 6.1 can clearly be divided into equal successive

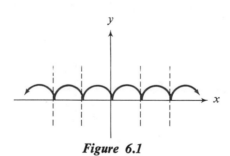

Figure 6.1

subintervals in such a way that the graph over each subinterval is a repetition of the graph over every other equal subinterval. Functions that display this property are said to be "periodic," and the length of each of the equal subintervals is called a **period** of the function.

Definition 6.1 *If f is a function with domain $D \subset R$, such that for some $p \in R$, $p \neq 0$, the values $x + p$ and $x - p$ are in D for each $x \in D$, and*

$$f(x + p) = f(x),$$

then f is **periodic** *with period p.*

If there is a least positive number p for which the function is periodic (see Figure 6.2), then p is called the **fundamental period** of the function. Of course, if a function is periodic with period p, it is also periodic with period $2p$, $3p$, $-p$, and,

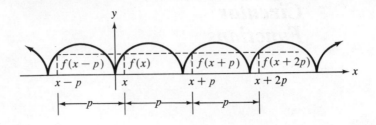

Figure 6.2

in general, kp, $k \in J$, $k \neq 0$; but we are primarily concerned only with the fundamental period, and when we refer to the period of a function we shall ordinarily mean its fundamental period. The most common periodic functions are the circular functions (defined using arc lengths on a circle) and trigonometric functions (defined using angles). In this chapter, we shall study circular functions.

Cosine and sine functions

Periodic functions can be defined using the **unit circle** with equation $x^2 + y^2 = 1$. Consider, intuitively, a point moving steadily in a counterclockwise direction around the circle. At any given time, the moving point occupies a position on the circle; the point, in turn, is associated with an ordered pair (x, y). If the distance along the circle from the point $(1, 0)$ to the point (x, y) is designated by s, then we can associate the real number s with the ordered pair (x, y) (Figure 6.3-a). We shall

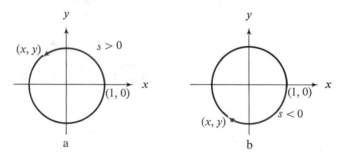

Figure 6.3

assume that every arc of a circle has a length and that there is a one-to-one correspondence between the members of the set of nonnegative real numbers and the lengths of all arcs of the unit circle measured in a *counterclockwise* direction from the point $(1, 0)$ to points (x, y) on the circle. Let us agree that values of $s < 0$ denote lengths of arc measured from $(1, 0)$ in a *clockwise* direction to points (x, y) on the circle (Figure 6.3-b). This extends the one-to-one correspondence of the set of lengths of arcs on the unit circle to the entire set R of real numbers.

Since the circumference of a circle is $2\pi r$ and the radius of the unit circle is 1, the distance once around the unit circle is 2π, twice around is 4π, and so on. Furthermore, the distance halfway around is π, one-fourth of the way around is $\pi/2$, and so forth. The endpoints of selected arcs as viewed in a counterclockwise direction from the point $(1, 0)$ and their corresponding arc lengths are shown in Figure 6.4.

Especially useful are the associations of the members of R with the members of the set of all first components of the ordered pairs (x, y), and of the members of R with the members of the set of all second components of the ordered pairs. Before making this association, let us first rename these components.

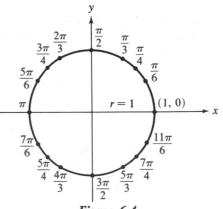

Figure 6.4

Definition 6.2   *If (x, y) is the point at arc length s from $(1, 0)$ on the unit circle with equation $x^2 + y^2 = 1$, then x is the* **cosine of s,** *and y is the* **sine of s.** *We denote this by writing*

$$x = \cos s \quad and \quad y = \sin s.$$

In other words, the first component of the point (x, y) located at arc length s from $(1, 0)$ on the unit circle is called the cosine of s, and the second component is called the sine of s; and we denote these by $\cos s$ and $\sin s$, respectively. Thus, $(x, y) = (\cos s, \sin s)$. Having given meaning to the symbolism $\cos s$ and $\sin s$, we can now define two new functions.

Definition 6.3   *If $s \in R$, then*

     I   **cosine** $= \{(s, x) | x = \cos s\}$,

     II   **sine** $= \{(s, y) | y = \sin s\}$.

Domain and range of sine and cosine

Since s is the length of an arc on the unit circle, the domain of each of these functions is the set R. The range of the cosine function is the set of all first components of the ordered pairs corresponding to points on the unit circle, and hence is the set $\{x \mid |x| \le 1\}$. Similarly, the range of the sine function is $\{y \mid |y| \le 1\}$.

    Since, for $k \in J$,

$$\cos (s + 2k\pi) = \cos s \tag{1}$$

and

$$\sin (s + 2k\pi) = \sin s, \tag{2}$$

the function cosine and sine are periodic, with fundamental period 2π (see Exercises 25 and 26, page 141).

Signs of cosine and sine

Because $\cos s$ and $\sin s$ are simply the coordinates of points on the unit circle, they are positive or negative in accord with the values of x and y in the various quadrants. Table 6.1 summarizes in a convenient way the sign associated with $x = \cos s$ and $y = \sin s$ in each quadrant.

Table 6.1

Quadrant II	Quadrant I
x or $\cos s < 0$ y or $\sin s > 0$	x or $\cos s > 0$ y or $\sin s > 0$
Quadrant III	Quadrant IV
x or $\cos s < 0$ y or $\sin s < 0$	x or $\cos s > 0$ y or $\sin s < 0$

cos s and sin s
in terms of
each other

By Definition 6.2, $x = \cos s$ and $y = \sin s$ are subject to the condition that $x^2 + y^2 = 1$. Hence we have the following basic identity relating $\cos s$ and $\sin s$.

Theorem 6.1   *For every $s \in R$,*

$$\cos^2 s + \sin^2 s = 1. \tag{3}$$

Note that for convenience we write $\cos^2 s$ for $(\cos s)^2$ and $\sin^2 s$ for $(\sin s)^2$. Now Theorem 6.1 and Table 6.1 can be used to write

$$\sin s = \begin{cases} \sqrt{1 - \cos^2 s} & \text{in Quadrants I and II,} \quad (4a) \\ -\sqrt{1 - \cos^2 s} & \text{in Quadrants III and IV,} \quad (4b) \end{cases}$$

and

$$\cos s = \begin{cases} \sqrt{1 - \sin^2 s} & \text{in Quadrants I and IV,} \quad (5a) \\ -\sqrt{1 - \sin^2 s} & \text{in Quadrants II and III.} \quad (5b) \end{cases}$$

Therefore, if either $\sin s$ or $\cos s$ is known and the quadrant in which the terminal point of the arc of length s lies can be determined, then we can find the value for the other function.

Example

Given that $\cos s = -3/5$ and $\pi < s < 3\pi/2$, find $\sin s$.

Solution

Using (4b), we have

$$\sin s = -\sqrt{1 - \cos^2 s} = -\sqrt{1 - \left(-\frac{3}{5}\right)^2} = -\sqrt{\frac{16}{25}} = -\frac{4}{5}.$$

cos (−s) and
sin (−s)

Because the defining unit circle is symmetric to the horizontal axis, it follows that if $(\cos s, \sin s) = (x, y)$, then

$$(\cos(-s), \sin(-s)) = (x, -y).$$

We thus have the following theorem, which is illustrated in Figure 6.5 for the case $0 < s < \pi/2$.

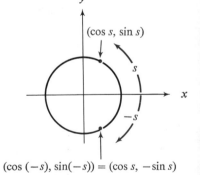

Figure 6.5

Theorem 6.2   *For every $s \in R$*

I $\cos(-s) = \cos s,$ (6)

II $\sin(-s) = -\sin s.$ (7)

Exercise 6.1

Use relationship (1) or (2) on page 133 to determine a value of s, $0 \le s < 2\pi$, for which the given equation is true. Then use Table 6.1 to state whether the given function value is positive or negative.

Example $\sin s = \sin \dfrac{11\pi}{4}$

Solution   Since $\dfrac{11\pi}{4}$ lies between 2π and 3π, we can write

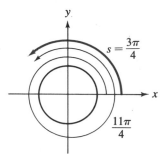

$$\sin \frac{11\pi}{4} = \sin \left(\frac{3\pi}{4} + 2\pi \right).$$

By relationship (2),

$$\sin \left(\frac{3\pi}{4} + 2\pi \right) = \sin \frac{3\pi}{4},$$

so that the desired value of s is $\dfrac{3\pi}{4}$. Since $\dfrac{3\pi}{4}$ lies between $\dfrac{\pi}{2}$ and π, we find from Table 6.1 that $\sin \dfrac{11\pi}{4}$ is positive.

1. $\cos s = \cos \dfrac{13\pi}{3}$ 2. $\cos s = \cos \dfrac{27\pi}{5}$

3. $\sin s = \sin \dfrac{17\pi}{7}$ 4. $\sin s = \sin \dfrac{29\pi}{4}$

5. $\cos s = \cos \dfrac{41\pi}{5}$ 6. $\sin s = \sin \dfrac{36\pi}{7}$

7. $\sin s = \sin \left(-\dfrac{4\pi}{3} \right)$ 8. $\cos s = \cos \left(-\dfrac{7\pi}{5} \right)$

9. $\cos s = \cos \left(-\dfrac{31\pi}{4} \right)$ 10. $\sin s = \sin \left(-\dfrac{17\pi}{3} \right)$

In Exercises 11–20, find the required function value and state the quadrant in which s terminates.

Example   Given that $\cos s = \dfrac{\sqrt{7}}{4}$ and $\sin s < 0$, find $\sin s$.

Solution   Since $\cos s = \dfrac{\sqrt{7}}{4}$ and $\sin s < 0$, s terminates in Quadrant **IV** and we see from (4b) on page 134 that

$$\sin s = -\sqrt{1 - \cos^2 s} = -\sqrt{1 - \left(\frac{\sqrt{7}}{4} \right)^2}$$

$$= -\sqrt{1 - \frac{7}{16}} = -\sqrt{\frac{9}{16}} = -\frac{3}{4}.$$

11. Given that $\sin s = 3/5$ and $\cos s > 0$,   find $\cos s$.
12. Given that $\cos s = 4/5$ and $\sin s < 0$,   find $\sin s$.
13. Given that $\cos s = 5/13$ and $\sin s < 0$,   find $\sin s$.
14. Given that $\sin s = 12/13$ and $\cos s < 0$,   find $\cos s$.

15. Given that $\sin s = -2/3$ and $\cos s > 0$,   find $\cos s$.
16. Given that $\cos s = -1/4$ and $\sin s > 0$,   find $\sin s$.
17. Given that $\cos s = -\sqrt{3}/2$ and $\sin s < 0$,   find $\sin s$.
18. Given that $\sin s = -\sqrt{3}/2$ and $\cos s > 0$,   find $\cos s$.
19. Given that $\sin s = -3/5$ and $\cos s > 0$,   find $\cos s$.
20. Given that $\cos s = -4/5$ and $\sin s > 0$,   find $\sin s$.

Use Theorem 6.2 to write the given expression in the form $\sin s$ *or* $\cos s$, *where* $s > 0$.

Example $\cos\left(-\dfrac{\pi}{3}\right)$

Solution $\cos\left(-\dfrac{\pi}{3}\right) = \cos\dfrac{\pi}{3}$

21. $\sin\left(-\dfrac{\pi}{3}\right)$ 22. $\cos\left(-\dfrac{\pi}{4}\right)$ 23. $\cos\left(-\dfrac{3\pi}{4}\right)$

24. $\sin\left(-\dfrac{11\pi}{6}\right)$ 25. $\sin\left(-\dfrac{13\pi}{6}\right)$ 26. $\cos\left(-\dfrac{20\pi}{3}\right)$

27. Use a figure similar to Figure 6.5 to argue that Theorem 6.2 is valid for $\pi/2 < s < \pi$.

6.2   *Values of Cosine and Sine for Special Arc Lengths*

In general, finding $\cos s$ and $\sin s$ for a given real number s is a difficult matter. However, some values of $\cos s$ and $\sin s$, corresponding to special values of s, are readily available.

Quadrantal function values

Figure 6.6 shows the coordinates of four points on the unit circle with which we can associate specific values of s. For $s = 0$,

$$\cos 0 = 1 \quad \text{and} \quad \sin 0 = 0. \tag{1}$$

Furthermore, since the arc of a circle included in a quadrant is one-fourth of the circumference of the circle, that is, $2\pi/4$, we know immediately that

$$\cos\frac{\pi}{2} = 0 \quad \text{and} \quad \sin\frac{\pi}{2} = 1, \tag{2}$$

$$\cos \pi = -1 \quad \text{and} \quad \sin \pi = 0, \tag{3}$$

$$\cos\frac{3\pi}{2} = 0 \quad \text{and} \quad \sin\frac{3\pi}{2} = -1. \tag{4}$$

Equations (1) and (2) on page 133 can be used to find values of cos s and sin s for values of s differing by $2k\pi$, $k \in J$, from those obtained above.

Example   Find $\cos \dfrac{7\pi}{2}$.

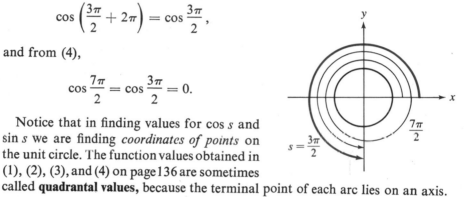

Figure 6.6

Solution   Expressing $7\pi/2$ as the sum of an integral multiple of 2π and a number between 0 and 2π, we have $7\pi/2 = 3\pi/2 + 2\pi$. A figure such as that shown at the right below is sometimes helpful. Since $\cos(s + 2k\pi) - \cos s$, we have

$$\cos\left(\frac{3\pi}{2} + 2\pi\right) = \cos \frac{3\pi}{2},$$

and from (4),

$$\cos \frac{7\pi}{2} = \cos \frac{3\pi}{2} = 0.$$

Notice that in finding values for cos s and sin s we are finding *coordinates of points* on the unit circle. The function values obtained in (1), (2), (3), and (4) on page 136 are sometimes called **quadrantal values,** because the terminal point of each arc lies on an axis.

We can find other special values for (cos s, sin s) by using the geometry of the unit circle and the distance formula,

$$d^2 = (x_2 - x_1)^2 + (y_2 - y_1)^2.$$

Values for multiples of $\pi/4$ First consider $s = \pi/4$. Figure 6.7 shows the unit circle and the designated value for s. Since (x, y) bisects the arc from $(1, 0)$ to $(0, 1)$, it follows from geometric considerations that $x = y$. Because $x^2 + y^2 = 1$, we have

$$x^2 + x^2 = 1,$$
$$2x^2 = 1,$$
$$x^2 = \frac{1}{2},$$

and

$$x = \frac{1}{\sqrt{2}} = \frac{\sqrt{2}}{2} \quad \text{or} \quad x = -\frac{1}{\sqrt{2}} = -\frac{\sqrt{2}}{2}.$$

Now, both x and y are positive in the first quadrant, so the desired value for x is $1/\sqrt{2}$. Since $x = y$, y is also equal to $1/\sqrt{2}$, and hence we have

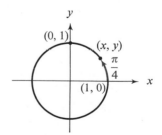

Figure 6.7

$$\cos \frac{\pi}{4} = \frac{1}{\sqrt{2}} = \frac{\sqrt{2}}{2} \quad \text{and} \quad \sin \frac{\pi}{4} = \frac{1}{\sqrt{2}} = \frac{\sqrt{2}}{2}.$$

Values for cos s and sin s for s equal to $3\pi/4$, $5\pi/4$, and $7\pi/4$ can be found using geometric symmetry as shown in Figure 6.8. These are listed in Table 6.2.

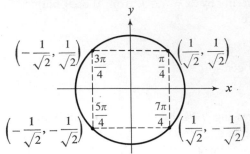

Figure 6.8

Values for multiples of $\pi/6$

Next, consider $s = \pi/6$, as pictured in Figure 6.9. If the ordered pair corresponding to $s = \pi/6$ is (x, y), then the ordered pair corresponding to $s = -\pi/6$ is $(x, -y)$. Now the arc from $(x, -y)$ to (x, y) is of length $\pi/6 + \pi/6 = \pi/3$, and so is the length of the arc from (x, y) to $(0, 1)$. Because equal arcs of a circle subtend equal chords, the distance from $(0, 1)$ to (x, y) is the same as the distance from (x, y) to

Table 6.2

s	$\cos s$	$\sin s$	s	$\cos s$	$\sin s$
0	1	0	π	-1	0
$\dfrac{\pi}{6}$	$\dfrac{\sqrt{3}}{2}$	$\dfrac{1}{2}$	$\dfrac{7\pi}{6}$	$-\dfrac{\sqrt{3}}{2}$	$-\dfrac{1}{2}$
$\dfrac{\pi}{4}$	$\dfrac{1}{\sqrt{2}}$	$\dfrac{1}{\sqrt{2}}$	$\dfrac{5\pi}{4}$	$-\dfrac{1}{\sqrt{2}}$	$-\dfrac{1}{\sqrt{2}}$
$\dfrac{\pi}{3}$	$\dfrac{1}{2}$	$\dfrac{\sqrt{3}}{2}$	$\dfrac{4\pi}{3}$	$-\dfrac{1}{2}$	$-\dfrac{\sqrt{3}}{2}$
$\dfrac{\pi}{2}$	0	1	$\dfrac{3\pi}{2}$	0	-1
$\dfrac{2\pi}{3}$	$-\dfrac{1}{2}$	$\dfrac{\sqrt{3}}{2}$	$\dfrac{5\pi}{3}$	$\dfrac{1}{2}$	$-\dfrac{\sqrt{3}}{2}$
$\dfrac{3\pi}{4}$	$-\dfrac{1}{\sqrt{2}}$	$\dfrac{1}{\sqrt{2}}$	$\dfrac{7\pi}{4}$	$\dfrac{1}{\sqrt{2}}$	$-\dfrac{1}{\sqrt{2}}$
$\dfrac{5\pi}{6}$	$-\dfrac{\sqrt{3}}{2}$	$\dfrac{1}{2}$	$\dfrac{11\pi}{6}$	$\dfrac{\sqrt{3}}{2}$	$-\dfrac{1}{2}$
π	-1	0	2π	1	0

$(x, -y)$. Using the distance formula, we therefore have

$$(x - 0)^2 + (y - 1)^2 = (x - x)^2 + (-y - y)^2,$$

or

$$x^2 + y^2 - 2y + 1 = 4y^2.$$

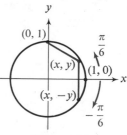

Figure 6.9

Since $x^2 + y^2 = 1$, we substitute 1 for $x^2 + y^2$ in this equation and obtain

$$1 - 2y + 1 = 4y^2,$$
$$4y^2 + 2y - 2 = 0,$$
$$2(2y - 1)(y + 1) = 0,$$

so that

$$y = \frac{1}{2} \quad \text{or} \quad y = -1.$$

Because (x, y) is in the first quadrant, we must select the value $1/2$ as the y-coordinate. Next, we use $x^2 + y^2 = 1$ to find a value for x. Thus

$$x^2 + \left(\frac{1}{2}\right)^2 = 1,$$

from which

$$x = \pm \frac{\sqrt{3}}{2}.$$

Because (x, y) is in the first quadrant, we have $x = \sqrt{3}/2$, and it follows that

$$\cos \frac{\pi}{6} = \frac{\sqrt{3}}{2} \quad \text{and} \quad \sin \frac{\pi}{6} = \frac{1}{2}. \quad (5)$$

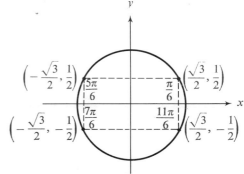

With this information, values for $\cos s$ and $\sin s$ for s equal to $5\pi/6$, $7\pi/6$, and $11\pi/6$ can be obtained using geometric symmetry, as shown in Figure 6.10. These values are also listed in Table 6.2.

Figure 6.10

From Equation (5), we can quickly find values for $\cos (\pi/3)$ and $\sin (\pi/3)$ by symmetry. Figure 6.11 shows the point (x, y) associated with s equal to $\pi/3$;

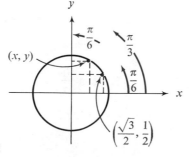

Figure 6.11

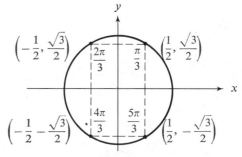

Figure 6.12

clearly, the abscissa of (x, y) is the ordinate of $(\sqrt{3}/2, 1/2)$, and the ordinate of (x, y) is the abscissa of $(\sqrt{3}/2, 1/2)$. Hence

$$\cos \frac{\pi}{3} = \frac{1}{2} \quad \text{and} \quad \sin \frac{\pi}{3} = \frac{\sqrt{3}}{2} \,.$$

Again, values for cos s and sin s for s equal to $2\pi/3$, $4\pi/3$, and $5\pi/3$ can be obtained from geometric considerations, as shown in Figure 6.12. These values are also listed in Table 6.2.

Table 6.2, in conjunction with Equations (1) and (2), page 133, can be used to find values of cos s and sin s for values of s differing by $2k\pi$, $k \in J$, from those listed in the table.

Example   Find $\cos \dfrac{8\pi}{3}$ and $\sin \dfrac{8\pi}{3}$.

Solution   We first observe that $\dfrac{8\pi}{3} = \dfrac{2\pi}{3} + \dfrac{6\pi}{3} = \dfrac{2\pi}{3} + 2\pi \,.$

Then, from (1) and (2) on page 133 and Table 6.2,

$$\cos \frac{8\pi}{3} = \cos \left(\frac{2\pi}{3} + 2\pi \right) = \cos \frac{2\pi}{3} = -\frac{1}{2}$$

and

$$\sin \frac{8\pi}{3} = \sin \left(\frac{2\pi}{3} + 2\pi \right) = \sin \frac{2\pi}{3} = \frac{\sqrt{3}}{2}$$

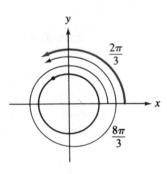

Exercise 6.2

In Exercises 1–16, use relationships 1 and 2 from page 133, together with Table 6.2, to find the given function value.

Example $\sin \dfrac{11\pi}{4}$

Solution $\sin \left(\dfrac{3\pi}{4} + 2\pi \right) = \sin \dfrac{3\pi}{4} = \dfrac{1}{\sqrt{2}}$

1. $\cos \dfrac{9\pi}{4}$ 2. $\sin \dfrac{9\pi}{4}$ 3. $\cos \dfrac{10\pi}{3}$ 4. $\sin \dfrac{5\pi}{3}$

5. $\sin \dfrac{15\pi}{6}$ 6. $\cos \dfrac{15\pi}{6}$ 7. $\sin \dfrac{11\pi}{2}$ 8. $\cos \dfrac{11\pi}{2}$

9. $\sin 8\pi$ 10. $\cos 12\pi$ 11. $\sin(-5\pi)$ 12. $\cos(-7\pi)$

13. $\cos \left(-\dfrac{7\pi}{4}\right)$ 14. $\sin \left(-\dfrac{9\pi}{4}\right)$ 15. $\cos \left(-\dfrac{8\pi}{3}\right)$ 16. $\sin \left(-\dfrac{11\pi}{2}\right)$

In Exercises 17–24, assume $0 \leq s < 2\pi$ and find the value of s for which both conditions are true. Use Table 6.2 as necessary.

17. $\sin s = 1/\sqrt{2}$ and $\cos s > 0$ 18. $\cos s = 1/\sqrt{2}$ and $\sin s < 0$

19. $\cos s = -\sqrt{3}/2$ and $\sin s > 0$ 20. $\sin s = -1/2$ and $\cos s < 0$

21. $\sin s = \sqrt{3}/2$ and $\cos s < 0$ 22. $\cos s = -1/\sqrt{2}$ and $\sin s > 0$

23. $\cos s = -1/2$ and $\sin s < 0$ 24. $\sin s = -\sqrt{3}/2$ and $\cos s > 0$

25. Use the fact that $(1, 0)$ is the only point on the unit circle with first coordinate 1 to show that there is no value a, $0 < a < 2\pi$, that is a period of the cosine function.

26. Usc the fact that $(0, 1)$ is the only point on the unit circle with second coordinate 1 to show that there is no value a, $0 < a < 2\pi$, that is a period of the sine function.

27. Prove that $f = \{(s, y) \mid y = \cos s + \sin s\}$ is periodic with period 2π.

6.3   *Properties of the Cosine Function*

Figure 6.13 shows selected points on the unit circle together with their coordinates. Since the arc length from the point $(1, 0)$ to $(\cos (s_1 + s_2), \sin (s_1 + s_2))$ is $s_1 + s_2$ and from $(\cos s_1, -\sin s_1)$ to $(\cos s_2, \sin s_2)$ the arc length is also $s_1 + s_2$, thc respective chords have equal lengths. Using the distance formula to express this fact, we have

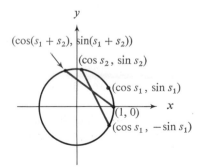

Figure 6.13

$$\sqrt{[\cos (s_1 + s_2) - 1]^2 + [\sin (s_1 + s_2) - 0]^2}$$
$$= \sqrt{(\cos s_2 - \cos s_1)^2 + [\sin s_2 - (-\sin s_1)]^2},$$

or, squaring both members,

$$[\cos (s_1 + s_2) - 1]^2 + [\sin (s_1 + s_2)]^2$$
$$= (\cos s_2 - \cos s_1)^2 + (\sin s_2 + \sin s_1)^2.$$

If we now perform the operations indicated, we have

$$\cos^2 (s_1 + s_2) - 2 \cos (s_1 + s_2) + 1 + \sin^2 (s_1 + s_2)$$
$$= \cos^2 s_2 - 2 \cos s_1 \cos s_2 + \cos^2 s_1 + \sin^2 s_2 + 2 \sin s_1 \sin s_2 + \sin^2 s_1.$$

Regrouping terms, we obtain

$$[\cos^2 (s_1 + s_2) + \sin^2 (s_1 + s_2)] - 2 \cos (s_1 + s_2) + 1$$
$$= (\cos^2 s_2 + \sin^2 s_2) + (\cos^2 s_1 + \sin^2 s_1) - 2 \cos s_1 \cos s_2 + 2 \sin s_1 \sin s_2,$$

and since $\cos^2 s + \sin^2 s = 1$, it follows that

$$1 - 2 \cos (s_1 + s_2) + 1 = 1 + 1 - 2 \cos s_1 \cos s_2 + 2 \sin s_1 \sin s_2.$$

This simplifies to the following result.

Theorem 6.3   *For each s_1, $s_2 \in R$,*

$$\cos (s_1 + s_2) = \cos s_1 \cos s_2 - \sin s_1 \sin s_2. \tag{1}$$

By replacing s_2 with $-s_2$ in the sum formula (1) and using formulas (6) and (7) in Theorem 6.2, we obtain the following companion result.

Theorem 6.4   *For each s_1, $s_2 \in R$,*

$$\cos (s_1 - s_2) = \cos s_1 \cos s_2 + \sin s_1 \sin s_2. \tag{2}$$

Reduction formulas for cosine

The extremely important relationships (1) and (2) are called, respectively, the **sum formula** and the **difference formula** for the cosine function. These are used in proving the following results.

Theorem 6.5   *For each $s \in R$,*

$$\text{I} \quad \cos (\pi - s) = -\cos s,$$

$$\text{II} \quad \cos (\pi + s) = -\cos s,$$

$$\text{III} \quad \cos (2\pi - s) = \cos s.$$

We shall prove only Part I here and leave Parts II and III as exercises.

Proof of 6.5-I   In relationship (2), replace s_1 with π and s_2 with s to obtain

$$\cos (\pi - s) = \cos \pi \cos s + \sin \pi \sin s.$$

From Table 6.2, $\cos \pi = -1$ and $\sin \pi = 0$, so that this equation becomes

$$\cos (\pi - s) = (-1)(\cos s) + 0 \sin s = -\cos s,$$

and the proof is complete.

The formulas in Theorem 6.5 are called **reduction formulas** for the cosine function. Geometric interpretations of I, II, and III are shown in Figure 6.14. In each

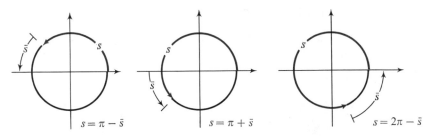

Figure 6.14

case, it is apparent that if s is the length of an arc terminating in Quadrants II, III, or IV, then the arc can be visualized either as the sum or as the difference of two arcs, one of measure $n\pi$, $n \in J$, and the other measuring $\bar{s}$, where $0 < \bar{s} < \pi/2$. We refer to the arc measuring $\bar{s}$ as the **reference arc** for s. For example, the reference arc of an arc measuring $5\pi/4$ is one measuring $\pi/4$, because $5\pi/4 = \pi + \pi/4$, while the reference arc for an arc with measure $-8\pi/3$ measures $\pi/3$ because $-8\pi/3 = -3\pi + \pi/3$.

Use of tables for cosine

In Section 6.2, we found function values (elements in the range) for the cosine function for selected elements in the domain by using the distance formula and certain geometric considerations. It is necessary, however, that we be able to find—or, more precisely, to approximate—a function value for any given real number. To do this, we use tables that provide such approximations to the desired number of decimal places. The means by which the tables are constructed will not concern us at present.

Because the circular functions are periodic and reduction formulas are available, we need tables only over the interval $0 \le s \le \pi/2$. Table V in the Appendix gives approximate function values at intervals of 0.01 for the cosine function and other circular functions to be discussed in the following sections.

Examples

a. $\cos 0.73 \approx 0.7452$ b. $\cos 1.29 \approx 0.2771$

Notice that in Table V the letter x is used to represent an element in the domain. This is customary in many tables. In this usage, of course, x can be thought of as representing an arc length along the unit circle, just as s did. In fact, because x is ordinarily the variable used to represent an element in the domain of a function, *we shall hereafter be using x where we heretofore used s.*

Use of reduction formulas

An appropriate reduction formula enables us to approximate function values of x outside the interval $0 \le x \le \pi/2$. In such cases, note that

$$\pi/2 \approx 1.57, \quad \pi \approx 3.14, \quad 3\pi/2 \approx 4.71, \quad \text{and} \quad 2\pi \approx 6.28.$$

Also note that $\bar{x}$ is used in the same way as $\bar{s}$.

Example

Find $\cos 3.03$.

Solution

Since $1.57 < 3.03 < 3.14$, we can use the reduction formula

$$\cos x = -\cos(\pi - x).$$

Thus,

$$\cos 3.03 \approx -\cos (3.14 - 3.03)$$
$$= -\cos 0.11$$
$$\approx -0.9940.$$

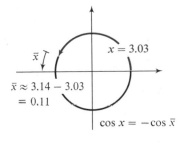

A sketch of the unit circle, including the reference arc, helps interpret the effect of the reduction formula.

To find approximations to function values for numbers with four-digit numerals, we can use the method of linear interpolation explained in Section 5.4 for tables of logarithms.

Finding elements in the domain

Table V can also be used to find elements in the domain of the cosine function for specified elements in the range. For the time being, we shall restrict our attention to finding such elements only in the interval $\{x \mid 0 \le x \le \pi/2\}$.

Example   Approximate the member of $\{x \mid 0 < x < 1.57\}$ for which

$$\cos x = 0.9664.$$

Solution   From Table V, we see that x is approximately equal to 0.26.

Exercise 6.3

Reduce the given expression to an equivalent one either of the form $\cos \bar{x}$ or of the form $-\cos \bar{x}$, where $0 \leq \bar{x} \leq \pi/2$.

Example $\cos \dfrac{7\pi}{5}$

Solution   We first note that $7\pi/5$ lies in Quadrant III, since $7\pi/5 = \pi + (2\pi/5)$. Thus the reference arc is $2\pi/5$. Using Theorem 6.5, Part II, we have

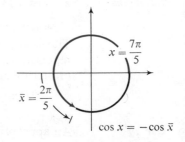

$$\cos \frac{7\pi}{5} = \cos \left(\pi + \frac{2\pi}{5} \right) = -\cos \frac{2\pi}{5}.$$

1. $\cos \dfrac{3\pi}{5}$ 2. $\cos \dfrac{7\pi}{8}$ 3. $\cos \dfrac{9\pi}{8}$ 4. $\cos \dfrac{11\pi}{9}$

5. $\cos \dfrac{13\pi}{7}$ 6. $\cos \dfrac{13\pi}{16}$ 7. $\cos \left(-\dfrac{7\pi}{6} \right)$ 8. $\cos \left(-\dfrac{11\pi}{6} \right)$

Use the fact that cosine is periodic, along with an appropriate reduction formula, to write an equivalent expression either of the form $\cos \bar{x}$ or of the form $-\cos \bar{x}$, where $0 \leq \bar{x} \leq \pi/2$.

Example $\cos \dfrac{24\pi}{5}$

Solution   We first write $\dfrac{24\pi}{5} = \dfrac{4\pi}{5} + \dfrac{20\pi}{5} = \dfrac{4\pi}{5} + 4\pi$. Then

$$\cos \frac{24\pi}{5} = \cos \left(\frac{4\pi}{5} + 4\pi \right) = \cos \frac{4\pi}{5}.$$

Next, because $4\pi/5$ is in Quadrant II, we note that $4\pi/5 = \pi - (\pi/5)$. Thus, the reference arc is $\pi/5$. Using Part I of Theorem 6.5, we have

$$\cos \frac{4\pi}{5} = \cos \left(\pi - \frac{\pi}{5} \right) = -\cos \frac{\pi}{5}.$$

9. $\cos \dfrac{18\pi}{7}$ **10.** $\cos \dfrac{23\pi}{8}$ **11.** $\cos \dfrac{13\pi}{5}$ **12.** $\cos \dfrac{31\pi}{16}$

13. $\cos \dfrac{35\pi}{11}$ **14.** $\cos \dfrac{53\pi}{10}$ **15.** $\cos \left(-\dfrac{38\pi}{9}\right)$ **16.** $\cos \left(-\dfrac{27\pi}{5}\right)$

Use Table V in the Appendix to find an approximation for the given expression. Use linear interpolation as required.

17. $\cos 0.59$ **18.** $\cos 0.47$ **19.** $\cos 1.32$ **20.** $\cos 0.39$

21. $\cos 1.43$ **22.** $\cos 0.235$ **23.** $\cos 1.216$ **24.** $\cos 1.042$

Use the fact that the cosine function is periodic, along with the reduction formulas, to find an approximate value for the given expression.

Example $\cos 2$

Solution   Since $\pi \approx 3.14$ and $\pi/2 \approx 1.57$, an arc of length 2 terminates in the second quadrant. We have $\pi - 2 \approx 1.14$, so that $\bar{x} \approx 1.14$ is the length of the reference arc. Using Theorem 6.5-I, then, we obtain

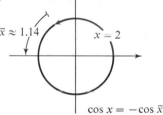

$$\cos 2 = -\cos (\pi - 2)$$

$$= -\cos 1.14$$

$$\approx -0.4176,$$

which is the function value we seek.

25. $\cos 2.01$ **26.** $\cos 4.52$ **27.** $\cos 5.21$ **28.** $\cos 2.24$

29. $\cos 8.41$ **30.** $\cos 10.32$ **31.** $\cos (-9.52)$ **32.** $\cos (-12.61)$

Use Table V to approximate element x in the domain of the function, $0 \leq x \leq 1.57$, for the specified element in the range.

33. $\cos x = 0.9664$ **34.** $\cos x = 0.2579$ **35.** $\cos x = 0.7038$

36. $\cos x = 0.1304$ **37.** $\cos x = 0.8600$ **38.** $\cos x = 0.2921$

Prove.

39. $\cos (x_1 - x_2) = \cos x_1 \cos x_2 + \sin x_1 \sin x_2$

40. $\cos (\pi + x) = -\cos x$

41. $\cos (2\pi - x) = \cos x$

42. $\cos \left(\dfrac{\pi}{2} - x\right) = \sin x$ *Hint:* Substitute $\dfrac{\pi}{2}$ for s_1 and x for s_2 in Theorem 6.4.

43. $\cos\left(\dfrac{\pi}{2} + x\right) = -\sin x$

44. $\cos\left(\dfrac{3\pi}{2} - x\right) = -\sin x$

45. $\cos\left(\dfrac{3\pi}{2} + x\right) = \sin x$

46. $\sin\left(\dfrac{\pi}{2} - x\right) = \cos x$ *Hint:* Substitute $\dfrac{\pi}{2} - x$ for x in Exercise 42.

6.4   *Properties of the Sine Function*

From Exercise 6.3-42 we have

$$\cos\left(\frac{\pi}{2} - x\right) = \sin x.$$

If now we set $x = x_1 + x_2$, we obtain

$$\cos\left(\frac{\pi}{2} - (x_1 + x_2)\right) = \sin(x_1 + x_2).$$

This can be rewritten as

$$\cos\left[\left(\frac{\pi}{2} - x_1\right) - x_2\right] = \sin(x_1 + x_2),$$

and if the left-hand member is expanded by means of Theorem 6.4, we find that

$$\cos\left(\frac{\pi}{2} - x_1\right)\cos x_2 + \sin\left(\frac{\pi}{2} - x_1\right)\sin x_2 = \sin(x_1 + x_2).$$

Since $\cos(\pi/2 - x_1) = \sin x_1$, and by Exercise 6.3-46. $\sin(\pi/2 - x_1) = \cos x_1$,

$$\cos\left(\frac{\pi}{2} - x_1\right)\cos x_2 + \sin\left(\frac{\pi}{2} - x_1\right)\sin x_2 = \sin x_1 \cos x_2 + \cos x_1 \sin x_2,$$

which establishes the following result.

Theorem 6.6   *If x_1, $x_2 \in R$, then*
$$\mathbf{\sin(x_1 + x_2) = \sin x_1 \cos x_2 + \cos x_1 \sin x_2.} \tag{1}$$

By replacing x_2 with $-x_2$ in Theorem 6.6, we obtain a formula for $\sin(x_1 - x_2)$.

Theorem 6.7   *If x_1, $x_2 \in R$, then*
$$\mathbf{\sin(x_1 - x_2) = \sin x_1 \cos x_2 - \cos x_1 \sin x_2.} \tag{2}$$

**Reduction
formulas
for sine**

Relationships (1) and (2), respectively called the **sum formula** and the **difference formula** for the sine function, can be used in the same way as the sum and difference formulas for the cosine function to establish the following reduction formulas.

Theorem 6.8   *For each* $x \in R$,

$$\text{I} \quad \sin (\pi - x) = \sin x,$$

$$\text{II} \quad \sin (\pi + x) = -\sin x,$$

$$\text{III} \quad \sin (2\pi - x) = -\sin x.$$

**Use of tables
for sine**

Table V in the Appendix also gives approximate function values for the sine function in the interval $0 \le x < \pi/2$. An appropriate reduction formula can be used to approximate function values of x outside this interval.

Example

Find sin 4.24.

Solution

Since we have $\pi \approx 3.14$ and $3\pi/2 \approx 4.71$, an arc of length 4.24 terminates in the third quadrant. The length of the reference arc, $\bar{x}$, therefore satisfies

$$\bar{x} \approx 4.24 - 3.14 = 1.10.$$

Using Part II of Theorem 6.8, we obtain

$$\sin 4.24 = \sin (3.14 + 1.10)$$
$$\approx -\sin 1.10$$
$$\approx -0.8912,$$

which is the function value we seek.

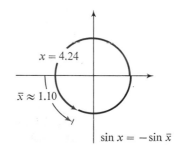

Exercise 6.4

Reduce to an equivalent expression either of the form $\sin \bar{x}$ *or of the form* $-\sin \bar{x}$, *where* $0 \le \bar{x} \le \pi/2$.

Example

$\sin \dfrac{5\pi}{7}$

Solution

We first observe that $5\pi/7$ lies in Quadrant II. Thus $5\pi/7 = \pi - (2\pi/7)$, and the reference arc is $2\pi/7$. Then, using Part I of Theorem 6.8, we have

$$\sin \frac{5\pi}{7} = \sin \left(\pi - \frac{2\pi}{7} \right) = \sin \frac{2\pi}{7}.$$

1. $\sin \dfrac{7\pi}{5}$ 2. $\sin \dfrac{7\pi}{8}$ 3. $\sin \dfrac{15\pi}{8}$ 4. $\sin \dfrac{12\pi}{13}$

5. $\sin \dfrac{15\pi}{11}$ 6. $\sin \dfrac{19\pi}{10}$ 7. $\sin \left(-\dfrac{8\pi}{7} \right)$ 8. $\sin \left(-\dfrac{7\pi}{8} \right)$

Use the fact that sine is periodic, along with an appropriate reduction formula, to write an equivalent expression of the form sin $\bar{x}$ *or* $-\sin \bar{x}$, *where* $0 \leq \bar{x} \leq \pi/2$.

Example $\sin \dfrac{38\pi}{5}$

Solution   We first write $\dfrac{38\pi}{5} = \dfrac{8\pi}{5} + \dfrac{30\pi}{5} = \dfrac{8\pi}{5} + 6\pi$. Then

$$\sin \frac{38\pi}{5} = \sin \left(\frac{8\pi}{5} + 6\pi\right) = \sin \frac{8\pi}{5}.$$

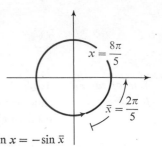

Next, because $8\pi/5$ is in Quadrant IV, we have $8\pi/5 = 2\pi - 2\pi/5$. Thus, the reference arc is $2\pi/5$. Using Part III of Theorem 6.8, we have

$$\sin \frac{8\pi}{5} = \sin \left(2\pi - \frac{2\pi}{5}\right) = -\sin \frac{2\pi}{5}.$$

9. $\sin \dfrac{25\pi}{6}$ **10.** $\sin \dfrac{27\pi}{7}$ **11.** $\sin \dfrac{38\pi}{11}$

12. $\sin \dfrac{35\pi}{17}$ **13.** $\sin \dfrac{43\pi}{5}$ **14.** $\sin \dfrac{58\pi}{11}$

15. $\sin \left(-\dfrac{24\pi}{5}\right)$ **16.** $\sin \left(-\dfrac{38\pi}{7}\right)$

Use Table V in the Appendix to find an approximation for the given expression. Use linear interpolation as required.

17. sin 0.31   **18.** sin 1.36   **19.** sin 1.21   **20.** sin 0.86

21. sin 1.50   **22.** sin 1.362   **23.** sin 1.042   **24.** sin 0.754

Use the fact that the sine function is periodic, along with the reduction formulas, to find an approximate value for the given expression.

25. sin 2.21   **26.** sin 3.68   **27.** sin 5.47   **28.** sin (−0.31)

29. sin (−1.92)   **30.** sin 12.24   **31.** sin (−9.61)   **32.** sin (−15.32)

Use Table V to approximate the element x $(0 \leq x \leq 1.57)$ *in the domain of the function for the specified element in the range.*

33. $\sin x = 0.9975$ **34.** $\sin x = 0.5396$ **35.** $\sin x = 0.8573$

36. $\sin x = 0.2280$ **37.** $\sin x = 0.4000$ **38.** $\sin x = 0.9230$

Prove.

39. $\sin (x_1 - x_2) = \sin x_1 \cos x_2 - \cos x_1 \sin x_2$

40. $\sin (\pi - x) = \sin x$ **41.** $\sin (\pi + x) = -\sin x$

42. $\sin(2\pi - x) = -\sin x$ **43.** $\sin\left(\dfrac{\pi}{2} + x\right) = \cos x$

44. $\sin\left(\dfrac{3\pi}{2} - x\right) = -\cos x$ **45.** $\sin\left(\dfrac{3\pi}{2} + x\right) = -\cos x$

6.5   *Graphs of Sine and Cosine Functions*

We have been using the variables s and x as real numbers associated with arc length on a unit circle. Figure 6.15 illustrates how such arc lengths, in turn, can be associated with line segments on the x-axis of a rectangular coordinate system. Here x

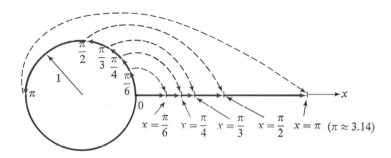

Figure 6.15

is the measure of the line segment on the x-axis that can be thought of as obtained by "unwinding" the arc that has the same measure.

     The periodic nature of the cosine and sine functions can be seen clearly by graphing these functions on a rectangular coordinate system in R^2, where values for s or x (elements in the domain) are associated with points on the horizontal axis.

Graph of sine   Although we have Table V available to find values for $\cos x$ and $\sin x$ for many values of x, the special values for 0, $\pi/6$, $\pi/4$, etc., in Table 6.2, page 138, will suffice for our purposes. Because the graphs of both the cosine function and the sine function are called *sine waves*, we shall look first at the graph of

$$\{(x, f(x)) \mid f(x) = \sin x\}, \quad \text{or simply} \quad \{(x, y) \mid y = \sin x\}.$$

Graphing some familiar ordered pairs in this function over the interval $0 \le x \le 2\pi$ gives us the graph in Figure 6.16 (note $1/\sqrt{2} = \sqrt{2}/2 \approx 1.414/2 \approx 0.7$). Assuming that sine is a continuous function—that is, that its graph contains no breaks or gaps (you will prove in calculus that this is true)—and that it increases as x increases from 0 to $\pi/2$, then decreases from $\pi/2$ to π, and so forth, we can join these points with a smooth curve to produce Figure 6.17. Then, because $\sin(x + 2\pi) = \sin x$, this pattern repeats itself over intervals of length 2π in both directions. Thus we have the pattern shown in Figure 6.18 for the graph of the sine function. Units on the horizontal axis representing integers (short marks) and

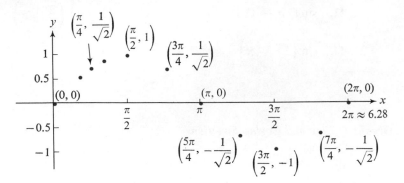

Figure 6.16

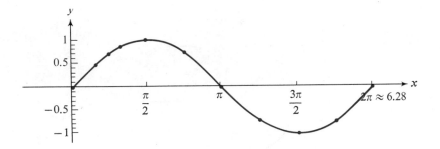

Figure 6.17

integral multiples of $\pi/2$ (long marks) are shown. The range of the function,

$$\{y \mid -1 \le y \le 1\},$$

is clearly evident from the graph. The zeros of the function, $k\pi$, where $k = 0, \pm 1,$ $\pm 2, \pm 3, \ldots$, namely the values of x associated with the points at which the curve crosses the x-axis, also appear on the graph. Graphs with this characteristic form are called **sine waves** (as remarked earlier), or **sinusoids.** The portion of the graph over any fundamental period of the function is called a **cycle** of the sine wave. Half the difference of the maximum and minimum ordinates on such a curve is called the **amplitude** of the wave. Thus, for the graph of $y = \sin x$, the amplitude of the wave is $(1/2)[1 - (-1)] = 1$.

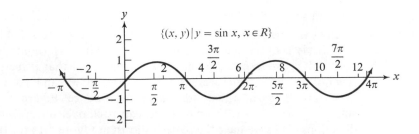

Figure 6.18

Graph of cosine

The graph of the cosine function can be obtained in the same manner as the graph of the sine function. From Table 6.2 on page 138, or from memory, we obtain the coordinates of several points on the graph, as shown in Figure 6.19. By connecting

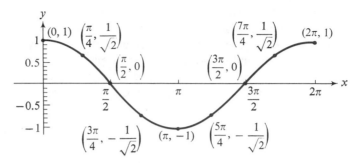

Figure 6.19

these points with a smooth curve, we obtain the graph of $\{(x, y) \mid y = \cos x\}$ over one period of the function. Duplicating this pattern over several more periods, we have a representative portion of the entire graph of the cosine function (Figure 6.20). Notice that the graph of the cosine function is a sinusoid, with amplitude 1.

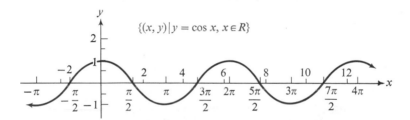

Figure 6.20

Furthermore, the zeros of the function, $\pi/2 + n\pi$, where $n = 0, \pm 1, \pm 2, \pm 3, \ldots$, and the range $\{y \mid -1 \le y \le 1\}$ are also evident.

Functions defined by equations of the form

$$y = A \sin (Bx + C) \quad \text{and} \quad y = A \cos (Bx + C),$$

where A, B, and C are constants and A and B are not 0, always have sine waves for graphs. With variations in the numbers A, B, and C, the graphs are variously situated with respect to the origin and have a variety of amplitudes and periods. To analyze such graphs, let us consider the effect of A on the graphs of the functions defined by $y = A \sin x$ and $y = A \cos x$.

Graphs of
$y = A \sin x$
and
$y = A \cos x$

For each value of x, each ordinate to the graph of $y = A \sin x$ is A times the ordinate to the graph of $y = \sin x$. Therefore, the amplitude of the graph of $y = A \sin x$ is $|A|$ times the amplitude of the graph of $y = \sin x$. Of course, the graph of $y = A \cos x$ is a similar modification of the graph of $y = \cos x$.

Example   Graph $y = 3 \sin x$, $-\pi \leq x \leq 4\pi$.

Solution   It may be helpful first to sketch $y = \sin x$, $0 \leq x \leq 2\pi$, as a reference. Then, since
the amplitude of $y = 3 \sin x$ is 3, we can sketch the desired graph on the same
coordinate system over the interval $0 \leq x \leq 2\pi$ by making each ordinate 3 times
the corresponding ordinate of the graph of $y = \sin x$.

We can then extend this cycle to include the entire interval $-\pi \leq x \leq 4\pi$, as
shown in the figure. In this figure, the first cycle is sketched with a heavier line for

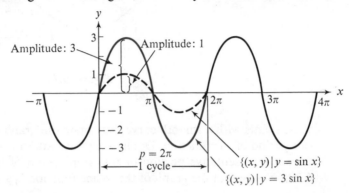

emphasis. Note also that, in this and some of the succeeding figures, different unit
lengths are used on the x- and y-axes. For $y = 3 \sin x$, the fundamental period p is
the same as for $y = \sin x$, namely $p = 2\pi$.

Graphs of   Next let us examine the way the graph of
$y = \sin Bx$
and $$y = \sin Bx, \quad B \neq 0,$$
$y = \cos Bx$ differs from the graph of $y = \sin x$. Note first that $\sin Bx$ has values between -1
and 1 inclusive, as has $\sin x$. Also,

$$\sin (Bx + 2\pi) = \sin Bx,$$

just as $\sin (x + 2\pi) = \sin x$. If we factor B from $Bx + 2\pi$, however, we have
$B(x + 2\pi/B)$, and hence the function defined by $y = \sin Bx$ is periodic with period

$$p = 2\pi/|B|.$$

We use $|B|$ instead of B to ensure a positive number for the period. It can be shown,
although it is not done here, that $2\pi/|B|$ is the fundamental period of the function.
Hence the graph of $y = \sin Bx$ is a sine wave with amplitude 1; it completes one
cycle over the interval $0 \leq x \leq 2\pi/|B|$.

Example   Graph $y = \cos 2x$, $\dfrac{-3\pi}{2} \leq x \leq 2\pi$.

Solution   Let us first sketch a cycle of $y = \cos x$, $0 \leq x \leq 2\pi$, as a reference. Since

$$p = \frac{2\pi}{|B|} = \frac{2\pi}{2} = \pi,$$

we next sketch a cycle of the graph of $y = \cos 2x$ over the interval $0 \le x \le \pi$ on the same coordinate system, and extend the cycle obtained over the interval $-3\pi/2 \le x \le 2\pi$.

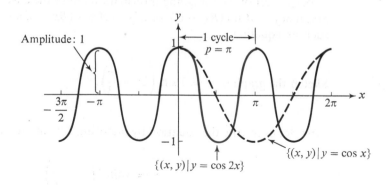

Example

Graph $y = -4 \sin \dfrac{1}{2} x, \quad -2\pi \le x \le 4\pi$.

Solution

We first sketch the graph of $y = \sin x$ as a reference. Since $A = -4$, each ordinate of the graph of $y = -4 \sin x/2$ is the negative of the ordinate of the graph of $y = 4 \sin (x/2)$. Since

$$p = \frac{2\pi}{|B|} = \frac{2\pi}{\dfrac{1}{2}} = 4\pi,$$

there is one cycle in the interval $0 \le x \le 4\pi$. Hence, we sketch one cycle, and extend it over the interval $-2\pi \le x \le 4\pi$.

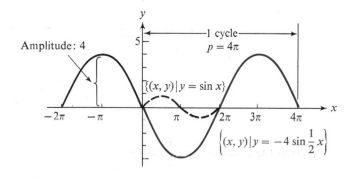

Graphs of
$y = \sin(x + C)$
and
$y = \cos(x + C)$

Finally, let us look at the difference between the graphs of

$$y = \sin (x + C) \quad \text{and} \quad y = \sin x,$$

where $C > 0$. For $x = -C$, we have

$$\sin (x + C) = \sin 0 = 0.$$

Similarly, for any real number x_1, the ordinate of $\sin (x + C)$ at $x_1 - C$ will be the same as the ordinate of $\sin x$ at x_1. Hence, the graph of $y = \sin (x + C)$ is said to

lead the graph of $y = \sin x$ by C. The number C, itself, is called the **phase shift** of the wave. If $C < 0$, then the graph of $y = \sin (x + C)$ is shifted $|C|$ units to the right of the graph of $y = \sin x$ and is said to **lag** the graph of $y = \sin x$.

We use all of the foregoing information about the effect of A, B, and C on the graph of $y = A \sin (Bx + C)$, or of $y = A \cos (Bx + C)$, to help sketch the graph of such an equation.

Example   Sketch the graph of $y = 3 \sin \left(2x + \dfrac{\pi}{3}\right)$.

Solution   First, let us rewrite the equation by factoring 2 from the expression in parentheses:

$$y = 3 \sin 2\left(x + \frac{\pi}{6}\right).$$

By inspecting this equation, we note the following things about the graph:

1. It is a sine wave.
2. It has amplitude 3.
3. It has period $2\pi/2 = \pi$.
4. It leads the graph of $y = 3 \sin 2x$ by $\pi/6$.

With these facts, we can quickly sketch the graph shown.

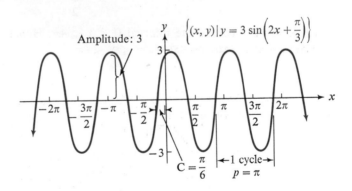

Example   Sketch the graph of $y = \dfrac{1}{2} \sin \dfrac{\pi x}{2}$.

Solution   By inspection, we note the following things concerning the graph:

1. It is a sine wave.
2. It has amplitude 1/2.
3. It has period $\dfrac{2\pi}{\pi/2} = 4$.

Since the period is 4, we use integers as elements of the domain. Scaling the x-axis in integral units facilitates sketching the graph as shown on the next page.

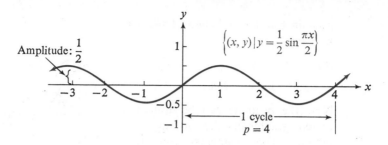

Exercise 6.5

Sketch the graph of the given equation over the interval $-2\pi \le x \le 2\pi$.

1. $y = 2 \sin x$ 2. $y = 3 \cos x$ 3. $y = \dfrac{1}{2} \cos x$

4. $y = \dfrac{1}{3} \cos x$ 5. $y = -4 \sin x$ 6. $y = -\dfrac{1}{2} \cos x$

7. $y = \sin 2x$ 8. $y = \cos 3x$ 9. $y = \cos \dfrac{1}{3} x$

10. $y - \cos \dfrac{1}{2} x$ 11. $y - -3 \sin 2x$ 12. $y = -\dfrac{1}{2} \sin 3x$

13. $y = \sin (x + \pi)$ 14. $y = \cos \left(x - \dfrac{\pi}{2} \right)$

15. $y = 2 \cos \left(x - \dfrac{\pi}{4} \right)$ 16. $y - 3 \sin \left(x + \dfrac{\pi}{6} \right)$

17. $y = 3 \sin 2 \left(x - \dfrac{\pi}{3} \right)$ 18. $y = 2 \cos 3 \left(x + \dfrac{\pi}{4} \right)$

19. $y = 2 \sin \pi x$ 20. $y = -3 \cos \dfrac{\pi}{2} x$

21. $y = -\dfrac{1}{2} \cos \dfrac{\pi}{3} x$ 22. $y = \dfrac{1}{4} \sin \dfrac{\pi}{4} x$

From the respective graph, determine the zeros (over the specified domain) of the function defined by the equation in the given exercise.

23. Exercise 7   24. Exercise 8   25. Exercise 9
26. Exercise 10   27. Exercise 19   28. Exercise 20

Example   Sketch the graph of $y = \sin x + 2 \cos x$ over the interval $0 \le x \le 2\pi$.

Solution   First, sketch the graphs of $y = \sin x$ and $y = 2 \cos x$ on the same coordinate system over the given interval. The ordinate of the graph of $y = \sin x + 2 \cos x$ at each point x on the x-axis is the *algebraic* sum of the corresponding ordinates of

$y = \sin x$ and $y = 2 \cos x$. Thus for $x \doteq \pi/6$,

$$\sin x = 0.5, \quad 2 \cos x = 2\sqrt{3}/2 \approx 1.7,$$
$$\text{and} \quad \sin x + 2 \cos x \approx 2.2.$$

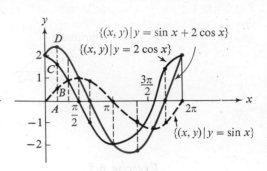

Now the ordinate can be approximated graphically by adding the directed line segments from the x-axis to the curves at $x = \pi/6$:

$$\overline{AB} + \overline{AC} = \overline{AB} + \overline{BD} = \overline{AD}.$$

If this is done for a few selected values of x, we obtain a good approximation for the curve.

Sketch the graph of the given equation over the interval $-2\pi \leq x \leq 2\pi$.

29. $y = \sin x + \cos x$ **30.** $y = 3 \sin x + \cos x$

31. $y = \sin 2x + \dfrac{1}{2} \cos x$ **32.** $y = \sin 3x + 2 \cos \dfrac{1}{2} x$

33. $y = \sin x - 2 \cos x$ **34.** $y = 3 \cos x - \sin 2x$

35. $y = 3 \sin x + 1$ **36.** $y = 4 \sin 2x - 3$

37. $y = x + \cos x$ **38.** $y = 2x - \cos x$

6.6   Additional Properties of the Cosine and Sine Functions

Formulas for cos 2x and sin 2x

The sum and difference formulas for cosine and sine that we developed in Sections 6.3 and 6.4 can be used to generate still other relationships between values of these functions.

Theorem 6.9   *If* $x \in R$, *then*

$$\text{I} \quad \cos 2x = \cos^2 x - \sin^2 x, \tag{1}$$

$$\text{II} \quad \sin 2x = 2 \sin x \cos x. \tag{2}$$

Proof   Let $s_1 = s_2 = x$ in Equation (1) on page 142. Then

$$\cos (x + x) = \cos x \cos x - \sin x \sin x,$$
$$\cos 2x = \cos^2 x - \sin^2 x.$$

Let $x_1 = x_2 = x$ in Equation (1) on page 146. Then

$$\sin (x + x) = \sin x \cos x + \cos x \sin x,$$
$$\sin 2x = 2 \sin x \cos x,$$

as was to be shown.

Formulas for cos x/2 and sin x/2

Alternative forms of the equation for $\cos 2x$ can be obtained by using the relationship $\cos^2 x + \sin^2 x = 1$ to obtain

$$\cos^2 x = 1 - \sin^2 x \quad \text{and} \quad \sin^2 x = 1 - \cos^2 x,$$

and then replacing the appropriate term in the right-hand member of (1). The results are the formulas:

$$\cos 2x = 1 - 2 \sin^2 x, \tag{3}$$

$$\cos 2x = 2 \cos^2 x - 1. \tag{4}$$

Relationships (3) and (4) lead directly to the following formulas:

Theorem 6.10 *If $x \in R$, then*

$$\text{I} \quad \cos \frac{x}{2} = \begin{cases} \sqrt{\dfrac{1 + \cos x}{2}} & \text{when } \dfrac{x}{2} \text{ terminates in Quadrant I or IV,} \tag{5} \\[3ex] -\sqrt{\dfrac{1 + \cos x}{2}} & \text{when } \dfrac{x}{2} \text{ terminates in Quadrant II or III;} \tag{5'} \end{cases}$$

$$\text{II} \quad \sin \frac{x}{2} = \begin{cases} \sqrt{\dfrac{1 - \cos x}{2}} & \text{when } \dfrac{x}{2} \text{ terminates in Quadrant I or II,} \tag{6} \\[3ex] -\sqrt{\dfrac{1 - \cos x}{2}} & \text{when } \dfrac{x}{2} \text{ terminates in Quadrant III or IV.} \tag{6'} \end{cases}$$

Proof We shall prove only the formula for $\sin (x/2)$ here and leave the proof of the formula for $\cos (x/2)$ as an exercise. If we replace x in (3) with $x/2$, we have

$$\cos 2 \left(\frac{x}{2} \right) = 1 - 2 \sin^2 \frac{x}{2},$$

from which

$$\cos x = 1 - 2 \sin^2 \frac{x}{2},$$

$$\sin^2 \frac{x}{2} = \frac{1 - \cos x}{2},$$

and

$$\sin \frac{x}{2} = \pm \sqrt{\frac{1 - \cos x}{2}}.$$

Now, because $\sin (x/2) > 0$ for $x/2$ in Quadrant I or II, we take the positive square root in these quadrants; and for a similar reason in the remaining quadrants we take the negative square root. Thus our proof is complete.

Example

Find $\sin \dfrac{\pi}{12}$.

(Solution overleaf)

Solution   Since $\dfrac{\pi}{12} = \dfrac{1}{2}\left(\dfrac{\pi}{6}\right)$, from (6) we have

$$\sin\frac{\pi}{12} = \sin\frac{1}{2}\left(\frac{\pi}{6}\right)$$

$$= \sqrt{\frac{1 - \cos(\pi/6)}{2}}$$

$$= \sqrt{\frac{1 - \sqrt{3}/2}{2}} = \sqrt{\frac{2 - \sqrt{3}}{4}} = \frac{1}{2}\sqrt{2 - \sqrt{3}}.$$

Exercise 6.6

Use Formulas (1)–(6) and Table 6.2 to find the value for the given expression.

1. $\sin\dfrac{\pi}{8}$ 2. $\cos\dfrac{\pi}{8}$ 3. $\cos\left(-\dfrac{\pi}{8}\right)$ 4. $\sin\left(-\dfrac{\pi}{8}\right)$

5. $\cos\dfrac{3\pi}{8}$ 6. $\sin\dfrac{3\pi}{8}$ 7. $\sin\dfrac{5\pi}{24}$ 8. $\cos\dfrac{5\pi}{24}$

Example   Given that $\sin x = 4/5$,   find $\sin 2x$, $\pi/2 \le x \le \pi$.

Solution   Since $\sin x = 4/5$ and x terminates in Quadrant II,

$$\cos x = -\sqrt{1 - \sin^2 x} = -\sqrt{1 - \left(\frac{4}{5}\right)^2} = -\sqrt{\frac{9}{25}} = -\frac{3}{5}.$$

From Formula (2) on page 156, we have

$$\sin 2x = 2 \sin x \cos x = 2\left(\frac{4}{5}\right)\left(-\frac{3}{5}\right) = -\frac{24}{25}.$$

9. Given that $\cos x = -\dfrac{3}{5}$, find:

    a. $\cos 2x$, $\dfrac{\pi}{2} \le x \le \pi$ b. $\sin 2x$, $\pi \le x \le \dfrac{3\pi}{2}$

    c. $\sin\dfrac{1}{2}x$, $3\pi \le x \le \dfrac{7\pi}{2}$ d. $\cos\dfrac{1}{2}x$, $\dfrac{\pi}{2} \le x \le \pi$

10. Given that $\sin x = \dfrac{\sqrt{5}}{3}$, find:

    a. $\cos 2x$, $\dfrac{\pi}{2} \le x \le \pi$ b. $\sin 2x$, $\dfrac{\pi}{2} \le x \le \pi$

    c. $\sin\dfrac{1}{2}x$, $0 \le x \le \dfrac{\pi}{2}$ d. $\cos\dfrac{1}{2}x$, $2\pi \le x \le \dfrac{5\pi}{2}$

By means of the formulas in this section, evaluate the given expression using the relationships in Theorems 6.9 and 6.10.

Example $2 \sin \dfrac{\pi}{6} \cos \dfrac{\pi}{6}$

Solution   From Theorem 6.9-II, we have

$$2 \sin \frac{\pi}{6} \cos \frac{\pi}{6} = \sin 2\left(\frac{\pi}{6}\right)$$

$$= \sin \frac{\pi}{3} = \frac{\sqrt{3}}{2}.$$

Hence, the value of the expression is $\sqrt{3}/2$.

11. $2 \sin \dfrac{\pi}{12} \cos \dfrac{\pi}{12}$ **12.** $\cos^2 \dfrac{\pi}{12} - \sin^2 \dfrac{\pi}{12}$

13. $2 \cos^2 \dfrac{5\pi}{12} - 1$ **14.** $1 - 2 \sin^2 \dfrac{5\pi}{12}$

15.   Express $\sin^2 x$ in terms of $\cos 2x$. *Hint*: Use Formula (3) on page 157.

16.   Express $\cos^2 x$ in terms of $\cos 2x$.

17.   Show that for all $x \in R$, $\cos 3x = 4 \cos^3 x - 3 \cos x$. *Hint:* $\cos 3x = \cos (2x + x)$.

18.   Show that for all $x \in R$, $\sin 3x = 3 \sin x - 4 \sin^3 x$.

19.   Show that $\sin 2x$ is periodic with fundamental period π.

20.   Show that $\cos 2x$ is periodic with fundamental period π.

21.   Show that $\sin (x/2)$ is periodic with fundamental period 4π.

22.   Show that $\cos (x/2)$ is periodic with fundamental period 4π.

23.   Prove Theorem 6.10-I.

6.7   Other Circular Functions

The functions sine and cosine can be used to define other periodic functions. First, however, let us give names to some ratios of sine and cosine function values.

Definition 6.4   *Let $x \in R$.*

    I   *The **tangent of x** is denoted by* **tan x,** *and*

$$\tan x = \frac{\sin x}{\cos x} \left(x \neq \frac{\pi}{2} + k\pi, \, k \in J \right).$$

                                                    (Definition continued)

II *The **cotangent** of x is denoted by* **cot x**, *and*

$$\cot x = \frac{\cos x}{\sin x} \quad (x \neq k\pi,\ k \in J).$$

III *The **secant** of x is denoted by* **sec x**, *and*

$$\sec x = \frac{1}{\cos x} \quad \left(x \neq \frac{\pi}{2} + k\pi,\ k \in J\right).$$

IV *The **cosecant** of x is denoted by* **csc x**, *and*

$$\csc x = \frac{1}{\sin x} \quad (x \neq k\pi,\ k \in J).$$

Example   If $\sin x = 3/5$ and $\pi/2 \leq x \leq \pi$, find $\cos x$, $\tan x$, $\cot x$, $\sec x$, and $\csc x$.

Solution   By Equation (5b) on page 134, for $\pi/2 \leq x \leq \pi$,

$$\cos x = -\sqrt{1 - \sin^2 x}.$$

Then, since $\sin x = 3/5$,

$$\cos x = -\sqrt{1 - \left(\frac{3}{5}\right)^2} = -\frac{4}{5}.$$

By Definition 6.4,

$$\tan x = \frac{\sin x}{\cos x} = \frac{3/5}{-4/5} = -\frac{3}{4}, \qquad \sec x = \frac{1}{\cos x} = \frac{1}{-4/5} = -\frac{5}{4},$$

$$\cot x = \frac{\cos x}{\sin x} = \frac{-4/5}{3/5} = -\frac{4}{3}, \qquad \csc x = \frac{1}{\sin x} = \frac{1}{3/5} = \frac{5}{3}.$$

Now, let us take the ratios in Definition 6.4 one at a time and examine the function associated with each.

Tangent   We can use Definition 6.4 together with Table 6.2, page 138, to find $\tan x$ for all values of x included in the table. For example,

$$\tan 0 = \frac{\sin 0}{\cos 0} = \frac{0}{1} = 0,$$

$$\tan \frac{\pi}{6} = \frac{\sin (\pi/6)}{\cos (\pi/6)} = \frac{1/2}{\sqrt{3}/2} = \frac{1}{\sqrt{3}},$$

and so forth. The values of $\tan x$ obtained in this way are tabulated in Table 6.3. Just as we do for $\sin x$ and $\cos x$, we can use Table V to obtain additional values for $\tan x$.

Definition 6.5   *If $x \in R$, and $x \neq \pi/2 + k\pi$, $k \in J$, then*

$$\textbf{tangent} = \{(x,y) | y = \tan x\}.$$

Because $\tan x = \sin x/\cos x$, the domain of the tangent function is R, with the exception of the real numbers x for which $\cos x = 0$. In other words, we except real

Table 6.3

x	$\tan x$	x	$\tan x$
0	0	π	0
$\dfrac{\pi}{6}$	$\dfrac{1}{\sqrt{3}}$	$\dfrac{7\pi}{6}$	$\dfrac{1}{\sqrt{3}}$
$\dfrac{\pi}{4}$	1	$\dfrac{5\pi}{4}$	1
$\dfrac{\pi}{3}$	$\sqrt{3}$	$\dfrac{4\pi}{3}$	$\sqrt{3}$
$\dfrac{\pi}{2}$	not defined	$\dfrac{3\pi}{2}$	not defined
$\dfrac{2\pi}{3}$	$-\sqrt{3}$	$\dfrac{5\pi}{3}$	$-\sqrt{3}$
$\dfrac{3\pi}{4}$	-1	$\dfrac{7\pi}{4}$	-1
$\dfrac{5\pi}{6}$	$-\dfrac{1}{\sqrt{3}}$	$\dfrac{11\pi}{6}$	$-\dfrac{1}{\sqrt{3}}$
π	0	2π	0

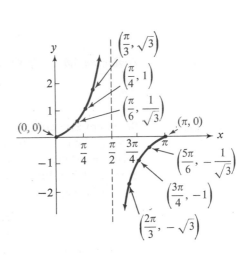

Figure 6.21

numbers of the form $\pi/2 + k\pi$, $k \in J$. The range of the tangent function is R. Since the sine and cosine functions have period 2π, so has the tangent function. We note, however, that

$$\tan x = \frac{\sin x}{\cos x} = \frac{-\sin(x + \pi)}{-\cos(x + \pi)} = \tan(x + \pi),$$

so that this function actually has period π also. Clearly, $\tan x$ alternates in sign from quadrant to quadrant, being positive in Quadrants I and III and negative in Quadrants II and IV. The graph of the tangent function over the interval $0 \le x \le \pi$ is shown in Figure 6.21. Its graph over several cycles is shown in Figure 6.22. Observe that the asymptotes to the curve are the graphs of $x = \pi/2 + k\pi$, $k \in J$, the zeros of the function are the real numbers $k\pi$, and the range of the function is R.

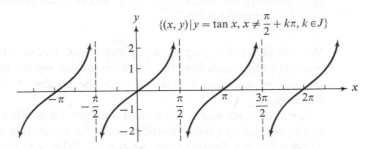

Figure 6.22

From Definition 6.4, we have

$$\tan(-x) = \frac{\sin(-x)}{\cos(-x)}.$$

From Theorem 6.2, $\sin(-x) = -\sin x$ and $\cos(-x) = \cos x$. Hence,

$$\tan(-x) = \frac{-\sin x}{\cos x} = -\tan x.$$

Definition 6.6   *If $x \in R$, and $x \neq k\pi$, $k \in J$, then*

$$\textbf{cotangent} = \{(x, y) \mid y = \cot x\}.$$

Cotangent

Since $\cot x = \cos x/\sin x$, the domain of the cotangent function is R, excepting those real numbers for which $\sin x = 0$, that is, all real numbers except those of the form $k\pi$, where $k \in J$. The range of cotangent is R, and, like the tangent function, cotangent has fundamental period π. Note that because $\cot x = \cos x/\sin x$ and $\tan x = \sin x/\cos x$,

$$\cot x = \frac{1}{\tan x} \quad \text{and} \quad \tan x = \frac{1}{\cot x}$$

for those values of x for which each expression is defined.

Definition 6.7   *If $x \in R$, and $x \neq \pi/2 + k\pi$, $k \in J$, then*

$$\textbf{secant} = \{(x, y) \mid y = \sec x\}.$$

Secant

Since $\sec x$ is the reciprocal of $\cos x$, the domain of the secant function is R, except for those values of x for which $\cos x = 0$, namely, $x = \pi/2 + k\pi$, $k \in J$. Because $|\cos x| \leq 1$ for all $x \in R$, $\sec x$ is never less than 1 in absolute value; the range of the secant function is $\{y \mid |y| \geq 1\}$. Since the cosine function has fundamental period 2π, the secant function also has fundamental period 2π.

Definition 6.8   *If $x \in R$, and $x \neq k\pi$, $k \in J$, then*

$$\textbf{cosecant} = \{(x, y) \mid y = \csc x\}.$$

Cosecant

Because $\csc x$ is the reciprocal of $\sin x$, the domain of the cosecant function contains all real numbers except those for which $\sin x = 0$; that is, the domain is all of R except the numbers $k\pi$, $k \in J$. The cosecant function has range $\{y \mid |y| \geq 1\}$ and fundamental period 2π.

Graphs of reciprocal functions

We can sketch the graphs of the cotangent, secant, and cosecant functions by plotting some points over $0 < x < 2\pi$ and joining the points with a curve. These are shown in Figure 6.23 along with the graphs of their respective reciprocal functions.

Since $\cot x$, $\sec x$, and $\csc x$ are the reciprocals of $\tan x$, $\cos x$, and $\sin x$, respectively, we can find these function values for commonly used values of x by using Table 6.2 or 6.3. Furthermore, we can use Table V in the Appendix for additional values.

Examples

a. $\sec \dfrac{7\pi}{6}$ b. $\csc 1.32$

Solutions

a. By using Definition 6.4-III and Table 6.2, we have

$$\sec \frac{7\pi}{6} = \frac{1}{\cos\left(\dfrac{7\pi}{6}\right)} = \frac{1}{-\sqrt{3}/2} = -\frac{2}{\sqrt{3}}.$$

b. By using Table V, we obtain

$$\csc 1.32 = 1.032.$$

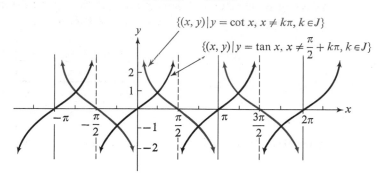

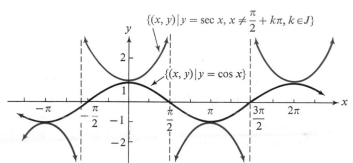

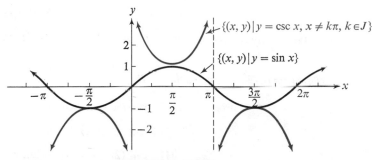

Figure 6.23

Other formulas for tangent

Each of the functions discussed in this section has its associated sum, difference, and reduction formulas, comparable to and derived from those of sine and cosine.

Theorem 6.11   *If $x_1, x_2 \in R$, and $x_1, x_2, x_1 + x_2 \neq \pi/2 + k\pi, k \in J$, then*

$$\tan(x_1 + x_2) = \frac{\tan x_1 + \tan x_2}{1 - \tan x_1 \tan x_2}.$$

Proof   By definition, for $x_1 + x_2 \in R$, $x_1 + x_2 \neq \pi/2 + k\pi$, $k \in J$,

$$\tan (x_1 + x_2) = \frac{\sin (x_1 + x_2)}{\cos (x_1 + x_2)}.$$

By Theorems 6.6 and 6.3, we have

$$\tan (x_1 + x_2) = \frac{\sin x_1 \cos x_2 + \cos x_1 \sin x_2}{\cos x_1 \cos x_2 - \sin x_1 \sin x_2}.$$

Dividing numerator and denominator of the right-hand member by $\cos x_1 \cos x_2$ for $x_1, x_2 \in R$, and $x_1, x_2 \neq \pi/2 + k\pi$, $k \in J$, we find

$$\tan (x_1 + x_2) = \frac{\dfrac{\sin x_1 \cos x_2}{\cos x_1 \cos x_2} + \dfrac{\cos x_1 \sin x_2}{\cos x_1 \cos x_2}}{\dfrac{\cos x_1 \cos x_2}{\cos x_1 \cos x_2} - \dfrac{\sin x_1 \sin x_2}{\cos x_1 \cos x_2}}.$$

After simplifying each term in the right-hand member, from Definition 6.4-I we obtain

$$\tan (x_1 + x_2) = \frac{\tan x_1 + \tan x_2}{1 - \tan x_1 \tan x_2},$$

as was to be shown.

Also, by substituting $x_1 - x_2$ for x in the definition of $\tan x$, we can establish a difference formula for tangents in a similar way.

Theorem 6.12   *If $x_1, x_2 \in R$, and $x_1, x_2, x_1 - x_2 \neq \pi/2 + k\pi$, $k \in J$, then*

$$\tan (x_1 - x_2) = \frac{\tan x_1 - \tan x_2}{1 + \tan x_1 \tan x_2}.$$

The following reduction formulas follow directly from Theorems 6.11 and 6.12.

Theorem 6.13   *For each $x \in R$,*

$$\text{I} \quad \tan (\pi - x) = -\tan x,$$

$$\text{II} \quad \tan (\pi + x) = \tan x,$$

$$\text{III} \quad \tan (2\pi - x) = -\tan x.$$

Their proofs are left as exercises. Formulas for $\tan 2x$ and $\tan \dfrac{x}{2}$ also follow immediately from formulas we have already developed.

Theorem 6.14   *If $x \in R$, $x \neq \pi/2 + k\pi$, $x \neq \pi/4 + k\pi/2$, $k \in J$, then*

$$\tan 2x = \frac{2 \tan x}{1 - \tan^2 x}.$$

Proof   Replacing x_1 and x_2 by x in Theorem 6.11, we obtain

$$\tan (x + x) = \frac{\tan x + \tan x}{1 - \tan x \tan x}, \qquad \tan 2x = \frac{2 \tan x}{1 - \tan^2 x}.$$

Theorem 6.15   *If $x \in R$ and $x \neq k\pi$, $k \in J$, then*

$$\text{I} \quad \tan \frac{x}{2} = \frac{1 - \cos x}{\sin x}, \qquad \text{II} \quad \tan \frac{x}{2} = \frac{\sin x}{1 + \cos x}.$$

Proof   We shall prove only Part I here. The proof of Part II, which is similar, is left as an exercise. By Definition 6.4,

$$\tan \frac{x}{2} = \frac{\sin \dfrac{x}{2}}{\cos \dfrac{x}{2}},$$

from which, by multiplying numerator and denominator by $2 \sin \dfrac{x}{2}$, we have

$$\tan \frac{x}{2} = \frac{2 \sin^2 \dfrac{x}{2}}{2 \sin \dfrac{x}{2} \cos \dfrac{x}{2}}.$$

Therefore, by Theorems 6.9 and 6.10,

$$\tan \frac{x}{2} = \frac{1 - \cos x}{\sin x},$$

as desired.

Relationships for cotangent, secant, and cosecant

Relationships similar to those developed for tangent in Theorems 6.11 through 6.15 can be derived for cotangent, secant, and cosecant. For practical purposes, however, the formulas for cosine, sine, and tangent are the only ones necessary in view of the fact that

$$\sec x = \frac{1}{\cos x}, \qquad \csc x = \frac{1}{\sin x}, \quad \text{and} \quad \cot x = \frac{1}{\tan x}.$$

Exercise 6.7

Using the definitions and theorems stated in this section, find values for the other five circular functions from the one function value and quadrant given.

Example $\tan x = \dfrac{4}{3};$ x terminates in Quadrant III.

                                              (Solution overleaf)

Solution   From Definition 6.4, $\tan x = \dfrac{\sin x}{\cos x}$, so $\dfrac{\sin x}{\cos x} = \dfrac{4}{3}$. In Quadrant III, we have

$\cos x = -\sqrt{1 - \sin^2 x}$. It follows that

$$\frac{\sin x}{-\sqrt{1 - \sin^2 x}} = \frac{4}{3}, \quad \text{or} \quad 3 \sin x = -4\sqrt{1 - \sin^2 x}.$$

Squaring each member and simplifying, we find that

$$9 \sin^2 x = 16(1 - \sin^2 x), \qquad 25 \sin^2 x = 16, \qquad \sin^2 x = \frac{16}{25}.$$

Therefore either $\sin x = 4/5$ or $\sin x = -4/5$. Since x terminates in Quadrant III, we have $\sin x = -4/5$. Using $\sin^2 x + \cos^2 x = 1$, we find that

$$\frac{16}{25} + \cos^2 x = 1,$$

from which either

$$\cos x = \frac{3}{5} \quad \text{or} \quad \cos x = -\frac{3}{5}.$$

In Quadrant III, $\cos x < 0$, and hence $\cos x = -3/5$. From these results, and using Definition 6.4, we find

$$\sin x = -\frac{4}{5}, \quad \cos x = -\frac{3}{5}, \quad \sec x = -\frac{5}{3}, \quad \csc x = -\frac{5}{4}, \quad \cot x = \frac{3}{4}.$$

1. $\sin x = -8/17$; x terminates in Quadrant III.
2. $\sec x = -5/3$; x terminates in Quadrant II.
3. $\tan x = 5/12$; x terminates in Quadrant I.
4. $\tan x = 8/15$; x terminates in Quadrant III.
5. $\cot x = -3$; x terminates in Quadrant IV.
6. $\csc x = -17/15$; x terminates in Quadrant IV.

Prove the given formula. In each case, state restrictions on x.

7. $\tan (\pi - x) = -\tan x$ 8. $\tan (2\pi - x) = -\tan x$

9. $\tan \left(\dfrac{\pi}{2} - x\right) = \cot x$ 10. $\tan \left(\dfrac{\pi}{2} + x\right) = -\cot x$

Use the fact that the period of tangent is π to express the function value in the form

$\tan \bar{x}$ *or* $-\tan \bar{x}$, *where* $0 \le \bar{x} < \dfrac{\pi}{2}$.

Example $\tan \dfrac{19\pi}{5}$

Solution   Since the period of tangent is π, we express $\dfrac{19\pi}{5}$ as

$$\frac{15\pi}{5} + \frac{4\pi}{5}, \quad \text{or} \quad 3\pi + \frac{4\pi}{5}$$

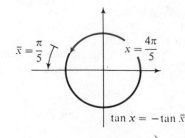

Then we have

$$\tan\left(\frac{19\pi}{5}\right) = \tan\left(3\pi + \frac{4\pi}{5}\right) = \tan\left(\frac{4\pi}{5}\right).$$

Since $\dfrac{\pi}{2} < \dfrac{4\pi}{5} < \pi$, we can use the reduction formula I in Theorem 6.13 to obtain

$$\tan\frac{4\pi}{5} = -\tan\frac{\pi}{5}.$$

11. $\tan\dfrac{8\pi}{7}$ **12.** $\tan\dfrac{13\pi}{5}$ **13.** $\tan\dfrac{21\pi}{11}$ **14.** $\tan\dfrac{23\pi}{12}$

15. $\tan\dfrac{47\pi}{13}$ **16.** $\tan\dfrac{39\pi}{5}$ **17.** $\tan\left(\dfrac{-11\pi}{7}\right)$ **18.** $\tan\left(\dfrac{-15\pi}{7}\right)$

Use appropriate reduction formulas and Tables 6.2 and 6.3 to find the value of each expression.

Example $\tan\dfrac{5\pi}{4}$

Solution   Since $\dfrac{5\pi}{4} = \pi + \dfrac{\pi}{4}$, $\tan\dfrac{5\pi}{4} = \tan\left(\pi + \dfrac{\pi}{4}\right) = \tan\dfrac{\pi}{4}$.

From Table 6.3, $\tan\dfrac{\pi}{4} = 1$. Hence, $\tan\dfrac{5\pi}{4} = 1$.

19. $\tan\dfrac{5\pi}{3}$ **20.** $\sec\dfrac{4\pi}{3}$ **21.** $\tan\dfrac{11\pi}{6}$

22. $\csc\dfrac{13\pi}{3}$ **23.** $\cot\left(-\dfrac{7\pi}{6}\right)$ **24.** $\sec\left(-\dfrac{13\pi}{3}\right)$

Use the definitions and periodicity of the circular functions, together with Table V to obtain an approximation for the given function value.

Example $\tan 3.31$

Solution $\tan 3.31 \approx \tan(3.31 - 3.14)$

           $= \tan 0.17$

           ≈ 0.1717

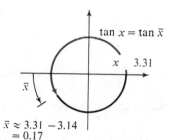

$\bar{x} \approx 3.31 - 3.14$
$= 0.17$

25. $\tan 4.52$ **26.** $\tan 5.61$ **27.** $\tan 7.45$ **28.** $\tan 8.30$

29. $\tan 12.50$ **30.** $\tan 27.30$ **31.** $\tan(-9.32)$ **32.** $\tan(-6.41)$

33. $\sec 6.92$ **34.** $\cot 8.31$ **35.** $\csc(-7.09)$ **36.** $\sec(-9.42)$

Use Table V to approximate the element x, $0 \leq x \leq 1.57$, in the domain of the function, for the specified element in the range. Interpolate as required.

37. $\tan x = 1.459$ **38.** $\cot x = 0.6563$ **39.** $\sec x = 2.448$

40. $\csc x = 1.010$ **41.** $\cot x = 0.0208$ **42.** $\tan x = 0.5000$

Chapter Review

[6.1]   *Determine a value of s, $0 \leq s \leq 2\pi$, for which the given equation is true. State whether the function value is positive or negative.*

1. $\cos s = \cos \dfrac{7\pi}{3}$ **2.** $\sin s = \sin \left(\dfrac{21\pi}{4}\right)$

3.   Given $\cos s = \dfrac{12}{13}$ and $\sin s < 0$, find $\sin s$.

4.   Given $\sin s = -\dfrac{8}{17}$ and $\cos s < 0$, find $\cos s$.

Write the expression in the form sin s or cos s, where $s > 0$.

5. $\cos \left(-\dfrac{2\pi}{3}\right)$ **6.** $\sin \left(-\dfrac{3\pi}{4}\right)$

[6.2]   *Find the function value.*

7. $\cos \dfrac{5\pi}{3}$ **8.** $\sin \dfrac{8\pi}{3}$ **9.** $\sin \left(\dfrac{5\pi}{4}\right)$ **10.** $\cos \left(-\dfrac{9\pi}{2}\right)$

Assume $0 \leq s < 2\pi$ and find the value of s for which both conditions are true.

11. $\cos s = \dfrac{1}{2}$ and $\sin s < 0$ **12.** $\sin s = -\dfrac{1}{\sqrt{2}}$ and $\cos s > 0$

[6.3]   *Write the expression as an equivalent one of the form $\cos \bar{x}$ or of the form $-\cos \bar{x}$, where $0 \leq \bar{x} \leq \dfrac{\pi}{2}$.*

13. $\cos \dfrac{4\pi}{5}$ **14.** $\cos \left(-\dfrac{11\pi}{8}\right)$ **15.** $\cos \dfrac{19\pi}{7}$ **16.** $\cos \left(-\dfrac{25\pi}{11}\right)$

Find an approximation for the given expression. Use linear interpolation as required.

17. $\cos 0.43$ **18.** $\cos 1.355$ **19.** $\cos 8.31$ **20.** $\cos(-4.40)$

Approximate x, for $0 \leq x \leq 1.57$.

21. $\cos x = 0.4357$ **22.** $\cos x = 0.7640$

[6.4] *Reduce the given expression to an equivalent one of the form* sin $\bar{x}$ *or of the form* $-\sin \bar{x}$, *where* $0 \le \bar{x} \le \dfrac{\pi}{2}$.

23. $\sin \dfrac{9\pi}{8}$

24. $\sin \left(-\dfrac{13\pi}{7} \right)$

Write an equivalent expression either of the form sin $\bar{x}$ *or of the form* $-\sin \bar{x}$, *where* $0 \le x \le \dfrac{\pi}{2}$.

25. $\sin \dfrac{28\pi}{3}$

26. $\sin \left(-\dfrac{42\pi}{11} \right)$

Find an approximation for the given expression. Use linear interpolation as required.

27. $\sin 1.49$ 28. $\sin 0.324$ 29. $\sin 3.57$ 30. $\sin (-6.94)$

Approximate x, for $0 \le x \le 1.57$.

31. $\sin x = 0.4259$

32. $\sin x = 0.6400$

[6.5] *Sketch the graph of the given equation over* $-2\pi \le x \le 2\pi$, *except as noted.*

33. $y = 3 \sin 2x$

34. $y = 2 \cos \left(x + \dfrac{\pi}{3} \right)$

35. $y = 2 \cos \pi x, \ -2 \le x \le 2$

36. $y = 3 \sin x + \cos 2x$

[6.6] *Given that* $\sin x = 0.4$ *and* $\dfrac{\pi}{2} \le x \le \pi$, *find an approximation for the given expression.*

37. $\cos x$ 38. $\cos 2x$ 39. $\sin 2x$ 40. $\sin \dfrac{x}{2}$

[6.7] *Find values for the other five of the six circular functions if:*

41. $\tan x = -\dfrac{3}{4}$ and x terminates in Quadrant IV.

42. $\sin x = \dfrac{5}{13}$ and x terminates in Quadrant II.

Find an approximation for the given function value.

43. $\tan 2.08$ 44. $\tan 11.42$

Find an approximation for x, where $0 \le x \le 1.57$. *Interpolate as required.*

45. $\sec x = 2.448$ 46. $\cot x = 4.150$

7 *Trigonometric Functions*

7.1   *Angles and Their Measure*

In studying geometry, we learn that an angle is the union of two rays with a common endpoint (Figure 7.1-a) and that an angle can be designated by naming a point on each ray together with the common endpoint of the rays, or else by simply assigning a single symbol, say α, to the angle. The common endpoint of the rays is called the **vertex** of the angle, and the rays are called the **sides** of the angle.

Now each angle in the plane is congruent ($\cong$) to an angle with one side along the positive x-axis and vertex at the origin (Figure 7.1-b). Such an angle is said to be in **standard position.** The side $\overrightarrow{OC'}$ of $\angle A'OC'$ in Figure 7.1-b is called the **initial side** of the angle, the side $\overrightarrow{OA'}$ is called the **terminal side** and the angle can be visualized as being formed by a rotation from the initial side $\overrightarrow{OC'}$ into the terminal side $\overrightarrow{OA'}$. In trigonometry, the *amount of rotation* is considered along with the angle itself. If the terminal side of an angle in standard position lies in a given quadrant, we say that the angle is *in* that quadrant.

Angle measurement   A measure is assigned to an angle by means of a circle with center at the vertex of the angle. For convenience, we shall restrict this discussion to the set of angles in

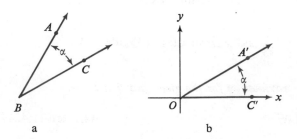

a   b

Figure 7.1

standard position, although the process described is perfectly general. If the circumference of a circle of radius $r > 0$ is divided into p arcs of equal length, then each of the arcs will have length $2\pi r/p$. Starting at the point $(r, 0)$, we can scale the circumference of the circle in both the counterclockwise and clockwise directions from this point, in terms of this arc length as a unit. In doing this, it is customary to assign *negative* numbers in the *clockwise* direction, and *positive* numbers in the *counterclockwise* direction. As an example, consider the circle in Figure 7.2, where the circumference is divided into 16 equal parts and the arc length of each part is $2\pi r/16$. The terminal side of any angle α in standard position intercepts the circumference at a point, and one of the scale numbers (arc lengths) assigned to that point becomes the measure of the angle, depending on the amount of rotation involved.

Note that although angles such as α_1, α_2, and α_3 in Figure 7.3 have the same initial side and the same terminal side, their measures are different because the amounts of rotation involved are different. Such angles are called **coterminal.** The measures of α_1 and α_3 are positive and the measure of α_2 is negative.

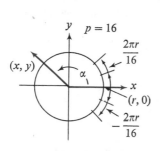

Figure 7.2

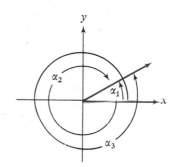

Figure 7.3

The two most commonly encountered units of angle measure are the **degree** and the **radian.** In degree measure, the circumference of a circle is divided into 360 arcs of equal length, and hence the unit arc is of length

$$\frac{2\pi r}{360} = \frac{\pi}{180}\, r,$$

where r is the radius of the circle (see Figure 7.4-a). For this measure, we use the notation $m°(\alpha)$.

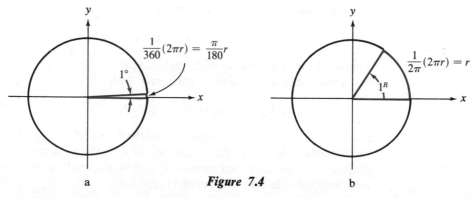

a   **Figure 7.4**   b

In radian measure, the circumference is divided into 2π arcs of equal length, and hence the length of the unit arc (see Figure 7.4-b) is

$$\frac{2\pi r}{2\pi} = r.$$

For this measure we use the notation $m^R(\alpha)$. Thus, in radian measure, the length of the unit arc is the length of the radius of the circle, and the circumference contains just 2π of these units. Recall from geometry that the lengths of arc intercepted by a given central angle on concentric circles are proportional to the circumferences of the circles. *It follows that the measure assigned to an angle in terms of a given unit is independent of the radius of the measuring circle.* In particular, if the *unit* circle is used as the basis for $m^R(\alpha)$, then the length of the unit arc is 1. For this reason, the unit circle is customarily used as a reference for the measure of an angle in radians.

Comparison of degree and radian measure

Because the measures of a given angle in different units are proportional to the circumference of the measuring circle in these same units, we have

$$\frac{m^\circ(\alpha)}{360} = \frac{m^R(\alpha)}{2\pi},$$

so that

$$m^\circ(\alpha) = \frac{180}{\pi}\, m^R(\alpha) \tag{1}$$

and

$$m^R(\alpha) = \frac{\pi}{180}\, m^\circ(\alpha). \tag{2}$$

Equations (1) and (2) are **conversion formulas** for expressing the relationships between degrees and radians.

Symbolism for angles and their measure

Note that m° and m^R are functions, each with the set of angles as domain and R as range. Several conventions regarding the use of symbolism for angles and their measure are customarily made in mathematics. For one thing, to express a relation such as $m^\circ(\alpha) = 40$ or $m^R(\beta) = \pi/4$, we usually write $\alpha = 40^\circ$ or $\beta = (\pi/4)^R$. Thus. for example, if α, β, and γ are the angles of a triangle, then

$$m^R(\alpha) + m^R(\beta) + m^R(\gamma) = \pi,$$

and this equation is generally abbreviated

$$\alpha + \beta + \gamma = \pi^R.$$

This notation is consistent with the fact that if α and β are two angles, and $\alpha + \beta$ denotes their **sum** then

$$m^R(\alpha + \beta) = m^R(\alpha) + m^R(\beta).$$

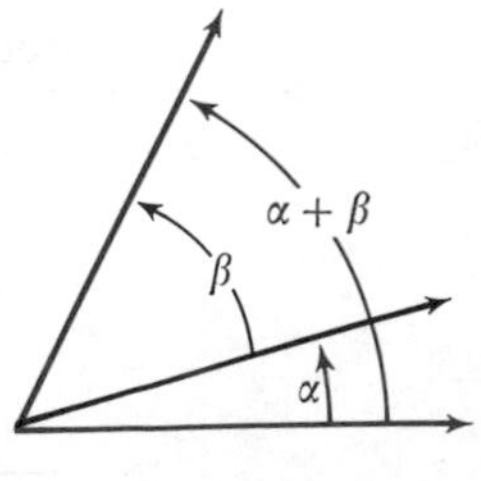

Figure 7.5

Here the sum $\alpha + \beta$ is construed to be the result of taking the terminal side of α as the initial side of β, with the same point for vertex, and then viewing $\alpha + \beta$ as the angle with initial side that of α and terminal side that of β, as in Figure 7.5.

Following another convention, we sometimes write, for example,

$$m^\circ(\alpha) = 60^\circ, \quad \text{or} \quad m^R(\alpha) = \frac{\pi^R}{3}.$$

Here, the appearance of the degree or radian symbol in the right-hand member of the equation should be interpreted as a clarifying redundancy.

Examples   Find the degree measure, to the nearest tenth of a degree, of the angle whose radian measure is given.

a. $\dfrac{\pi^R}{3}$ b. 0.62^R

Solutions   a. $\dfrac{\pi^R}{3} = \left(\dfrac{180}{\pi} \cdot \dfrac{\pi}{3}\right)^\circ = 60^\circ$ b. $0.62^R = \left(\dfrac{180}{\pi} \cdot 0.62\right)^\circ$

$$= \frac{111.6^\circ}{\pi} \approx 35.5^\circ$$

Examples   Find the radian measure, to the nearest hundredth of a radian, of the angle whose degree measure is given.

a. 30° b. 300°

Solutions   a. $30^\circ = \left(\dfrac{\pi}{180} \cdot 30\right)^R = \dfrac{\pi^R}{6} \approx 0.52^R$ b. $300^\circ = \left(\dfrac{\pi}{180} \cdot 300\right)^R = \dfrac{5\pi^R}{3} \approx 5.24^R$

Note that, in these examples, the number of units in one measure does not equal the number of units in another measure; for example, $30 \neq \pi/6$. But an angle whose measure is 30° is congruent to an angle whose measure is $(\pi/6)^R$, and we express this fact by writing $30^\circ = (\pi/6)^R$. This is analogous to 36 inches $=$ 3 feet, 6 feet $=$ 2 yards, and so forth, when indicating lengths of line segments in different units.

If the measure of an angle is given in radians, then the length s of the intercepted arc of a circle with given radius r can be found directly. That is,

$$s = r \cdot m^R(\alpha). \tag{3}$$

If the measure of an angle is given in degrees, it can first be changed to radian measure and then the length of the intercepted arc can be found directly from (3).

Exercise 7.1

Find the degree measure of the angle whose radian measure is as given.

1. a. 0^R **b.** $\dfrac{\pi^R}{2}$ **c.** π^R **d.** $\dfrac{3\pi^R}{2}$ **e.** $2\pi^R$

2. By filling in the blank spaces, complete the following table, which compares the radian measure and degree measure of angles that are of frequent occurrence.

$m°(\alpha)$	30°	45°		120°	135°	
$m^R(\alpha)$			$\dfrac{\pi^R}{3}$			$\dfrac{5\pi^R}{6}$

$m°(\alpha)$	210°				315°	330°
$m^R(\alpha)$		$\dfrac{5\pi^R}{4}$	$\dfrac{4\pi^R}{3}$	$\dfrac{5\pi^R}{3}$		

Find the degree measure, to the nearest tenth of a degree, of the angle whose radian measure is as given.

3. $\dfrac{2\pi^R}{9}$ **4.** $\dfrac{3\pi^R}{5}$ **5.** $\dfrac{7\pi^R}{5}$ **6.** $\dfrac{5\pi^R}{8}$

7. 0.30^R **8.** 1.25^R **9.** 3.62^R **10.** 9.14^R

Find the radian measure, to the nearest hundredth of a radian, of the angle whose degree measure is as given.

11. 20°   **12.** 50°   **13.** 130°   **14.** 310°

15. 420°   **16.** 580°   **17.** 750°   **18.** 800°

19. What is the degree measure, to the nearest hundredth of a degree, of an angle whose radian measure is 1^R?

20. What is the radian measure, to the nearest thousandth of a radian, of an angle whose degree measure is 1°?

Find all angles α satisfying $-360° \leq \alpha \leq 720°$ that are coterminal with the angle whose measure is given. Write a representation for the measures of all angles coterminal with the angle.

Examples   **a.** 24°   **b.** −184°

Solutions   **a.** $24° + 360° = 384°$ **b.** $-184° + 360° = 176°$
            $24° - 360° = -336°$ $-184° + 720° = 536°$
            $24° + 360°k,\ k \in J$ $-184° + 360°k,\ k \in J$

21. 30°   **22.** −150°   **23.** −240°   **24.** 190°

25. 420°   **26.** 683°   **27.** −330°   **28.** −271°

On a circle with given radius, find the length of the arc intercepted by the angle whose measure is as given.

Example $r = 3$ in.; $m°(\alpha) = 120°$

Solution   The measure of the angle is first changed to radian measure. Thus

$$m^R(\alpha) = \left(\frac{\pi}{180} \cdot 120\right)^R = \frac{2\pi^R}{3}.$$

Then, by Equation (3) on page 173,

$$s = r \cdot m^R(\alpha) = 3 \cdot \frac{2\pi}{3} = 2\pi \approx 6.28.$$

Therefore, the desired arc length is approximately 6.28 in.

29. $r = 4$ cm; $(\pi/6)^R$ **30.** $r = 5.2$ in.; π^R **31.** $r = 1.2$ m; 0.60^R

32. $r = 3.6$ ft; 1^R **33.** $r = 2$ m; $135°$ **34.** $r = 4.3$ yd; $180°$

7.2   *Functions of Angles*

Historically, interest in the circular functions arose from the study of angles and triangles. In defining the circular functions, we used the notion of arc length on a unit circle to associate a real number (the arc length from the point $(1, 0)$ to a point on the circle) with another real number, the first or second coordinate of the point. Thus we defined functions with real numbers for *both* domain and range.

Geometric ratios in the coordinate plane

Now, consider any angle α in standard position and let (x_1, y_1) and (x_2, y_2) be any two distinct points (except the origin) in its terminal side or ray (Figure 7.6). We can show that the following equalities of ratios hold:

$$\frac{x_1}{y_1} = \frac{x_2}{y_2} \quad (y_1, y_2 \neq 0),$$

$$\frac{x_1}{\sqrt{x_1^2 + y_1^2}} = \frac{x_2}{\sqrt{x_2^2 + y_2^2}},$$

and

$$\frac{y_1}{\sqrt{x_1^2 + y_1^2}} = \frac{y_2}{\sqrt{x_2^2 + y_2^2}}.$$

The proofs of these are left as exercises.

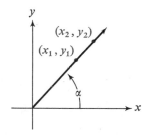

Figure 7.6

Although the figure shows the terminal side of α in the first quadrant, the ratios are equal for coordinates of points on the terminal side of an angle in any quadrant. Because these ratios do not depend on the choice of a point in the terminal side of α, we can use them to define some new functions, each with *the set of angles in standard position* as domain, and with sets of real numbers as ranges. Traditionally, six such assignments are made, and the names given the resulting functions are those we have used to name the circular functions.

Definition 7.1   *If* α *is an angle in standard position and if* $(x, y) \neq (0, 0)$ *is any point on the terminal side of* α, *then*

$$\text{I}\qquad \textbf{cosine} = \left\{ (\alpha, \cos \alpha) \mid \cos \alpha = \frac{x}{\sqrt{x^2 + y^2}} \right\},$$

$$\text{II}\qquad \textbf{sine} = \left\{ (\alpha, \sin \alpha) \mid \sin \alpha = \frac{y}{\sqrt{x^2 + y^2}} \right\},$$

$$\text{III}\qquad \textbf{tangent} = \left\{ (\alpha, \tan \alpha) \mid \tan \alpha = \frac{y}{x}, \quad x \neq 0 \right\},$$

$$\text{IV}\qquad \textbf{cotangent} = \left\{ (\alpha, \cot \alpha) \mid \cot \alpha = \frac{x}{y}, \quad y \neq 0 \right\},$$

$$\text{V}\qquad \textbf{secant} = \left\{ (\alpha, \sec \alpha) \mid \sec \alpha = \frac{\sqrt{x^2 + y^2}}{x}, \quad x \neq 0 \right\},$$

$$\text{VI}\qquad \textbf{cosecant} = \left\{ (\alpha, \csc \alpha) \mid \csc \alpha = \frac{\sqrt{x^2 + y^2}}{y}, \quad y \neq 0 \right\}.$$

These functions are called **trigonometric.** Observe that in each place where the expression $\sqrt{x^2 + y^2}$ occurs, the *positive* root is used.

Example   Find the value of each of the six trigonometric functions of α, if the terminal side of α contains the point $(-3, 5)$.

Solution   By Definition 7.1,

$$\cos \alpha = \frac{x}{\sqrt{x^2 + y^2}} \qquad\qquad \sec \alpha = \frac{\sqrt{x^2 + y^2}}{x}$$

$$= \frac{-3}{\sqrt{9 + 25}} = -\frac{3}{\sqrt{34}}, \qquad\qquad = \frac{\sqrt{9 + 25}}{-3} = -\frac{\sqrt{34}}{3},$$

$$\sin \alpha = \frac{y}{\sqrt{x^2 + y^2}} \qquad\qquad \csc \alpha = \frac{\sqrt{x^2 + y^2}}{y}$$

$$= \frac{5}{\sqrt{9 + 25}} = \frac{5}{\sqrt{34}}, \qquad\qquad = \frac{\sqrt{9 + 25}}{5} = \frac{\sqrt{34}}{5},$$

$$\tan \alpha = \frac{y}{x} = \frac{5}{-3} = -\frac{5}{3}, \qquad\qquad \cot \alpha = \frac{x}{y} = \frac{-3}{5} = -\frac{3}{5}.$$

Because every angle in the plane is congruent to an angle in standard position, the definitions of the trigonometric functions can be extended to assign the same numbers to every angle congruent to a given angle α. Thus, while these functions are defined in terms of angles in standard position, they can be viewed as applying to the set of all angles in the plane. Moreover, since congruent angles have the same measure, we can identify angles in the domain of each function with a particular unit of measure. Thus, we write

$$\sin 30° \quad \text{and} \quad \sin \frac{\pi^R}{6}$$

as abbreviations for "the sine of an angle whose measure is 30 degrees" and "the sine of an angle whose measure is $\pi/6$ radians," respectively.

Relationship between circular and trigonometric functions

The fact that the circular functions are defined using the unit circle, together with the fact that the unit circle can be used to assign measures to angles, makes it reasonable to expect a very close relationship to exist between the circular and trigonometric functions. Such is indeed the case.

Since the trigonometric functions of an angle α have been defined in terms of the coordinates of *any* point other than the origin on the terminal side of α, we can arbitrarily choose the point $P(\cos x, \sin x)$ where the terminal side of the angle intersects the unit circle. Thus,

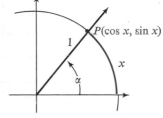

$$\cos \alpha = \frac{\cos x}{1} = \cos x,$$

and, similarly,

$$\sin \alpha = \frac{\sin x}{1} = \sin x$$

(Figure 7.7). From this, we see that the elements in the ranges of the trigonometric functions are equal to the corresponding elements in the ranges of the analogous circular functions. Thus, if we denote any of the six trigonometric functions by T, and the circular functions of the same name by C, then

$$T(\alpha) = C(x), \tag{1}$$

Figure 7.7

where x is the length of the arc intercepted by α on the unit circle. Notice that elements x in the domain of C are *real numbers*, while elements α in the domain of T are *angles*. The elements in the range of each function are real numbers. Moreover, the correspondence between the values of the circular functions and those of the trigonometric functions renders all of the relationships thus far studied for circular functions equally applicable to the trigonometric functions. Because the elements in the domains of the trigonometric functions are angles, an equation such as

$$\sin 2\alpha = 2 \sin \alpha \cos \alpha$$

is sometimes called a **double-angle formula,** while an equation such as

$$\sin \frac{\alpha}{2} = \pm\sqrt{\frac{1 - \cos \alpha}{2}}$$

is called a **half-angle formula.**

In Section 7.1 we agreed to name an angle by its measure. In particular, if $m^R(\alpha) = x$, then we write $\alpha = x^R$, and if $m°(\alpha) = t$, then we write $\alpha = t°$. Thus by replacing α by x^R and $t°$ in (1), we obtain

$$T(x^R) = C(x) \quad \text{and} \quad T(t°) = C(x).$$

For example,

$$\cos \frac{\pi^R}{3} = \cos \frac{\pi}{3} \quad \text{and} \quad \cos 60° = \cos \frac{\pi}{3},$$

where in each equation the left-hand member is an element in the range of the *trigonometric* function and the right-hand member is an element in the range of the analogous *circular* function.

Using relationship (1), we obtain the entries of Table 7.1 directly from Table 6.2, page 138, Definition 6.4, and the relationship between degree and radian measures of angles. These are left for you to verify.

Table 7.1

$m°(\alpha)$	$m^R(\alpha)$	$\sin \alpha$	$\csc \alpha$	$\cos \alpha$	$\sec \alpha$	$\tan \alpha$	$\cot \alpha$
$0°$	0^R	0	not defined	1	1	0	not defined
$30°$	$\dfrac{\pi^R}{6}$	$\dfrac{1}{2}$	2	$\dfrac{\sqrt{3}}{2}$	$\dfrac{2}{\sqrt{3}}$	$\dfrac{1}{\sqrt{3}}$	$\sqrt{3}$
$45°$	$\dfrac{\pi^R}{4}$	$\dfrac{1}{\sqrt{2}}$	$\sqrt{2}$	$\dfrac{1}{\sqrt{2}}$	$\sqrt{2}$	1	1
$60°$	$\dfrac{\pi^R}{3}$	$\dfrac{\sqrt{3}}{2}$	$\dfrac{2}{\sqrt{3}}$	$\dfrac{1}{2}$	2	$\sqrt{3}$	$\dfrac{1}{\sqrt{3}}$
$90°$	$\dfrac{\pi^R}{2}$	1	1	0	not defined	not defined	0
$180°$	π	0	not defined	-1	-1	0	not defined
$270°$	$\dfrac{3\pi^R}{2}$	-1	-1	0	not defined	not defined	0

Examples　　a.　$\cos 45° = \dfrac{1}{\sqrt{2}}$　　　　　　b.　$\tan \pi^R = 0$

Tables for trigonometric functions　　Trigonometric function values that are not listed in Table 7.1 can be obtained from Table V in the Appendix if $m^R(\alpha)$ is known. Table VI in the Appendix gives trigonometric function values of angles with given degree measure. The table is graduated in intervals of 10 minutes ($10'$), one minute being equal to one-sixtieth of a degree. We can interpolate as necessary to find, to four significant figures, values between those listed in the table. Observe that the table reads from top to bottom for $0° \leq \alpha \leq 45°$, where the function values are identified at the top of the page, and from bottom to top for $45° \leq \alpha \leq 90°$, where the function values are identified at the bottom of the page.

Example   Find $\tan 149°$.

Solution   From the reduction formula

$$\tan \alpha = -\tan (180° - \alpha),$$

we have

$$\tan 149° = -\tan (180° - 149°)$$

$$= -\tan 31°.$$

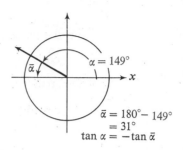

$\bar{\alpha} = 180° - 149°$
$= 31°$
$\tan \alpha = -\tan \bar{\alpha}$

From Table VI in the Appendix, we find that $\tan 31° \approx 0.6009$. Thus

$$\tan 149° \approx -0.6009.$$

Observe that in the preceding example we again used the symbol $\approx$, because the function values obtained from the table are approximations to irrational numbers. Note also how a reference angle can be used in the same way that we used a reference arc in Chapter 6 to provide a geometric interpretation of a reduction formula.

Exercise 7.2

Find the value of each of the six trigonometric functions of α for the terminal side of α containing the given point.

1. $(3, 4)$ **2.** $(5, -12)$ **3.** $(\sqrt{2}, -\sqrt{2})$ **4.** $(-1, \sqrt{3})$

5. $(-6, -8)$ **6.** $(-4, -3)$ **7.** $(0, 3)$ **8.** $(-7, 0)$

Determine in which quadrant the terminal side of the angle α lies.

Example $\sin \alpha > 0$, $\cos \alpha < 0$

Solution   Since $\sin \alpha > 0$, the terminal side of α lies in Quadrant I or II; since $\cos \alpha < 0$, the terminal side of α lies in Quadrant II or III. Therefore, the terminal side of α lies in Quadrant II.

9. $\sin \alpha < 0$, $\cos \alpha > 0$ **10.** $\cos \alpha > 0$, $\tan \alpha < 0$

11. $\sec \alpha > 0$, $\sin \alpha > 0$ **12.** $\tan \alpha < 0$, $\sin \alpha > 0$

13. $\sec \alpha > 0$, $\csc \alpha < 0$ **14.** $\cot \alpha > 0$, $\sin \alpha < 0$

Use Table 7.1 and reduction formulas to find the value of the given expression.

Example $\cos 150°$

Solution   From the reduction formula,

$$\cos \alpha = -\cos (180° - \alpha),$$

of Theorem 6.5-I, we have

$$\cos 150° = -\cos (180 - 150)°$$

$$= -\cos 30° = -\frac{\sqrt{3}}{2}.$$

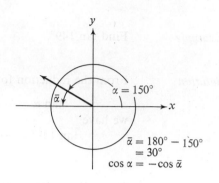

15. $\tan 330°$ 16. $\cot 120°$ 17. $\cos 225°$

18. $\tan 150°$ 19. $\sin 330°$ 20. $\cos 240°$

21. $\sin (-30°)$ 22. $\tan (-60°)$ 23. $\sin (-240°)$

24. $\cos (-135°)$ 25. $\tan (-300°)$ 26. $\tan (-750°)$

Use Table V or Table VI, as appropriate, to find function values to four significant figures for the given expression.

27. $\sin 32°$ 28. $\cos 49°$ 29. $\tan 33°$

30. $\cot 51°$ 31. $\sec 17°$ 32. $\csc 84°$

33. $\cos 0.63^R$ 34. $\sin 0.42^R$ 35. $\cot 1.42^R$

36. $\tan 1.03^R$ 37. $\csc \dfrac{5\pi^R}{9}$ 38. $\sec \dfrac{5\pi^R}{12}$

39. $\sin 132°$ 40. $\cos 153°$ 41. $\tan 320°$

42. $\sin 312°$ 43. $\cos (-130°)$ 44. $\tan (-605°)$

45. $\sin 2.07^R$ 46. $\cos 3.51^R$ 47. $\tan 4.21^R$

48. $\tan 6.00^R$ 49. $\cos (-1.63^R)$ 50. $\sin (-12.32^R)$

51. Show (geometrically) that if (x_1, y_1) and (x_2, y_2) are the coordinates of any two points (except the origin) on the terminal side (ray) of an angle in the first quadrant, then the following equalities of ratios hold:

$$\frac{x_1}{y_1} = \frac{x_2}{y_2} \quad (y_1, y_2 \neq 0),$$

$$\frac{x_1}{\sqrt{x_1^2 + y_1^2}} = \frac{x_2}{\sqrt{x_2^2 + y_2^2}}, \quad \text{and} \quad \frac{y_1}{\sqrt{x_1^2 + y_1^2}} = \frac{y_2}{\sqrt{x_2^2 + y_2^2}}.$$

52. Under the same conditions as in Exercise 51, explain how the equalities of these ratios hold for the coordinates of two points on the terminal side of an angle whose terminal side is in the second, third, and fourth quadrants.

7.3   *Right Triangles*

An examination of Figure 7.8-a makes it evident that in the first quadrant, any point (x, y) on the terminal side of an angle α determines a right triangle with sides

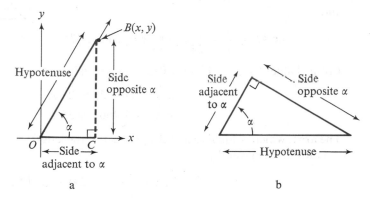

Figure 7.8

measuring x and y and with hypotenuse of length $\sqrt{x^2 + y^2}$. Therefore, as special cases of trigonometric function values, we can consider the following **trigonometric ratios** of the sides of a right triangle:

$$\sin \alpha = \frac{\text{length of side opposite } \alpha}{\text{length of hypotenuse}},$$

$$\cos \alpha = \frac{\text{length of side adjacent to } \alpha}{\text{length of hypotenuse}},$$

$$\tan \alpha = \frac{\text{length of side opposite } \alpha}{\text{length of side adjacent to } \alpha}.$$

Similar statements can be made about csc α, sec α, and cot α. It is not necessary that the right triangle be oriented in such a way that α is in standard position. The foregoing relations involving ratios of the lengths of the sides of a right triangle are equally applicable to a right triangle in any position, as suggested in Figure 7.8-b.

Solution of a triangle

If we are given some parts of a triangle, that is, the measures of some angles and the lengths of some sides, and asked to find the remaining measures, we are asked to **solve** the triangle. In general, we shall give solutions to the nearest tenth of a unit for lengths and to the nearest 10' for the degree measures of angles.

Example

If one angle of a right triangle measures 47° and the side adjacent to this angle has length 24, solve the triangle and determine its area.

(Solution overleaf)

Solution

We first make a sketch illustrating the situation. Generally the hypotenuse is labeled c, the legs a and b, and the angles opposite a, b, and c are labeled α, β, and γ, respectively, or A, B, and C, respectively. In this case, we note first that

$$\beta = 90° - \alpha = 90° - 47° = 43°.$$

Next we have

$$\tan 47° = \frac{a}{b} = \frac{a}{24}, \quad \text{or} \quad a = 24 \tan 47°,$$

and

$$\sec 47° = \frac{c}{b} = \frac{c}{24}, \quad \text{or} \quad c = 24 \sec 47°.$$

From Table VI, $\tan 47° \approx 1.072$ and $\sec 47° \approx 1.466$, so that

$$a \approx 24(1.072) \approx 25.7,$$
$$c \approx 24(1.466) \approx 35.2.$$

The area, which is equal to half the product of the length of the base by the length of the altitude, is given by

$$\mathscr{A} = \frac{1}{2}ab \approx \frac{1}{2}(25.7)(24) = 308.4.$$

Note that the ratio selected in each case in the foregoing example was one in which the length of the remaining side and the length of the hypotenuse appeared in the numerator. With such a selection of ratios, the operation in the computation is multiplication rather than division.

Logarithms can be used to perform the computations in the preceding example and in many of the problems that follow. Handbooks are available with the logarithms of the trigonometric ratios to various degrees of accuracy if you wish to use them, although devices such as the slide rule, desk calculator, and computer have largely replaced logarithmic tables for such computations.

Right triangles play a useful role in finding one trigonometric function value, given another.

Example

If $\tan \alpha = \dfrac{2}{3}$ and α is in Quadrant I, find $\sin \alpha$.

Solution

Sketch a right triangle and label the sides so that $\tan \alpha$ is as given. By the Pythagorean theorem, the hypotenuse of the triangle has length

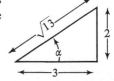

$$\sqrt{2^2 + 3^2} = \sqrt{13}, \text{ so that } \sin \alpha = \frac{2}{\sqrt{13}}.$$

Note in this example that once the length of the hypotenuse is determined, *all* of the values for trigonometric functions of α are available by inspection.

Right triangles are helpful also in quadrants other than the first if care is taken to assign *directed distances* as lengths to the sides, while viewing the length of the hypotenuse as always positive. This is simply a variation on our procedures for finding function values using either reduction formulas or reference angles.

Example   If $\cot \alpha = \dfrac{3}{7}$, and $\sin \alpha < 0$, find $\cos \alpha$.

Solution   Since $\cot \alpha > 0$ while $\sin \alpha < 0$, α can lie only in Quadrant III. Sketch a right triangle in Quadrant III and label the sides with directed lengths so that $\cot \bar{\alpha} = \dfrac{3}{7}$. By the Pythagorean theorem,

$$\text{length of } \overline{OA} = l(\overline{OA})$$
$$= \sqrt{(-3)^2 + (-7)^2}$$
$$= \sqrt{9 + 49} = \sqrt{58}.$$

By inspection, then, $\cos \alpha = \dfrac{-3}{\sqrt{58}}$.

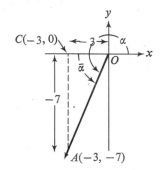

Exercise 7.3

In all exercises in this set, give lengths to the nearest tenth of a unit and angle measures to the nearest 10′.

Solve the right triangle ABC. Consider C — 90°.

1. $a = 6, b = 8$ 2. $c = 24, A = 32°$
3. $b = 120, B = 54°$ 4. $a = 3.5, B = 48°$
5. $c = 16, A = 22°$ 6. $a = 5, \quad c = 16$

7. In the right triangle ABC, find a if $\sin A = 4/5$ and $c = 20$.
8. In the right triangle ABC, find b if $\tan A = 3/4$ and $a = 12$.
9. Find the other trigonometric function values of θ if $\sin \theta = -1/2$ and the terminal side of θ is in the third quadrant.
10. Find the other trigonometric function values of θ if $\tan \theta = -7/24$ and the terminal side of θ is in the fourth quadrant.
11. Find the other trigonometric function values of θ if $\cos \theta = -3/7$ and the terminal side of θ is in the second quadrant.
12. Find the other trigonometric function values of θ if $\sin \theta = 2/3$ and the terminal side of θ is in the second quadrant.

Find the other trigonometric function values by using a sketch of the angle θ and an appropriate right triangle.

13. $\sin \theta = 3/5, \cos \theta < 0$ 14. $\cos \theta = -12/13, \sin \theta > 0$
15. $\tan \theta = 5/12, \sec \theta < 0$ 16. $\cot \theta = 4/3, \sin \theta < 0$
17. $\sec \theta = -2, \sin \theta < 0$ 18. $\csc \theta = 3, \sec \theta < 0$
19. Find the value of $\dfrac{\sin \theta + 2 \cos \theta - \tan \theta}{1 - \cot \theta + \sec \theta}$ if $\tan \theta = 1$ and the terminal side of θ is in the third quadrant.

20. Find the value of $\dfrac{\csc^2 \theta + \sin \theta - 1}{2 - \tan^2 \theta + \cot^2 \theta}$ if $\sin \theta = \dfrac{1}{2}$ and the terminal side of θ is in the second quadrant.

In Exercises 21 and 22, given the information pertaining to the figure, find the length of the line segment denoted by x.

21. Given: $l(\overline{BA}) = 25,$

$\qquad \qquad \alpha = 16°,$

$\qquad \qquad \beta = 12°,$

$\qquad \overleftrightarrow{BC}$ is a straight line.

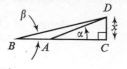

22. Given: $l(\overline{AB}) = 350,$

$\qquad \qquad \alpha = 21°,$

$\qquad \qquad \beta = 8°,$

$\qquad \overleftrightarrow{BD}$ is a straight line.

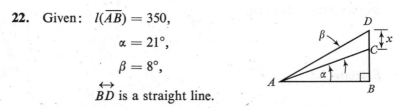

23. Find the area of a parallelogram if two of its adjacent sides are 22 and 28 inches in length and the measure of the included angle is 40°.

24. In a circle, the length of the radius is 10 inches and the length of a chord is 12 inches. Find the measure of the angle made by two lines tangent to the circle at the ends of the chord.

25. Find the length of the base of an isosceles triangle if the length of one of the equal sides is 12 inches and the measure of one of the equal angles is 40°.

26. Find the perimeter of a regular pentagon inscribed in a circle of radius 8 inches.

27. A rectangle is 80 feet long and 64 feet wide. Find the measures of the angles between a diagonal and the sides.

28. The sides of an isosceles triangle have lengths 6, 6, and 8. Find a measure for each angle in the triangle.

29. The angle of elevation (the angle between the line of sight and the horizontal) from a point 200 feet from the base of a building to the base of a flagpole on top of the building is 60°. The angle of elevation from the same spot to the top of the flagpole is 65°. How tall is the flagpole?

30. Each of two surveyors is located 200 feet from a flagpole. If the angle between the flagpole and one surveyor, when measured by the other surveyor, is 36°, how far apart are the two surveyors?

7.4   *The Law of Sines*

The fact that the area $\mathscr{A}$ of a triangle is equal to one-half the product of the length of its base and the length of its altitude gives us an immediate expression for its area

in terms of the lengths of two sides of the triangle
and the measure of the included angle. Thus, from
Figure 7.9 and the fact that $\sin(180° - \gamma) = \sin \gamma$,

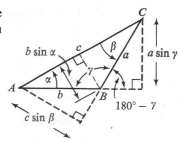

$$\mathscr{A} = \frac{1}{2} c(b \sin \alpha) = \frac{1}{2} bc \sin \alpha,$$

$$\mathscr{A} = \frac{1}{2} a(c \sin \beta) = \frac{1}{2} ac \sin \beta,$$

$$\mathscr{A} = \frac{1}{2} b(a \sin \gamma) = \frac{1}{2} ab \sin \gamma.$$

Figure 7.9

Because each triangle has only one area, we can equate the right-hand members
of these equations to obtain

$$\frac{1}{2} bc \sin \alpha = \frac{1}{2} ac \sin \beta = \frac{1}{2} ab \sin \gamma.$$

Multiplying each member here by $2/(abc)$, we obtain the following result, called the
law of sines.

Theorem 7.1   *If α, β, and γ are the angles of a triangle and a is the length of the
side opposite α, b is the length of the side opposite β, and c is the length of the side
opposite γ, then*

$$\frac{\sin \alpha}{a} = \frac{\sin \beta}{b} = \frac{\sin \gamma}{c}.$$

***Use of the
law of sines
in solving
triangles***

The law of sines can be used to solve certain triangles. Let us first consider the case
in which two angles and one side of a triangle are given.

Example

Solve the triangle for which

$$\alpha = 45°, \quad \beta = 60°, \quad \text{and} \quad a = 10.$$

Solution

It is helpful first to make a sketch. Then, from Theorem 7.1, we have

$$\frac{\sin 45°}{10} = \frac{\sin 60°}{b},$$

from which

$$b = \frac{10 \sin 60°}{\sin 45°} = \frac{10 \dfrac{\sqrt{3}}{2}}{\dfrac{1}{\sqrt{2}}} \approx 12.245 \approx 12.2.$$

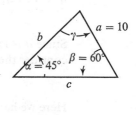

Next, we observe that

$$\gamma = 180° - \alpha - \beta = 180° - 45° - 60° = 75°.$$

(Solution continued)

From Theorem 7.1, we have

$$\frac{\sin 45°}{10} = \frac{\sin 75°}{c}.$$

Then, from Table VI,

$$c = \frac{10 \sin 75°}{\sin 45°} \approx 10\left(\frac{0.9659}{0.7071}\right) \approx 10(1.37) = 13.7.$$

Thus, we have

$$b \approx 12.2, \quad c \approx 13.7, \quad \text{and} \quad \gamma = 75°.$$

Ambiguous case

If the lengths of two sides of a triangle, say a and b, and the measure of an angle opposite one of them, say α, are given, we may encounter ambiguity, depending on the value of a in relation to those of b and α. In Figure 7.10, we hold b and α constant and observe the possible situations as a assumes different values.

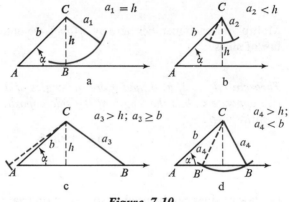

Figure 7.10

First we determine the length h of the altitude of a triangle with the given measures for the angle α and the adjacent side b. Since $\sin \alpha = h/b$,

$$h = b \sin \alpha.$$

Now consider the case (Figure 7.10-a) in which the length a of the given side satisfies the equation

$$a = b \sin \alpha.$$

We have a unique solution—a right triangle. Next consider the case (Figure 7.10-b) in which the length a of the given side satisfies the inequality

$$a < b \sin \alpha.$$

Since a is less than h, the given values are such that no triangle is possible. Next consider the case (Figure 7.10-c) in which the length a of the given side is such that

$$a > b \sin \alpha \quad \text{and} \quad a \geq b.$$

Here we have a unique solution. For the fourth and last possibility (Figure 7.10-d), in which the length a of the given side is such that

$$a > b \sin \alpha \quad \text{and} \quad a < b,$$

we have two triangles possible, $\triangle ABC$ and $\triangle AB'C$.

Of course, if the given angle α is obtuse (Figure 7.11), then any specified lengths of the sides and measures of the angles permit only two possibilities:

1. $a \le b$, no triangle (as shown in Figure 7.11-a);
2. $a > b$, one oblique triangle (as shown in Figure 7.11-b).

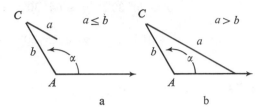

Figure 7.11

Example   Solve the triangle for which $a = 4$, $b = 3$, and $\beta = 45°$.

Solution   We first sketch a figure and observe that this gives rise to the ambiguous case in which we are given the lengths of two sides and the measure of an angle opposite one of them. We then check for the number of possible solutions and observe that $b > a \sin \beta$ and $b < a$. Therefore, there are two triangles with the given measurements. Using Theorem 7.1, we next determine a value for $\sin \alpha$. We have

$$\frac{\sin \alpha}{4} = \frac{\sin 45°}{3},$$

$$\sin \alpha = \frac{4}{3} \cdot \frac{\sqrt{2}}{2} = \frac{2}{3} \sqrt{2}$$

$$\approx \frac{2}{3}(1.414) \approx 0.9426.$$

Because $\sin \alpha > 0$ in Quadrants I and II, from Table VI we have either

$$\alpha \approx 70° \, 30' \quad \text{or} \quad \alpha' \approx 180° - 70° \, 30'$$

$$= 109° \, 30'.$$

If $\alpha \approx 70° \, 30'$, then $\gamma \approx 180° - 45° - 70° \, 30' = 64° \, 30'$, and we therefore have

$$\frac{\sin 45°}{3} \approx \frac{\sin 64° \, 30'}{c},$$

so that

$$c \approx 3\left(\frac{0.9026}{0.7071}\right) \approx 3.8.$$

If $\alpha' \approx 109° \, 30'$, then $\gamma' \approx 180° - 45° - 109° \, 30' = 25° \, 30'$, and we have

$$\frac{\sin 45°}{3} \approx \frac{\sin 25° \, 30'}{c},$$

(Solution continued)

so that

$$c' \approx 3\left(\frac{0.4305}{0.7071}\right) \approx 1.8.$$

Thus, the two solutions for the triangle are

$$\alpha \approx 70° 30', \quad \gamma \approx 64° 30', \quad c \approx 3.8$$

and

$$\alpha' \approx 109° 30', \quad \gamma' \approx 25° 30', \quad c' \approx 1.8.$$

Exercise 7.4

Solve the triangle. In all exercises in this set, give lengths to the nearest tenth and angle measures to the nearest 10'.

1. $b = 10, B = 30°, A = 80°$
2. $a = 64, B = 36° 10', C = 82° 20'$
3. $c = 78.1, A = 58° 10', C = 63° 10'$
4. $b = 1.02, B = 41° 10', C = 80° 20'$
5. $a = 84.2, A = 110°, C = 22° 20'$
6. $c = 0.94, A = 41° 10', B = 96° 50'$

Determine the number of triangles that satisfy the given conditions.

Example $b = 34, \quad c = 12, \quad C = 30°$

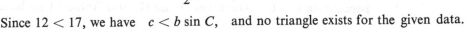

Solution   Sketch a figure and determine

$$h = 34 \sin 30° = 34 \cdot \frac{1}{2} = 17.$$

Since $12 < 17$, we have $c < b \sin C$, and no triangle exists for the given data.

7. $a = 5.4, b = 7.0, B = 30°$
8. $b = 4.9, c = 3.2, C = 30°$
9. $a = 31.1, c = 41.3, C = 30°$
10. $a = 42.3, b = 20.7, B = 30°$
11. $b = 16.2, c = 14.3, C = 20°$
12. $a = 141, b = 182, B = 20°$
13. $a = 4.6, b = 2.3, B = 30°$
14. $b = 68.1, c = 41.3, C = 30°$

Solve the triangle.

15. $a = 4.8, c = 3.9, A = 113°$
16. $a = 3.2, b = 2.6, B = 54°$
17. $b = 6.21, c = 4.39, B = 42° 40'$
18. $a = 179, b = 212, B = 114° 10'$
19. $b = 1.8, c = 1.5, C = 32° 30'$
20. $a = 9.4, b = 8.6, B = 54° 20'$
21. $a = 8.4, b = 6.9, B = 62° 10'$
22. $b = 0.42, c = 0.21, B = 31° 50'$

23. $a = 13.84$, $b = 6.92$, $A = 60°$ **24.** $a = 4.72$, $c = 9.44$, $A = 30°$

25. $b = 420$, $c = 610$, $B = 33° 20'$ **26.** $a = 5.42$, $b = 6.82$, $A = 43° 30'$

27. In a given triangle $a = 6.4$, $b = 8.2$, and $B = 30°$. Find the triangle's area to the nearest whole unit.

28. Find the area of a triangle to the nearest tenth where $c = 4.5$, $A = 24°$, and $C = 61°$.

29. From a window in a tower, 85 feet above the ground, the angle of elevation to the top of a nearby building measures $34° 30'$. From a point on the ground directly below the window, the angle of elevation to the top of the same building measures $50° 20'$. Find the height of the building.

30. Two men, 500 feet apart, observe a balloon between them that is in the same vertical plane with them. The respective angles of elevation of the balloon are observed by the men to measure $80° 10'$ and $52° 50'$. Find the height of the balloon above the ground.

Show that, in any triangle ABC, the given equality is true.

31. $\dfrac{a + b}{b} = \dfrac{\sin \alpha + \sin \beta}{\sin \beta}$ **32.** $\dfrac{a - b}{b} = \dfrac{\sin \alpha - \sin \beta}{\sin \beta}$

33. $\dfrac{a - b}{a + b} = \dfrac{\tan \frac{1}{2}(\alpha - \beta)}{\tan \frac{1}{2}(\alpha + \beta)}$ (This is called the **law of tangents**.)

7.5   The Law of Cosines

The x- and y-coordinates of a point on the terminal side of an angle β in standard position, when the point is located b units from the origin, are

$$b \cos \beta \quad \text{and} \quad b \sin \beta,$$

respectively (Figure 7.12). We can use this fact to derive a very useful formula. Figure 7.13 shows points $A(x_1, y_1)$ and $B(x_2, y_2)$ lying on the terminal sides of

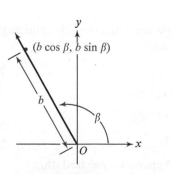

Figure 7.12

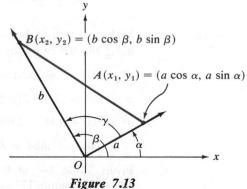

Figure 7.13

angles α and β, respectively. Now, if $B(x_2, y_2)$ is located b units from the origin, while $A(x_1, y_1)$ is located a units from the origin, then

$$[l(\overline{AB})]^2 = (x_2 - x_1)^2 + (y_2 - y_1)^2$$

$$= (b \cos \beta - a \cos \alpha)^2 + (b \sin \beta - a \sin \alpha)^2$$

$$= b^2 \cos^2 \beta - 2ab \cos \beta \cos \alpha + a^2 \cos^2 \alpha$$

$$+ b^2 \sin^2 \beta - 2ab \sin \alpha \sin \beta + a^2 \sin^2 \alpha$$

$$= b^2(\cos^2 \beta + \sin^2 \beta) + a^2(\cos^2 \alpha + \sin^2 \alpha)$$

$$- 2ab(\cos \beta \cos \alpha + \sin \beta \sin \alpha).$$

Now from Theorems 6.1 and 6.4, $\cos^2 \beta + \sin^2 \beta = 1$, $\cos^2 \alpha + \sin^2 \alpha = 1$, and $\cos \beta \cos \alpha + \sin \beta \sin \alpha = \cos (\beta - \alpha)$. Making the appropriate substitutions, we have the following formula for the square of the distance between two points A and B in the plane:

$$[l(\overline{AB})]^2 = a^2 + b^2 - 2ab \cos (\beta - \alpha).$$

Note that $\beta - \alpha$, or γ, is just the angle between the terminal sides of the angles β and α.

In view of the fact that the location of the axes in the plane is purely a matter of convenience, we have established the following theorem, which is known as the **law of cosines.**

Theorem 7.2   *If α, β, and γ are the angles of a triangle, and a, b, and c are the lengths of the sides opposite α, β, and γ, respectively, then*

$$c^2 = a^2 + b^2 - 2ab \cos \gamma \tag{1}$$

$$b^2 = a^2 + c^2 - 2ac \cos \beta \tag{2}$$

$$a^2 = b^2 + c^2 - 2bc \cos \alpha. \tag{3}$$

Solving triangles

This theorem has many applications, among them the solution of certain triangles. If we are given the measure of an angle and the lengths of the adjacent sides, we can use the law of cosines to help us find the remaining parts of the triangle.

Example

Solve the triangle for which

$$\alpha = 100°, \quad b = 10, \quad \text{and} \quad c = 12.$$

Solution

Make a sketch of the triangle and label the sides and angles. We can first find a by using the law of cosines; thus,

$$a^2 = b^2 + c^2 - 2bc \cos \alpha,$$

$$a^2 = 10^2 + 12^2 - 2(10)(12) \cos 100°$$

$$a^2 \approx 100 + 144 - 240(-0.1736)$$

$$= 244 + 41.664 \approx 285.7.$$

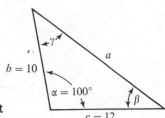

From Table I   in the Appendix, we find that $16^2 = 256$, while $17^2 = 289$; therefore, a is a little less than 17. We shall use 17. Next, to find β, we again use the law of cosines (we could use the law of sines because now we have the length of a side opposite a given angle) in the form

$$b^2 = a^2 + c^2 - 2ac \cos \beta,$$

or

$$\cos \beta = \frac{b^2 - a^2 - c^2}{-2ac}.$$

With $a^2 = 285.7$, $b^2 = 100$, and $c^2 = 144$, we obtain

$$\cos \beta \approx \frac{100 - 285.7 - 144}{-2(17)(12)} \approx 0.8.$$

From Table VI, $\beta \approx 36° 50'$. Since the sum of the angles of a triangle is 180°, we can find an approximation for γ by noting that

$$\gamma = 180° - \alpha - \beta \approx 180° - 100° - 36° 50' = 43° 10'.$$

We have, then, for the remaining parts of the triangle,

$$a \approx 17, \quad \beta \approx 36° 50', \quad \text{and} \quad \gamma \approx 43° 10'.$$

Of course, if greater precision were desired in the preceding example, we could resort to various other means of approximating $\sqrt{285.7}$, and dispense with rounding measures of angles to the nearest 10′.

Since Equations (1), (2), and (3) in Theorem 7.2 are relationships between three sides and one angle of a triangle, we can, as we did in the preceding example, use *any* of these forms to solve for the measure of an angle, given the lengths of the three sides.

Exercise 7.5

Find the remaining parts of the triangle. In all exercises in this set, give lengths and areas to the nearest tenth and angle measurements to the nearest 10′.

1. $a = 10$, $b = 4$, $\gamma = 30°$ 2. $b = 14.2$, $c = 7.9$, $\alpha = 64° 10'$
3. $a = 4.9$, $c = 6.8$, $\beta = 122° 20'$ 4. $a = 241$, $c = 104$, $\beta = 148° 10'$
5. $b = 14.6$, $c = 6.21$, $\alpha = 80° 40'$ 6. $b = 9.4$, $c = 10.2$, $\alpha = 100° 50'$
7. $a = 5$, $b = 8$, $c = 7$ 8. $a = 4.5$, $b = 5.3$, $c = 2.8$

9. Find the greatest angle of the triangle whose sides are 5.1, 4.2, and 4.5.
10. Find the least angle of the triangle whose sides are 29.5, 33.2, and 41.4.
11. Find the area of the triangle in Exercise 1.
12. Find the area of the triangle in Exercise 2.

13. Show that $1 + \cos \alpha = \dfrac{(b + c + a)(b + c - a)}{2bc}$.

14. Show that $1 - \cos \alpha = \dfrac{(a - b + c)(a + b - c)}{2bc}$.

15. Use the results of Exercises 13 and 14 to show that the area $\mathscr{A}$ of a triangle is given by

$$\mathscr{A} = \sqrt{s(s-a)(s-b)(s-c)},$$

where $s = \dfrac{a+b+c}{2}$. This formula is known as **Heron's formula.**

16. Use the results of Exercise 15 to find the area of the triangle in
a. Exercise 7;   **b.** Exercise 8.

17. Show that the Pythagorean theorem is a special case of the law of cosines.

18. Find the least angle of the triangle with vertices at the points $(0, 0)$, $(5, -2)$, and $(-7, -3)$.

19. Find the area of the triangle with vertices at the points $(1, 1)$, $(5, 5)$, and $(-2, 6)$. (See Exercise 15.)

20. Find the area of the triangle with vertices at the points $(2, -3)$, $(4, 6)$, and $(-3, 1)$.

Chapter Review

[7.1]   *Find the degree measure, to the nearest tenth of a degree, of the angle whose radian measure is as given.*

1. $\dfrac{3\pi^R}{8}$ **2.** 8.72^R

Find the radian measure, to the nearest hundredth of a radian, of the angle whose degree measure is as given.

3. $40°$ **4.** $610°$

Find all angles α, where $-360° \leq \alpha \leq 720°$, that are coterminal with the angle whose measure is given.

5. $-330°$ **6.** $518°$

On a circle with given radius, find the length of the arc intercepted by the angle whose measure is as given.

7. $r = 4.8''$; $\dfrac{\pi^R}{3}$ **8.** $r = 1.6'$; $150°$

[7.2]   *Determine in which quadrant the terminal side of the angle α lies.*

9. $\tan \alpha > 0$ and $\cos \alpha < 0$ **10.** $\cos \alpha < 0$ and $\sin \alpha < 0$

Find the value of each expression.

11. $\cos 210°$ **12.** $\sin 330°$ **13.** $\tan 225°$

14. $\cos 120°$ **15.** $\sin (-210°)$ **16.** $\cos (-315°)$

Find approximations for the given expression. Use any appropriate tables.

17. $\cos 23°$ **18.** $\tan 1.24^R$ **19.** $\sin 120°$ **20.** $\cos (-135°)$

21. $\sin 192°$ **22.** $\tan 422°$ **23.** $\cos 3.68^R$ **24.** $\sin (-5.12^R)$

[7.3]   *Solve the right triangle. Consider $C = 90°$.*

25. $b = 12$ and $A = 56°$ **26.** $c = 6.2$ and $B = 20°$

27. In the right triangle ABC, find b if $\cos A = \dfrac{5}{13}$ and $a = 18$.

28. Find the perimeter and area of a regular hexagon inscribed in a circle of radius 10 inches.

[7.4]   *Solve the triangle.*

29. $a = 4$, $B = 28°$, $A = 74°$ **30.** $c = 9.2$, $B = 32° \, 10'$, $C = 80° \, 40'$

Determine the number of triangles that satisfy the given conditions.

31. $b = 6$, $c = 5$, $C = 22°$ **32.** $a = 4$, $b = 12$, $A = 68°$

33. Solve the triangle(s) of Exercise 31.

34. Find the area of the triangle(s) of Exercise 31.

[7.5]   *Solve the triangle.*

35. $a = 6$, $b = 8$, $\gamma = 24°$ **36.** $b = 4.8$, $c = 5.4$, $\alpha = 32° \, 10'$

37. Find the angle of greatest measure in the triangle whose sides are 4.8, 3.7, and 4.4 in length.

38. Find the area of the triangle of Exercise 37.

8 More About Circular and Trigonometric Functions

The fact that all of the circular functions are defined either in terms of the unit circle or in terms of other circular functions suggests that these functions and the corresponding trigonometric functions are interrelated in a number of ways. These relationships are called **identities,** since they are true for every permissible replacement of any variables involved. Thus, for x any real number or any angle, sine and cosine are related by the identity

$$\sin^2 x + \cos^2 x = 1,$$

as we observed earlier. Also, the definitions of $\tan x$, $\cot x$, $\sec x$, and $\csc x$ are identities. Thus, the defining relations,

$$\tan x = \frac{\sin x}{\cos x}, \quad \sec x = \frac{1}{\cos x}, \quad \cot x = \frac{\cos x}{\sin x}, \quad \text{and} \quad \csc x = \frac{1}{\sin x},$$

are true for all real numbers x or angles with measure x for which the denominators in the right-hand members are not zero.

Proofs of identities

These four identities can be used to prove still other identities.

Example

Show that

$$\tan^2 x + 1 = \sec^2 x$$

is an identity.

Solution

We can "prove an identity" by showing that the given equation is equivalent to a true statement or to another equation known to be an identity. Since, for every real number x such that $x \neq (\pi/2) + k\pi, k \in J$, we have

$$\tan x = \frac{\sin x}{\cos x} \quad \text{and} \quad \sec x = \frac{1}{\cos x},$$

we can substitute appropriately in the given equation and obtain

$$\frac{\sin^2 x}{\cos^2 x} + 1 = \frac{1}{\cos^2 x}.$$

Because $\cos^2 x \neq 0$ for any x for which $\tan x$ and $\sec x$ are defined, we can multiply each member here by $\cos^2 x$ to obtain for all such x the equivalent equation

$$\sin^2 x + \cos^2 x = 1.$$

This is a known identity, and hence the given equation is also an identity.

When we prove an identity, we are actually proving a theorem. However, we shall present such relationships simply as exercises. We can also prove, for example, that

$$\cot^2 x + 1 = \csc^2 x$$

is also an identity.

In proving identities, it is sometimes more convenient to restrict manipulation to one member of the given equation, and sometimes more convenient to work with both members. In the foregoing example, both members were multiplied by $\cos^2 x$ to prove the identity. The following example restricts the transformations to the left-hand member.

Example   Show that $\dfrac{\sin \alpha}{1 - \cos \alpha} - \cot \alpha = \dfrac{1}{\sin \alpha}$ is an identity.

Solution   In proving identities, it is often helpful to rewrite a given equation in terms of $\sin \alpha$ and $\cos \alpha$ only. Writing $\cos \alpha / \sin \alpha$ for $\cot \alpha$, we have

$$\frac{\sin \alpha}{1 - \cos \alpha} - \frac{\cos \alpha}{\sin \alpha} = \frac{1}{\sin \alpha}.$$

Writing the left-hand member as a single fraction, we have

$$\frac{\sin^2 \alpha - (1 - \cos \alpha) \cos \alpha}{(1 - \cos \alpha) \sin \alpha} = \frac{1}{\sin \alpha}$$

from which

$$\frac{\sin^2 \alpha - \cos \alpha + \cos^2 \alpha}{(1 - \cos \alpha) \sin \alpha} = \frac{1}{\sin \alpha},$$

$$\frac{(\sin^2 \alpha + \cos^2 \alpha) - \cos \alpha}{(1 - \cos \alpha) \sin \alpha} = \frac{1}{\sin \alpha},$$

$$\frac{1 - \cos \alpha}{(1 - \cos \alpha) \sin \alpha} = \frac{1}{\sin \alpha}.$$

Since $1 - \cos \alpha$ is restricted from 0 in the original equation, we can rewrite the left-hand member here and arrive at the equivalent equation

$$\frac{1}{\sin \alpha} = \frac{1}{\sin \alpha},$$

which is clearly an identity, and the demonstration is complete.

The sum and difference formulas, which were discussed in Section 6.3, as well as the formulas derived from these, are often involved in identities.

Example   Show that

$$\frac{\sin 2\theta}{1 + \cos 2\theta} = \tan \theta$$

is an identity.

(*Solution overleaf*)

Solution   Replacing $\sin 2\theta$ and $\cos 2\theta$ in the given equation with $2 \sin \theta \cos \theta$ and $2 \cos^2 \theta - 1$, respectively, we have

$$\frac{2 \sin \theta \cos \theta}{1 + (2 \cos^2 \theta - 1)} = \tan \theta,$$

$$\frac{2 \sin \theta \cos \theta}{2 \cos^2 \theta} = \tan \theta,$$

$$\frac{\sin \theta \cos \theta}{\cos^2 \theta} = \tan \theta.$$

With the restriction that $\cos \theta \neq 0$, the left-hand member of this equation can be written $\dfrac{\sin \theta}{\cos \theta}$, and we have

$$\frac{\sin \theta}{\cos \theta} = \tan \theta,$$

which is true by definition.

For convenience, let us list again some important identities we have encountered. While no restrictions are given for variables here, it is important to keep such restrictions in mind when using any identity. Furthermore, the variable x should be viewed as representing an element in the set of real numbers or an element in the set of all angles.

Summary of Identities

1. $\sin^2 x + \cos^2 x = 1$ 2. $\cos(-x) = \cos x$ 3. $\sin(-x) = -\sin x$

4. $\cos(x_1 + x_2) = \cos x_1 \cos x_2 - \sin x_1 \sin x_2$

5. $\cos(x_1 - x_2) = \cos x_1 \cos x_2 + \sin x_1 \sin x_2$

6. $\sin(x_1 + x_2) = \sin x_1 \cos x_2 + \cos x_1 \sin x_2$

7. $\sin(x_1 - x_2) = \sin x_1 \cos x_2 - \cos x_1 \sin x_2$

8. a. $\cos 2x = \cos^2 x - \sin^2 x$ 9. $\sin 2x = 2 \sin x \cos x$

   b. $\cos 2x = 2 \cos^2 x - 1$

   c. $\cos 2x = 1 - 2 \sin^2 x$

10. $\cos \dfrac{x}{2} = \pm \sqrt{\dfrac{1 + \cos x}{2}}$ 11. $\sin \dfrac{x}{2} = \pm \sqrt{\dfrac{1 - \cos x}{2}}$

12. $\tan x = \dfrac{\sin x}{\cos x}$ 13. $\tan(-x) = -\tan x$

14. $\tan^2 x + 1 = \sec^2 x$ 15. $\cot^2 x + 1 = \csc^2 x$

16. $\sec x = \dfrac{1}{\cos x}$ 17. $\csc x = \dfrac{1}{\sin x}$

18. $\cot x = \dfrac{\cos x}{\sin x}$ 19. $\cot x = \dfrac{1}{\tan x}$

20. $\tan(x_1 + x_2) = \dfrac{\tan x_1 + \tan x_2}{1 - \tan x_1 \tan x_2}$ 21. $\tan(x_1 - x_2) = \dfrac{\tan x_1 - \tan x_2}{1 + \tan x_1 \tan x_2}$

22. $\tan 2x = \dfrac{2 \tan x}{1 - \tan^2 x}$ 23. $\tan \dfrac{x}{2} = \dfrac{1 - \cos x}{\sin x}$ 24. $\tan \dfrac{x}{2} = \dfrac{\sin x}{1 + \cos x}$

Exercise 8.1

Each of the following expressions may be written as a circular function or trig-
onometric function of kx, or k(x$_1$ ± x$_2$), where k is a positive integer, or as a
power of such a function with one or at most two steps. Write the answer directly.
Assume that x, x$_1$, and x$_2$ take on no value for which a denominator vanishes.

Examples

  a. $\cos(-x)$ b. $2 \sin x \cos x$ 2. $2 \cos^2 4x - 1$

Solutions

  a. $\cos x$ b. $\sin 2x$ c. $\cos 2(4x)$

                                             $\cos 8x$

1. $\tan(-x)$ **2.** $-\sin(-x)$

3. $\dfrac{1}{\cot x}$ **4.** $1 - 2\sin^2 x$

5. $\dfrac{\tan x_1 + \tan x_2}{1 - \tan x_1 \tan x_2}$ **6.** $\dfrac{2 \tan x}{1 - \tan^2 x}$

7. $\cos^2 x - \sin^2 x$ **8.** $\cos^2 3x - \sin^2 3x$

9. $\sin x_1 \cos x_2 - \cos x_1 \sin x_2$ **10.** $\sin 5x \cos 3x + \cos 5x \sin 3x$

11. $\dfrac{2 \tan 3x}{1 - \tan^2 3x}$ **12.** $\cot^2 x + 1$

13. $1 - \cos^2 x$ **14.** $1 - \sec^2 x$

15. $\cos(x_1 + x_2) \cos(x_1 - x_2) - \sin(x_1 + x_2) \sin(x_1 - x_2)$

16. $\sin(x_1 - x_2) \cos(x_1 + x_2) + \cos(x_1 - x_2) \sin(x_1 + x_2)$

Transform the first expression and show that it is identical to the second expression.
Assume suitable restrictions on x.

17. $\cos x \tan x;\quad \sin x$ **18.** $\sin x \sec x;\quad \tan x$

19. $\sin^2 x \cot^2 x;\quad \cos^2 x$ **20.** $\dfrac{\cos^2 x}{\cot^2 x};\quad \sin^2 x$

21. $\cos^2 \alpha(1 + \tan^2 \alpha);\quad 1$ **22.** $(\csc^2 \alpha - 1);\ \cot^2 \alpha$

23. $\sec \alpha \csc \alpha - \cot \alpha;\quad \tan \alpha$ **24.** $(1 - \cos^2 \alpha)(1 + \cot^2 \alpha);\quad 1$

25. $\dfrac{\sin \theta \sec \theta}{\tan \theta};\quad 1$ **26.** $\cos \theta \tan \theta \csc \theta;\quad 1$

27. $(\sec^2 \theta - 1)(\csc^2 \theta - 1);\quad 1$ **28.** $\dfrac{\cos \theta - \sin \theta}{\cos \theta};\quad 1 - \tan \theta$

29. $\dfrac{1}{1 + \sin \theta} + \dfrac{1}{1 - \sin \theta};\quad 2 \sec^2 \theta$

30. $\dfrac{1 + \tan^2 \theta}{\csc^2 \theta};\quad \tan^2 \theta$

Verify that the formula is an identity.

31. $\sin x \cot x = \cos x$ **32.** $\tan x \csc x = \sec x$

33. $\sec x - \cos x = \sin x \tan x$ **34.** $\sec x - \sin x \tan x = \cos x$

35. $\dfrac{1 + \tan^2 x}{\tan^2 x} = \csc^2 x$ **36.** $\dfrac{\sin^2 x}{1 - \cos x} = \dfrac{1 + \sec x}{\sec x}$

37. $\tan^2 \alpha - \sin^2 \alpha = \sin^2 \alpha \tan^2 \alpha$ **38.** $\cot^2 \alpha + \sec^2 \alpha = \tan^2 \alpha + \csc^2 \alpha$

39. $\tan \alpha + \sec \alpha = \dfrac{1}{\sec \alpha - \tan \alpha}$ **40.** $\dfrac{\sec \alpha + \csc \alpha}{1 + \tan \alpha} = \csc \alpha$

41. $\sin 2\alpha = \dfrac{2 \tan \alpha}{1 + \tan^2 \alpha}$ **42.** $\dfrac{2}{\sin 2\alpha} = \tan \alpha + \cot \alpha$

43. $\cot \theta - \cot 2\theta = \csc 2\theta$ **44.** $\dfrac{2}{1 + \cos 2\theta} = \sec^2 \theta$

45. $\dfrac{1 + \cos 2\theta}{\sin 2\theta} = \cot \theta$ **46.** $\cos 2\theta = \dfrac{1 - \tan^2 \theta}{1 + \tan^2 \theta}$

47. $\sin^2 x = \dfrac{1 - \cos 2x}{2}$ **48.** $\cos^2 x = \dfrac{1 + \cos 2x}{2}$

8.2   *Conditional Equations*

In Section 8.1 we observed that certain equations involving circular or trigonometric function values are *identities*; that is, they are true for all permissible values of any variables involved. In this section, we shall be concerned with *conditional* equations of a similar kind, namely, equations that involve circular or trigonometric function values but that are not satisfied by all permissible values of any variables involved.

Solution of conditional equations

Various procedures exist for solving conditional equations that involve circular or trigonometric function values. Of course, the methods applicable to algebraic equations are also valid for such equations, as you will see in the following examples.

Because the circular and trigonometric functions are periodic, we should expect equations involving $\sin x$, $\tan x$, or other circular or trigonometric function values to have infinite solution sets.

Example

Solve $\sin x = -1$ for:

a. $x \in R$,

b. $x \in \{\text{angles with measures given in radians}\}$,

c. $x \in \{\text{angles with measures given in degrees}\}$.

Solution   From the entries in Table 6.2, or using a unit circle where the point $(0, -1)$ is associated with an arc of length x, we observe that the solutions are:

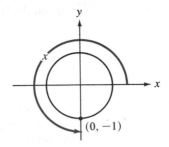

a. $\left\{ x \mid x = \dfrac{3\pi}{2} + 2k\pi, \quad k \in J \right\}$,

b. $\left\{ x \mid x = \left(\dfrac{3\pi}{2} + 2k\pi\right)^R, \quad k \in J \right\}$,

c. $\{ x \mid x = (270 + k \cdot 360)°, \quad k \in J \}$.

Next let us consider an example in which a reference arc proves helpful.

Example   Solve $\sin w = \sqrt{3}/2, \quad w \in R$.

Solution   As in the preceding example, we can use the unit circle to help us visualize the situation here. Since we know (from Table 6.2 or from memory) that $\sin(\pi/3) = \sqrt{3}/2$, we then know that w_1, the first-quadrant value for w, must be $\pi/3$. Using $\pi/3$ as the length of the reference arc, and the fact that $\sin w$ is positive only in the first and second quadrants, we observe that the only other value for w in the interval $0 \le w < 2\pi$ is $w_2 = 2\pi/3$. Hence, the solutions are all of the elements of

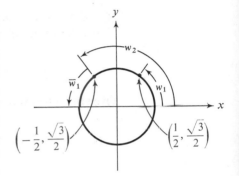

$$\left\{ w \mid w = \frac{\pi}{3} + 2k\pi, \quad k \in J \right\} \quad \text{or} \quad \left\{ w \mid w = \frac{2\pi}{3} + 2k\pi, \quad k \in J \right\}.$$

Since the set of all elements that belong to either of two sets A and B or to both can be designated by $A \cup B$, the solution set is

$$\left\{ w \mid w = \frac{\pi}{3} + 2k\pi, \quad k \in J \right\} \cup \left\{ w \mid w = \frac{2\pi}{3} + 2k\pi, \quad k \in J \right\}.$$

The identities we have developed for the circular and trigonometric functions are useful in solving some equations. We can use these identities to rewrite equations involving more than one function value as equivalent equations involving values of only one function.

Example   Solve $\sin \alpha = \cos \alpha$ over the interval $0^R \le \alpha \le \dfrac{\pi^R}{2}$.

Solution   Since $\sin \alpha \neq 0$ when $\cos \alpha = 0$, the equation is not satisfied if $\cos \alpha = 0$. We can accordingly assume that $\cos \alpha \neq 0$, and hence can multiply each member by $1/\cos \alpha$ to obtain

$$\frac{\sin \alpha}{\cos \alpha} = 1,$$

(Continued overleaf)

or, equivalently,

$$\tan \alpha = 1.$$

Therefore we have as our solution set $\left\{\dfrac{\pi}{4}^R\right\}$.

Some equations that involve circular or trigonometric function values are not of the first degree. In such cases, an algebraic factorization will sometimes lead to a solution.

Example   Solve $(2 \cos u - 1)(\sin u - 1) = 0$ over $0 \leq u < 2\pi$.

Solution   We seek

$$\{u \mid 2 \cos u - 1 = 0\} \cup \{u \mid \sin u - 1 = 0\}.$$

If $2 \cos u - 1 = 0$, then $\cos u = 1/2$, and if $\sin u - 1 = 0$, then $\sin u = 1$. Thus, the values of u satisfying these equations are of the form

$$\frac{\pi}{3} + 2k\pi, \qquad \frac{5\pi}{3} + 2k\pi, \qquad \text{or} \qquad \frac{\pi}{2} + 2k\pi, \quad k \in J.$$

Over the interval $0 \leq u < 2\pi$, the solution set is $\{\pi/3, 5\pi/3, \pi/2\}$.

Note that, in the foregoing example, if the equation had been presented in the form

$$2 \sin u \cos u - \sin u - 2 \cos u + 1 = 0,$$

it would first have been necessary to factor the equation into the given form before proceeding with the solution.

Exercise 8.2

Specify the given set in terms of rational multiples of π.

Example $\{w \mid \tan w = -1\}$.

Solution   From memory or the entries in Table 6.2, we obtain

$$\{w \mid \tan w = -1\} = \left\{w \mid w = \frac{3\pi}{4} + k\pi\right\}, \quad \text{where } k \in J.$$

1. $\left\{p \mid \sin p = \dfrac{1}{2}\right\}$ **2.** $\{q \mid \cot q = 1\}$ **3.** $\{y \mid \cos y = 2\}$

4. $\{y \mid \tan y = \sqrt{3}\}$ **5.** $\left\{r \mid \cos r = \dfrac{1}{\sqrt{2}}\right\}$ **6.** $\{s \mid \sec s = 2\}$

7. $\left\{x \mid \sin x = -\dfrac{\sqrt{3}}{2}\right\}$ **8.** $\left\{x \mid \cos x = -\dfrac{\sqrt{3}}{2}\right\}$

In all the following exercises in this set, use Table V or VI as necessary, and give results to the nearest reading in the table.

In Exercises 9–20, find the solution set of the given equation for
a. $x \in R$,   **b.** *x an angle whose measure is given in (1) radians, and (2) degrees.*

9. $\cos x = \dfrac{1}{2}$ 10. $\sin x = \dfrac{\sqrt{2}}{2}$ 11. $\tan x = \sqrt{3}$

12. $\cot x = -1$ 13. $\sec x - \sqrt{2} = 0$ 14. $\csc x + 1 = 0$

15. $4 \sin x - 1 = 0$ 16. $2 \tan x + 3 = 0$ 17. $3 \cot x - 1 = 0$

18. $3 \cos x + 1 = 0$ 19. $2 \sec x - 5 = 0$ 20. $3 \csc x + 8 = 0$

In Exercises 21–26, find the solution set of the given equation over R.

Example

$\sqrt{3} \sin x + \cos x = 0$

Solution

By first noting that $\cos x \neq 0$ in any solution, we can divide each member by $\cos x$ to produce

$$\sqrt{3} \, \frac{\sin x}{\cos x} + 1 = 0,$$

from which

$$\sqrt{3} \tan x + 1 = 0,$$

$$\tan x = -\frac{1}{\sqrt{3}}.$$

Therefore, the solution set is $\left\{ x \mid x = \dfrac{5\pi}{6} + k\pi, \quad k \in J \right\}$.

21. $\sin x - \sqrt{3} \cos x = 0$ 22. $3 \sin x + \cos x = 0$
23. $\tan^2 x - 1 = 0$ 24. $2 \sin^2 x - 1 = 0$
25. $\sin^2 x - \cos^2 x = 1$ 26. $\sin x + \cos x \tan x = 3$

In Exercises 27–32, find the solution set of the given equation for θ a member of the set of angles such that $0° \leq \theta < 360°$.

Example

$2 \cos^2 \theta \tan \theta - \tan \theta = 0$

Solution

Factoring $\tan \theta$ from the terms in the left-hand member, we have

$$\tan \theta \, (2 \cos^2 \theta - 1) = 0,$$

from which either

$$\tan \theta = 0 \quad \text{or} \quad 2 \cos^2 \theta - 1 = 0.$$

If $\tan \theta = 0$, either $\theta = 0°$ or $\theta = 180°$. If $2 \cos^2 \theta - 1 = 0$, then $\cos^2 \theta = 1/2$, $\cos \theta = \pm 1/\sqrt{2}$, and θ is an odd multiple of $45°$. Therefore, the solution set over $0° \leq \theta < 360°$ is

$$\{0°, \, 45°, \, 135°, \, 180°, \, 225°, \, 315°\}.$$

27. $(2 \sin \theta - 1)(2 \sin^2 \theta - 1) = 0$ **28.** $(\tan \theta - 1)(2 \cos \theta + 1) = 0$

29. $2 \sin \theta \cos \theta + \sin \theta = 0$ **30.** $\tan \theta \sin \theta - \tan \theta = 0$

In Exercises 31–38, find the solution set of the given equation over the interval $0^R \le \alpha < 2\pi^R$.

Example $\tan^2 \alpha + \sec \alpha - 1 = 0.$

Solution   Since $\tan^2 \alpha = \sec^2 \alpha - 1$, the given equation can be written as

$$\sec^2 \alpha - 1 + \sec \alpha - 1 = 0,$$
$$\sec^2 \alpha + \sec \alpha - 2 = 0,$$
$$(\sec \alpha + 2)(\sec \alpha - 1) = 0.$$

Now observe that the solution set is

$$\{\alpha \mid \sec \alpha + 2 = 0\} \cup \{\alpha \mid \sec \alpha - 1 = 0\}$$

over the interval $0^R \le \alpha < 2\pi^R$. If $\sec \alpha + 2 = 0$, then $\sec \alpha = -2$, and if $\sec \alpha - 1 = 0$, then $\sec \alpha = 1$. Over the interval $0^R \le \alpha < 2\pi^R$ the required solution set is

$$\left\{ \frac{2\pi^R}{3}, \frac{4\pi^R}{3} \right\} \cup \{0^R\} = \left\{ 0^R, \frac{2\pi^R}{3}, \frac{4\pi^R}{3} \right\}.$$

31. $\tan^2 \alpha - 2 \tan \alpha + 1 = 0$ **32.** $4 \sin^2 \alpha - 4 \sin \alpha + 1 = 0$

33. $\cos^2 \alpha + \cos \alpha = 2$ **34.** $\cot^2 \alpha = 5 \cot \alpha - 4$

35. $\sec^2 \alpha + 3 \tan \alpha - 11 = 0$ **36.** $\tan^2 \alpha + 4 = 2 \sec^2 \alpha$

37. $\sin^2 \alpha + \sin \alpha - 1 = 0$ *Hint*: Use the quadratic formula.

38. $\tan^2 \alpha = \tan \alpha + 3$

39. Approximate a solution of $x/2 - \sin x = 0$, $x \in R$, $x \ne 0$, by graphical methods. *Hint*: Graph $y_1 = x/2$ and $y_2 = \sin x$ on the same coordinate system and determine a value of x for which $y_1 = y_2$.

40. Approximate a solution of $\cos x = x^2$, $x \in R$, by graphical methods.

8.3   *Conditional Equations for Multiples*

Equations that contain circular or trigonometric function values such as $\sin 2x$, $\cos 3\alpha$, etc., need further consideration.

Example   Solve $\sqrt{2} \cos 3\alpha = 1$ over the interval $0° \le \alpha < 360°$.

Solution   The equation can be written equivalently as $\cos 3\alpha = \dfrac{1}{\sqrt{2}}$.

Therefore, we have

$$3\alpha = 45° + k \cdot 360° \quad \text{or} \quad 3\alpha = 315° + k \cdot 360°,$$

from which, by dividing each member by 3, we obtain

$$\alpha = 15° + k \cdot 120° \quad \text{or} \quad \alpha = 105° + k \cdot 120°, \quad k \in J.$$

Because we want solutions over the interval $0° \leq \alpha < 360°$, we consider 0, 1, 2 as replacements for k and obtain the solution set

$$\{15°, 135°, 255°, 105°, 225°, 345°\}.$$

A judicious selection of one or more of the identities encountered earlier can frequently help you solve certain kinds of equations. The following examples illustrate two such cases.

Example   Solve $\sin \alpha \cos \alpha = 1/4$ over the interval $0^R \leq \alpha < 2\pi^R$.

Solution   Multiplying each member of the given equation by 2, we obtain the equivalent equation

$$2 \sin \alpha \cos \alpha = \frac{1}{2}.$$

Since $2 \sin \alpha \cos \alpha = \sin 2\alpha$ for every α in the desired interval, this latter equation is equivalent to

$$\sin 2\alpha = \frac{1}{2}.$$

Therefore, we have

$$2\alpha = \left(\frac{\pi}{6} + 2k\pi\right)^R \quad \text{or} \quad 2\alpha = \left(\frac{5\pi}{6} + 2k\pi\right)^R, \quad k \in J,$$

so that

$$\alpha = \left(\frac{\pi}{12} + k\pi\right)^R \quad \text{or} \quad \alpha = \left(\frac{5\pi}{12} + k\pi\right)^R, \quad k \in J.$$

Because we want the solutions over the interval $0^R \leq \alpha < 2\pi^R$, we consider 0 and 1 as replacements for k and obtain the solution set

$$\left\{\frac{\pi^R}{12}, \frac{13\pi^R}{12}, \frac{5\pi^R}{12}, \frac{17\pi^R}{12}\right\}.$$

Example   Solve $\cos 2x = \sin x$, $x \in R$.

Solution   Since $\cos 2x = 1 - 2 \sin^2 x$, the equation $\cos 2x = \sin x$ can be written equivalently as

$$1 - 2 \sin^2 x = \sin x,$$
$$2 \sin^2 x + \sin x - 1 = 0,$$
$$(2 \sin x - 1)(\sin x + 1) = 0,$$

from which

$$\sin x = \frac{1}{2} \quad \text{or} \quad \sin x = -1.$$

(Solution continued)

Then as solution set we have

$$\left\{x \mid x = \frac{\pi}{6} + 2\pi k\right\} \cup \left\{x \mid x = \frac{5\pi}{6} + 2\pi k\right\} \cup \left\{x \mid x = \frac{3\pi}{2} + 2\pi k\right\}, \quad k \in J.$$

Exercise 8.3

In all exercises in this set, use Table V or VI as necessary.

In Exercises 1–12, solve the given equation over the interval $0° \leq \theta < 360°$.

1. $\cos 2\theta = \dfrac{\sqrt{2}}{2}$

2. $\tan 2\theta = \sqrt{3}$

3. $\sin \dfrac{1}{2} \theta = \dfrac{1}{2}$

4. $\cot \dfrac{1}{3} \theta = -1$

5. $\tan 3\theta = 0$

6. $\sin 4\theta = 1$

7. $\sin \theta \cos \theta = \dfrac{1}{2}$

8. $\cos^2 \theta - \sin^2 \theta = -1$

9. $\cos 2\theta + \sin 2\theta = 0$

10. $\sin \theta \cos \theta = \dfrac{\cos 2\theta}{2}$

11. $2 \cos^2 2\theta + \cos 2\theta - 1 = 0$

12. $\tan^2 2\theta + 2 \tan 2\theta + 1 = 0$

In Exercises 13–20, solve the given equation over R.

13. $\sin 2x - \cos x = 0$

14. $\cos 2x = \cos^2 x - 1$

15. $\cos 2x = \cos x - 1$

16. $\sin x = \sin 2x$

17. $\cos 2x \sin x + \sin x = 0$

18. $\sin 2x \cos x - \sin x = 0$

19. $\sin 4x - 2 \sin 2x = 0$ *Hint*: $\sin 4x = \sin 2(2x)$.

20. $\sin 3x + 4 \sin^2 x = 0$ *Hint*: $\sin 3x = \sin (x + 2x)$.

8.4   Inverses of Circular and Trigonometric Functions

Inverse relations

Each of the circular functions and each of the trigonometric functions has an inverse relation, but none of these inverses is a function. We recall from Sections 2.5 and 3.1 that the inverse of a function is the relation obtained by interchanging the components x and y of each ordered pair in the function and that the graph of the inverse of a function can be obtained by reflecting the graph of the function in the line with equation $y = x$. Furthermore, the resulting inverse is a function if and only if the function is one-to-one. Now consider the sine function

$$\{(x, y) \mid y = \sin x\}, \tag{1}$$

and the inverse of this function,

$$\{(x, y) \mid x = \sin y\}. \tag{2}$$

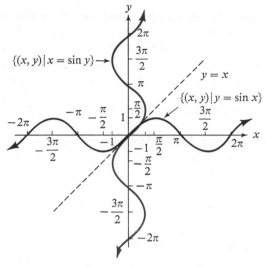

Figure 8.1

Both graphs are shown in Figure 8.1, where it is evident that each is the reflection of the other in the line $y = x$. Equation (2), $x = \sin y$, does not define a function, because for each element x in its domain, $\{x \mid -1 \le x \le 1, x \in R\}$, there are an unlimited number of elements in its range.

The inverse relation of the sine function is called the **arcsine relation**:

$$\text{arcsine} = \{(x, y) \mid x = \sin y\}.$$

Each circular and trigonometric function has an inverse relation. Thus we have

$$\text{arccosine} = \{(x, y) \mid x = \cos y\},$$

and

$$\text{arctangent} = \{(x, y) \mid x = \tan y\},$$

whose graphs are shown in Figure 8.2. The arccosecant, arcsecant, and arccotangent are less frequently used and their graphs are not shown.

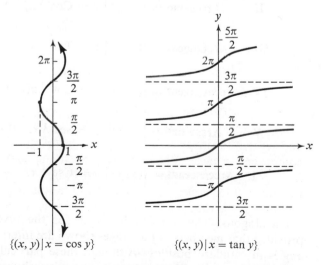

$\{(x, y) \mid x = \cos y\}$ $\{(x, y) \mid x = \tan y\}$

Figure 8.2

*Inverse
functions*

By suitably restricting the *domains* of the circular functions—that is, by suitably restricting the *ranges* of the respective inverse relations—we can define an **inverse function** for each circular and each trigonometric function. This is called the **principal-valued** inverse function. To distinguish it, a capital initial letter is often used. Thus for the sine function, we have the principal-valued inverse function

$$\text{Arcsine} = \left\{(x, y) \mid x = \sin y, -\frac{\pi}{2} \le y \le \frac{\pi}{2}\right\}.$$

The graph of Arcsine is shown in Figure 8.3.

Because there is now a unique element y in the range $\{y \mid -\pi/2 \le y \le \pi/2\}$ corresponding to each x in the domain $\{x \mid -1 \le x \le 1\}$, we can use inverse-function notation and write

$$y = \text{Arcsin } x$$

to mean

$$x = \sin y \quad \text{and} \quad -\frac{\pi}{2} \le y \le \frac{\pi}{2}.$$

Alternatively, we write

$$y = \text{Sin}^{-1} x$$

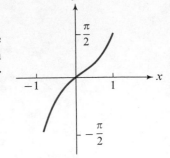

Figure 8.3

with this same meaning (but with the understanding, of course, that $\text{Sin}^{-1} x$ does *not* denote $1/\text{Sin } x$). Similar notation is used for the other inverse circular functions.

The restricted ranges for the inverse functions for the cosine and tangent functions are indicated in color in Figure 8.2. These ranges and those of the other inverse circular functions are specified in the following definition.

Definition 8.1   *The inverse circular functions corresponding to the six circular functions are*

I   **Arcsine** $= \left\{(x, y) \mid y = \text{Sin}^{-1} x, \quad -\frac{\pi}{2} \le y \le \frac{\pi}{2}\right\},$

II   **Arccosine** $= \{(x, y) \mid y = \text{Cos}^{-1} x, \quad 0 \le y \le \pi\},$

III   **Arctangent** $= \left\{(x, y) \mid y = \text{Tan}^{-1} x, \quad -\frac{\pi}{2} < y < \frac{\pi}{2}\right\},$

IV   **Arccotangent** $= \{(x, y) \mid y = \text{Cot}^{-1} x, \quad 0 < y < \pi\},$

V   **Arcsecant** $= \left\{(x, y) \mid y = \text{Sec}^{-1} x, \quad 0 \le y \le \pi, y \ne \frac{\pi}{2}\right\},$

VI   **Arccosecant** $= \left\{(x, y) \mid y = \text{Csc}^{-1} x, \quad -\frac{\pi}{2} \le y \le \frac{\pi}{2}, y \ne 0\right\}.$

*Choice
of ranges*

An analogous definition can be made for the inverse *trigonometric* functions provided that members of the ranges shown are interpreted as radian measures for angles in standard position. Although these particular ranges are the ones customarily chosen, the choices actually are quite arbitrary. They have generally been selected to involve small values of y, to have relatively simple graphs, and of course to yield a one-to-one correspondence between domain and range.

Example   Find Arctan 0.2236.

Solution   It may be helpful first to let $y = \text{Arctan } 0.2236$ and then rewrite the equation equivalently as

$$\tan y = 0.2236, \quad -\frac{\pi}{2} < y < \frac{\pi}{2}.$$

From Table V we obtain $y = 0.22$. Since $-\pi/2 < 0.22 < \pi/2$, we have

$$\text{Arctan } 0.2236 \approx 0.22.$$

Triangle interpretation   It is helpful to interpret symbolism such as $\cos(\text{Arcsin}(\sqrt{3}/4))$ by using triangles. If we let

$$\alpha = \text{Arcsin } \frac{\sqrt{3}}{4},$$

then equivalently

$$\sin \alpha = \frac{\sqrt{3}}{4}, \quad -\frac{\pi}{2} < \alpha < \frac{\pi}{2},$$

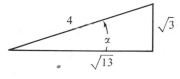

Figure 8.4

and $\sqrt{3}/4$ can be interpreted as the ratio of the length of the side of a triangle opposite an angle α to the length of the hypotenuse of the triangle, as shown in Figure 8.4. From the Pythagorean relationship, the remaining side of the triangle has length

$$\sqrt{4^2 - (\sqrt{3})^2} = \sqrt{16 - 3} = \sqrt{13},$$

which is also labeled. Then, since $\text{Arcsin}(\sqrt{3}/4) = \alpha$, by inspecting the triangle sketched we can see that

$$\cos\left(\text{Arcsin } \frac{\sqrt{3}}{4}\right) = \cos \alpha = \frac{\sqrt{13}}{4}.$$

Consider now a case in which the element in the domain is not specified.

Example   Write $\sin(\text{Cos}^{-1} x)$ as an equivalent expression without inverse notation.

Solution   Let $\text{Cos}^{-1} x = \alpha$. Then, $\cos \alpha = x$. Notice also that $0 \leq \text{Cos}^{-1} x \leq \pi$ for each x for which it is defined, and hence $\sin(\text{Cos}^{-1}x) \geq 0$. Therefore, using $\sin^2 \alpha + \cos^2 \alpha = 1$, we have

$$\sin(\text{Cos}^{-1} x) = \sin \alpha$$
$$= \sqrt{1 - \cos^2 \alpha} = \sqrt{1 - x^2}.$$

Using a sketch, we can visualize the relationship as shown at the right, where, by inspection,

$$\sin(\text{Cos}^{-1} x) = \sqrt{1 - x^2}.$$

Observe that it follows from the definitions of the inverse functions that $\sin(\text{Sin}^{-1} x) = x$, $\cos(\text{Cos}^{-1} x) = x$, etc. For example, if $\text{Sin}^{-1} x$ is a real number y, then

$$x = \sin y = \sin(\text{Sin}^{-1} x).$$

Exercise 8.4

Find the value of the given expression if it exists. Express the result in terms of rational multiples of π. You may be able to recall these values from memory, or you can use Table 7.1, page 178.

Examples   a. $\text{Arccos}\dfrac{1}{2}$ b. $\text{Tan}^{-1}(-1)$

Solutions

a. We seek a number y such that

$$\cos y = \frac{1}{2}, \quad 0 \le y \le \frac{\pi}{2}.$$

The number is $\dfrac{\pi}{3}$.

b. We seek a number y such that

$$\tan y = -1, \quad -\frac{\pi}{2} < y < \frac{\pi}{2}.$$

The number is $-\dfrac{\pi}{4}$.

1. $\text{Arcsin}\dfrac{1}{2}$ 2. $\text{Arctan}\sqrt{3}$ 3. $\text{Cot}^{-1} 1$ 4. $\text{Cos}^{-1}\dfrac{1}{\sqrt{2}}$

5. $\text{Cos}^{-1} 2$ 6. $\text{Sec}^{-1}(-2)$ 7. $\text{Arctan}\left(-\dfrac{1}{\sqrt{3}}\right)$ 8. $\text{Arcsin}\dfrac{\sqrt{3}}{2}$

Use Table V in the Appendix as necessary to find the value of the given expression.

9. $\text{Arctan}\ 0.1003$ 10. $\text{Arcsec}\ 1.053$ 11. $\text{Sin}^{-1}\ 0.3802$
12. $\text{Cos}^{-1}\ 0.6675$ 13. $\text{Arcsin}\ (-0.8624)$ 14. $\text{Arccos}\ (-0.3902)$
15. $\text{Tan}^{-1}\ 3.467$ 16. $\text{Csc}^{-1}\ 1.422$

Find the value for the given expression if it exists.

Examples   a. $\text{Cos}^{-1}(\tan \pi)$ b. $\sin\dfrac{1}{2}\left(\text{Cos}^{-1}\dfrac{1}{2}\right)$

Solutions

a. Since $\tan \pi = 0$,

$$\text{Cos}^{-1}(\tan \pi) = \text{Cos}^{-1} 0 = \frac{\pi}{2}.$$

b. Since $\text{Cos}^{-1}\dfrac{1}{2} = \dfrac{\pi}{3}$,

$$\sin\frac{1}{2}\left(\text{Cos}^{-1}\frac{1}{2}\right) = \sin\frac{1}{2}\left(\frac{\pi}{3}\right) = \frac{1}{2}.$$

17. $\mathrm{Sin}^{-1}\left(\cos\dfrac{\pi}{4}\right)$ **18.** $\mathrm{Cos}^{-1}\left(\sin\dfrac{\pi}{2}\right)$

19. $\mathrm{Tan}^{-1}\left(\tan\dfrac{\pi}{3}\right)$ **20.** $\mathrm{Sin}^{-1}\left(\sin\dfrac{3\pi}{2}\right)$

21. $\sin\left(\mathrm{Arccos}\,\dfrac{1}{2}\right)$ **22.** $\tan\left(\mathrm{Arcsin}\,\dfrac{\sqrt{3}}{2}\right)$

23. $\cos\left(\mathrm{Cot}^{-1}(-\sqrt{3})\right)$ **24.** $\sin\left(\mathrm{Tan}^{-1}(-1)\right)$

25. $\sin\left(2\,\mathrm{Arcsin}\,\dfrac{1}{2}\right)$ **26.** $\sin\left(2\,\mathrm{Cos}^{-1}\dfrac{3}{5}\right)$

27. $\tan\dfrac{1}{2}\left(\mathrm{Arcsin}\,\dfrac{12}{13}\right)$ **28.** $\cos\dfrac{1}{2}\left(\mathrm{Tan}^{-1}0\right)$

29. $\sin\left(\mathrm{Sin}^{-1}\dfrac{1}{2}+\mathrm{Cos}^{-1}\dfrac{3}{5}\right)$ **30.** $\cos\left(\mathrm{Sin}^{-1}\dfrac{1}{\sqrt{2}}+\mathrm{Cos}^{-1}\dfrac{4}{5}\right)$

31. $\mathrm{Arccos}\left(\sin\left(\mathrm{Arctan}(-1)\right)\right)$ **32.** $\sin\left(\mathrm{Cos}^{-1}(\tan 0)\right)$

Write the given expression without inverse notation.

Example

$\sin\left(\mathrm{Cos}^{-1}x+\mathrm{Sin}^{-1}y\right)$

Solution

From the formula for $\sin(x_1+x_2)$, page 196, we have

$\sin\left(\mathrm{Cos}^{-1}x+\mathrm{Sin}^{-1}y\right)$
$=\sin\left(\mathrm{Cos}^{-1}x\right)\cos\left(\mathrm{Sin}^{-1}y\right)$
$+\cos\left(\mathrm{Cos}^{-1}x\right)\sin\left(\mathrm{Sin}^{-1}y\right).$

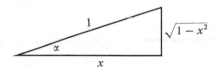

Using sketches, we can visualize the relationships in each factor of each term in the right-hand member. By inspection, we obtain

$\sin\left(\mathrm{Cos}^{-1}x+\mathrm{Sin}^{-1}y\right)$
$=\left(\sqrt{1-x^2}\right)\left(\sqrt{1-y^2}\right)+xy$
$=\sqrt{(1-x^2)(1-y^2)}+xy.$

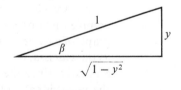

33. $\cos\left(\mathrm{Arcsin}\,x\right)$ **34.** $\cot\left(\mathrm{Arctan}\,x\right)$

35. $\tan\left(\mathrm{Sin}^{-1}y\right)$ **36.** $\sin\left(\mathrm{Tan}^{-1}y\right)$

37. $\cos\left(\dfrac{1}{2}\,\mathrm{Arccos}\,x\right)$ **38.** $\sin\left(\mathrm{Sin}^{-1}x-\mathrm{Sin}^{-1}y\right)$

Solve the given equation for x in terms of y.

Example

$y=2\,\mathrm{Arcsin}\,3x$

(Solution overleaf)

Solution   We have

$$\frac{y}{2} = \text{Arcsin } 3x,$$

and from the definition of Arcsine, we write the expression as

$$3x = \sin \frac{y}{2}, \quad \text{and then as} \quad x = \frac{1}{3} \sin \frac{y}{2},$$

with $-\dfrac{\pi}{2} \leq \dfrac{y}{2} \leq \dfrac{\pi}{2}$, or $-\pi \leq y \leq \pi$.

39. $y = 3 \text{ Arccos } 2x$ **40.** $y = 2 \text{ Sin}^{-1} \dfrac{x}{5}$

41. $y = \dfrac{1}{2} \text{ Tan}^{-1}(x + \pi)$ **42.** $y = \dfrac{2}{3} \text{ Cot}^{-1}(\pi x)$

43. Show that $\text{Arcsin } \dfrac{2}{5} = \text{Arctan } \dfrac{2}{\sqrt{21}}$.   *Hint*: Sketch a triangle.

44. Show that $\text{Arccos } \dfrac{1}{3} = \text{Arccot } \dfrac{1}{2\sqrt{2}}$.

45. Is $\text{Arccos }(\cos x) = x$? Why or why not?

46. Is $\text{Arcsin }(\sin x) = x$? Why or why not?

8.5   *Trigonometric Form of Complex Numbers*

Graphs of complex numbers

In Chapter 2, we used Cartesian coordinates to establish a one-to-one corre-spondence between the set of ordered pairs (a, b) in R^2 and the set of points P in the geometric plane. By pairing each complex number $a + bi$ with the ordered pair (a, b), we can establish a one-to-one correspondence between the set of all complex numbers and R^2, and hence each point in the plane can be viewed as the graph of a complex number (see Figure 8.5). Since the real part a of $a + bi$ is taken as the abscissa, or x-coordinate, of P, in this context the x-axis is called the **real axis.** Similarly, since the imaginary part b is taken as the ordinate or y-coordinate of P, the y-axis is called the **imaginary axis.** Just as we sometimes speak, for instance, of the point $(2, 3)$, meaning of course the point having coordinates 2 and 3, we

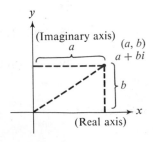

Figure 8.5

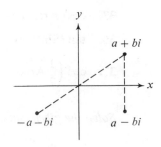

Figure 8.6

likewise speak of the point $2 + 3i$, meaning the point $(2, 3)$ representing the complex number $2 + 3i$.

A plane on which complex numbers are thus represented is often called a **complex plane.** It is also sometimes called an **Argand plane,** after the French mathematician Jean Robert Argand (1768–1822), who systematically used it, or a **Gauss plane,** after the great German mathematician Carl Friedrich Gauss (1777–1855).

A complex number $z = a + bi$, its conjugate $\bar{z} = a - bi$, and also its negative $-z = -a - bi$ are represented in Figure 8.6. It is evident that $\bar{z}$ is the reflection of z in the real axis, and $-z$ is the reflection of z in the origin as well as the reflection of $\bar{z}$ in the imaginary axis.

Modulus and argument

Since each nonzero complex number $z = a + bi$ lies on a ray with the origin as endpoint (Figure 8.7), we can associate with each z two useful concepts.

Definition 8.2   *The **absolute value**, or **modulus**, of the complex number $z = a + bi$ is denoted by r, $|z|$, or $|a + bi|$, and is given by*

$$r = |z| = |a + bi| = \sqrt{a^2 + b^2}.$$

Thus the modulus $|a + bi|$ is just the distance from the origin to the point $a + bi$.

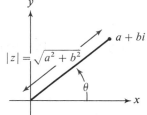

Figure 8.7

Definition 8.3   *An **argument**, or an **amplitude**, of the complex number $z = a + bi$ is an angle θ with initial side the positive x-axis and terminal side the ray from the origin containing $a + bi$.*

Note that if θ is an argument of $a + bi$, then so is $\theta + 2k\pi^R$, or $\theta + k360°$, for each $k \in J$. For $a + bi = 0$, that is, for $a^2 + b^2 = 0$, any angle θ might be used as an argument of $(a + bi)$. Also if θ is an argument of $a + bi$, then $b/a = \tan \theta$ when $a \neq 0$.

Trigonometric form

Because any ordered pair (a, b) can be written in the form $(r \cos \theta, r \sin \theta)$, where $r = \sqrt{a^2 + b^2}$ is the modulus of $a + bi$ and θ is an argument of $a + bi$, it follows that any complex number $a + bi$ can be written in the form

$$r \cos \theta + ir \sin \theta, \quad \text{or} \quad r(\cos \theta + i \sin \theta),$$

which is called the **trigonometric form,** or **polar form,** for a complex number (see Figure 8.8). More generally, using degree measure for angles, we have

$a + bi$
$= r[\cos (\theta + k360°) + i \sin (\theta + k360°)], k \in \mathbf{J}.$

A convenient abbreviation that is used for the expression $\cos \theta + i \sin \theta$ is **cis θ** (read "cosine θ plus i sine θ"), so that we can write

$$a + bi = r \operatorname{cis} (\theta + k360°), \quad k \in J.$$

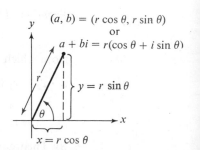

Figure 8.8

We ordinarily use for θ the angle of least nonnegative measure that is a solution of $a + bi = r \operatorname{cis} \theta$.

Example   Represent $1 + \sqrt{3}i$ in trigonometric form.

Solution $r = |1 + \sqrt{3}\,i| = \sqrt{1 + 3} = 2.$

Noting that the graph of the complex number is in Quadrant I and that

$$\tan \theta = \frac{\sqrt{3}}{1} = \sqrt{3},$$

we find that $\theta = 60°$. Hence,

$$1 + \sqrt{3}\,i = 2\,(\cos 60° + i \sin 60°) = 2 \operatorname{cis} 60°.$$

Example   Represent $4 \operatorname{cis} 225°$ graphically, and write the number in rectangular form.

Solution
$$a = r \cos \theta = 4\left(-\frac{\sqrt{2}}{2}\right) = -2\sqrt{2},$$

$$b = r \sin \theta = 4\left(-\frac{\sqrt{2}}{2}\right) = -2\sqrt{2},$$

and

$$a + bi = -2\sqrt{2} - 2\sqrt{2}i.$$

Products and quotients of complex numbers can be found quite easily when the complex numbers are in trigonometric form, as shown by the following results.

Theorem 8.1   If $z_1, z_2 \in C$, with $z_1 = r_1 \operatorname{cis} \theta_1$ and $z_2 = r_2 \operatorname{cis} \theta_2$, then

I $z_1 \cdot z_2 = r_1 r_2 \operatorname{cis}(\theta_1 + \theta_2),$

II $\dfrac{z_1}{z_2} = \dfrac{r_1}{r_2} \operatorname{cis}(\theta_1 - \theta_2) \quad (z_2 \neq 0 + 0i).$

We shall prove only Part I here and leave the proof of Part II as an exercise.

Proof of Theorem 8.1–I   We have
$$z_1 = r_1\,(\cos \theta_1 + i \sin \theta_1) \quad \text{and} \quad z_2 = r_2\,(\cos \theta_2 + i \sin \theta_2),$$

from which

$z_1 \cdot z_2 = r_1\,(\cos \theta_1 + i \sin \theta_1) \cdot r_2(\cos \theta_2 + i \sin \theta_2)$

$\quad = r_1 \cdot r_2 \cdot [\cos \theta_1 \cos \theta_2 + i \cos \theta_1 \sin \theta_2 + i \sin \theta_1 \cos \theta_2 + i^2 \sin \theta_1 \sin \theta_2]$

$\quad = r_1 \cdot r_2 \cdot [(\cos \theta_1 \cos \theta_2 - \sin \theta_1 \sin \theta_2) + i(\cos \theta_1 \sin \theta_2 + \sin \theta_1 \cos \theta_2)].$

By Theorems 6.3 and 6.6, the right-hand member can be written

$$r_1 r_2\,[\cos(\theta_1 + \theta_2) + i \sin(\theta_1 + \theta_2)],$$

so that

$$z_1 \cdot z_2 = r_1 r_2 \text{ cis } (\theta_1 + \theta_2),$$

as was to be shown.

Example   Write the product $3 \text{ cis } 80° \cdot 5 \text{ cis } 40°$ in the form $a + bi$.

Solution   By Theorem 8.1-I,

$$3 \text{ cis } 80° \cdot 5 \text{ cis } 40° = 15 \text{ cis } 120° = 15 (\cos 120° + i \sin 120°).$$

Since $\cos 120° = -\dfrac{1}{2}$ and $\sin 120° = \dfrac{\sqrt{3}}{2}$, we have

$$15(\cos 120° + i \sin 120°) = 15\left(-\frac{1}{2} + \frac{\sqrt{3}}{2} i\right) = -\frac{15}{2} + \frac{15\sqrt{3}}{2} i.$$

Example   Write the quotient $\dfrac{8 \text{ cis } 540°}{2 \text{ cis } 225°}$ as a complex number in the form $a + bi$.

Solution   By Theorem 8.1-II,

$$\frac{8 \text{ cis } 540°}{2 \text{ cis } 225°} = \frac{8}{2} \text{ cis } (540° - 225°) = 4 \text{ cis } 315° = 4(\cos 315° + i \sin 315°).$$

Since $\cos 315° = \dfrac{1}{\sqrt{2}} = \dfrac{\sqrt{2}}{2}$ and $\sin 315° = \dfrac{-1}{\sqrt{2}} = \dfrac{-\sqrt{2}}{2}$, we have

$$4(\cos 315° + i \sin 315°) = 4\left(\frac{\sqrt{2}}{2} - \frac{\sqrt{2}}{2} i\right) = 2\sqrt{2} - 2\sqrt{2} \, i.$$

Exercise 8.5

Graph the given complex number, its conjugate, its negative, and the negative of its conjugate. Draw line segments joining each pair of these four points.

1. $2 + 3i$ **2.** $-3 + 4i$ **3.** $4 - i$

4. $-2 - i$ **5.** $4i$ **6.** $-3i$

Write without absolute-value notation.

Examples   a. $|-3|$ b. $|2 + 5i|$ c. $|(-3, 2)|$

Solutions   By Definition 8.2,

       a. $\sqrt{(-3)^2} = 3$ b. $\sqrt{2^2 + 5^2} = \sqrt{29}$ c. $\sqrt{(-3)^2 + (2)^2} = \sqrt{13}$

7. $|4|$ **8.** $|-2|$ **9.** $|3 + 2i|$

10. $|4 - i|$ **11.** $|2|$ **12.** $|3 - 5i|$

13. $|-2 - i|$ **14.** $|-7 - i|$

Write the complex number in the form r cis θ.

Example $3\sqrt{3} - 3i$

Solution $r = |3\sqrt{3} - 3i| = \sqrt{27 + 9} = 6.$

Noting that the graph of the complex number is in Quadrant IV and also that

$\tan \theta = \dfrac{-3}{3\sqrt{3}} = \dfrac{-1}{\sqrt{3}}$, we find that $\theta = 330°$. Hence

$$3\sqrt{3} - 3i = 6 (\cos 330° + i \sin 330°)$$
$$= 6 \text{ cis } 330°.$$

15. $3 + 3i$ **16.** $2 - 2i$ **17.** 5

18. $-7i$ **19.** $2\sqrt{3} - 2i$ **20.** $-3\sqrt{3} - 3i$

Write the complex number in the form $a + bi$.

Example 2 cis $120°$

Solution   Since $r = 2$ and $\theta = 120°$, we have

$$a = r \cos \theta = 2 \cos 120° = 2\left(-\frac{1}{2}\right) = -1,$$

and

$$b = r \sin \theta = 2 \sin 120° = 2\left(\frac{\sqrt{3}}{2}\right) = \sqrt{3}.$$

Hence, $a + bi = -1 + \sqrt{3}\, i.$

21. 4 cis $240°$ **22.** 3 cis $300°$ **23.** 6 cis $(-30°)$

24. 5 cis $180°$ **25.** 12 cis $420°$ **26.** 10 cis $(-480°)$

For the given pair of complex numbers z_1 and z_2, find (a) $z_1 \cdot z_2$, and (b) z_1/z_2. Express each result in the form $a + bi$. Use Table VI as needed.

27. $z_1 = 3$ cis $90°$ and $z_2 = \sqrt{2}$ cis $45°$

28. $z_1 = 4$ cis $30°$ and $z_2 = 2$ cis $60°$

29. $z_1 = 6$ cis $150°$ and $z_2 = 18$ cis $570°$

30. $z_1 = 14$ cis $210°$ and $z_2 = 2$ cis $120°$

31. $z_1 = -3 + i$ and $z_2 = -2 - 4i$

32. $z_1 = 2 + 3i$ and $z_2 = 2 + 3i$

33. Prove that the sum and product of two conjugate complex numbers are both real.

34. Show that if $a + bi = r$ cis θ, then $(a + bi)^2 = r^2$ cis 2θ.

35. Use the result of Exercise 34 to show that if $a + bi = r$ cis θ, then

$$(a + bi)^3 = r^3 \text{ cis } 3\theta.$$

8.6   De Moivre's Theorem—Powers and Roots

Powers of complex numbers

Since $a + bi = r$ cis θ, an application of Theorem 8.1-I to $(a + bi)^2$ results in

$$(a + bi)^2 = (r \text{ cis } \theta)(r \text{ cis } \theta) = r^2 \text{ cis } 2\theta. \tag{1}$$

Because

$$(a + bi)^3 = (a + bi)^2(a + bi),$$

from (1) we have

$$(a + bi)^3 = (r^2 \text{ cis } 2\theta)(r \text{ cis } \theta) = r^3 \text{ cis } 3\theta.$$

In a similar way, we can show that

$$(a + bi)^4 = (r^3 \text{ cis } 3\theta)(r \text{ cis } \theta) = r^4 \text{ cis } 4\theta,$$

and it seems plausible to make the following assertion, known as **De Moivre's theorem.** The proof is omitted.

Theorem 8.2   *If $z \in C$, $z = r$ cis θ and $n \in N$, then*

$$z^n = r^n \text{ cis } n\theta.$$

Example

Write $(\sqrt{3} + i)^7$ in the form $a + bi$.

Solution

For the modulus, we have $r = \sqrt{(\sqrt{3})^2 + 1^2} = 2$. For an argument, $\tan \theta = 1/\sqrt{3}$ and from the fact that the graph of $\sqrt{3} + i$ is in Quadrant I, we obtain $\theta = 30°$. Thus, $(\sqrt{3} + i)^7 = (2 \text{ cis } 30°)^7$. Then, by De Moivre's theorem,

$$(2 \text{ cis } 30°)^7 = 2^7 \text{ cis } (7 \cdot 30)° = 128 \text{ cis } 210°.$$

Converting to the form $a + bi$, we find

$$128(\cos 210° + i \sin 210°) = 128\left(-\frac{\sqrt{3}}{2} - \frac{1}{2}i\right)$$

$$= -64\sqrt{3} - 64i,$$

so

$$(\sqrt{3} + i)^7 = -64\sqrt{3} - 64i.$$

By appropriately defining z^0 and z^{-n}, we can extend De Moivre's theorem to include as exponents all $n \in J$.

Definition 8.4   *If $z \neq 0 + 0i$, then*

$$\text{I}\quad z^0 = 1 + 0i, \qquad \text{II}\quad z^{-n} = \frac{1}{z^n}, \quad for\ n \in J.$$

Theorem 8.3   *If $z \in C$, $z \neq 0 + 0i$, $z = r\operatorname{cis}\theta$ and $n \in J$, then*

$$z^n = r^n \operatorname{cis} n\theta.$$

Example   Write $(1 + i)^{-6}$ in the form $a + bi$.

Solution $r = \sqrt{1^2 + 1^2} = \sqrt{2}.$

Noting that the graph of $1 + i$ is in Quadrant I and $\tan\theta = 1/1$, we have $\theta = 45°$. Hence,

$$(1 + i)^{-6} = (\sqrt{2}\operatorname{cis}45°)^{-6}.$$

By Theorem 8.3,

$$(\sqrt{2}\operatorname{cis}45°)^{-6} = (\sqrt{2})^{-6}\operatorname{cis}(-6 \cdot 45°)$$

$$= \frac{1}{8}\operatorname{cis}(-270°)$$

$$= \frac{1}{8}[\cos(-270°) + i\sin(-270°)].$$

Since $\cos(-270°) = 0$ and $\sin(-270°) = 1$, we obtain

$$(1 + i)^{-6} = \frac{1}{8}(0 + i) = \frac{1}{8}i.$$

Roots of complex numbers   Yet another extension of De Moivre's theorem is possible if we make the following definition.

Definition 8.5   *For $z \in C$, $n \in N$, w is an nth root of z provided*

$$w^n = z.$$

Theorem 8.4   *If $z \in C$, $z = r\operatorname{cis}\theta$ and $n \in N$, then*

$$w = r^{1/n}\operatorname{cis}\left(\frac{\theta}{n}\right)$$

is an nth root of z.

This theorem follows directly from De Moivre's theorem. The fact that

$$\operatorname{cis}\theta = \operatorname{cis}(\theta + k360°),$$

for $k \in J$, enables us to find n distinct complex nth roots for each $z \in C$, $z \neq 0 + 0i$ as illustrated in the following example.

Example

Write each of the four fourth roots of $z = 2 + 2\sqrt{3}\,i$ in the form r cis θ.

Solution

$r = \sqrt{2^2 + (2\sqrt{3}\,)^2} = 4$.

The graph of $2 + 2\sqrt{3}\,i$ is in Quadrant I and $\tan\theta = \sqrt{3}$. Hence, $\theta = 60°$ and

$$z = 2 + 2\sqrt{3}\,i = 4 \text{ cis } 60°.$$

From Theorem 8.4 and the periodic property of cosine and sine, each number

$$4^{1/4} \text{ cis }\left(\frac{60° + k360°}{4}\right),$$

for $k \in J$, is a fourth root of z. Taking $k = 0$, 1, 2, and 3, in turn, gives the roots

$$w_0 = \sqrt{2} \text{ cis } 15°, \quad w_1 = \sqrt{2} \text{ cis } 105°, \quad w_2 = \sqrt{2} \text{ cis } 195°, \quad w_3 = \sqrt{2} \text{ cis } 285°.$$

The substitution of any other integer for k will produce one of these four complex numbers.

Exercise 8.6

Use Theorem 8.2 or 8.3 as appropriate to write the given expression as a complex number of the form $a + bi$. Use Table VI as necessary.

1. $[2 \text{ cis } (-30°)]^7$

2. $(4 \text{ cis } 36°)^5$

3. $-\left(\frac{1}{2} + \frac{1}{2}\sqrt{3}\,i\right)^3$

4. $(1 + i)^{12}$

5. $(\sqrt{3} \text{ cis } 5°)^{12}$

6. $(\sqrt{2} \text{ cis } 30°)^{-7}$

7. $(\sqrt{3} - i)^{-5}$

8. $(1 - i)^{-6}$

9. $\dfrac{(1 + i)^3}{(1 + \sqrt{3}\,i)^5(1 - i)^2}$

10. $\dfrac{4(\sqrt{3} + i)^3}{(1 - i)^3}$

11. $\dfrac{(1 - i)^5}{(1 + i)^6}$

12. $\dfrac{(1 + \sqrt{3}\,i)^{-4}}{(\sqrt{3} + i)^{-6}}$

Find the nth roots of z by applying Theorem 8.4. Leave the results in trigonometric form and list all n of the nth roots.

13. $z = 32 \text{ cis } 45°, \quad n = 5$

14. $z = 27, \quad n = 3$

15. $z = -16\sqrt{3} + 16i, \quad n = 5$

16. $z = 1 - i, \quad n = 4$

17. $z = -i, \quad n = 6$

18. $z = 2 + 2\sqrt{3}\,i, \quad n = 3$

Solve the given equation over C.

19. $x^5 = 16 - 16\sqrt{3}\,i$

20. $x^3 + 4i = 4\sqrt{3}$

21. $x^7 + 1 = 0$

22. $x^7 - 1 = 0$

23. Factor $x^4 + 16$ into linear factors.

24. Factor $x^5 - 1$ into linear factors.

25. Show that the sum of the four fourth roots of 1 is $0 + 0i$.

26. Explain why the sum of the *n*th roots of any complex number is zero for all $n > 1$.

8.7   *Polar Coordinates*

The Cartesian coordinates that we have been using specify the location of a point in the plane by giving the directed distances of the point from a pair of fixed perpendicular lines, the axes. There is an alternative coordinate system that is frequently used in the plane, in which the location of a point is specified in a different manner.

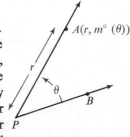

In the plane, consider a fixed ray $\overrightarrow{PB}$ and any point A. We can describe the location of A by giving the distance r from P to A and specifying the angle BPA (Figure 8.9), which is customarily designated by θ. By stating the ordered pair $(r, m°(\theta))$ or $(r, m^R(\theta))$, we clearly identify the location of A. We ordinarily write (r, θ) for either of these ordered pairs, where the meaning should be clear from the context. The components of such an ordered pair are called **polar coordinates** of A. The fixed ray $\overrightarrow{PB}$ is called the **polar axis**, and the initial point P of the polar axis is called the **pole** of the system.

Figure 8.9

Notice that while there is a one-to-one correspondence between the set of ordered pairs in a Cartesian coordinate system and the points in the geometric plane, each point in the plane has infinitely many pairs of polar coordinates. In the first place, if (r, θ) are polar coordinates of A, then so are

$$(r, \theta + k360°), \quad k \in J$$

(Figure 8.10-a). In the second place, if we let $\overrightarrow{PA'}$ denote the ray in a direction opposite that of $\overrightarrow{PA}$ (Figure 8.10-b), then we see that $-r \leq 0$ denotes the directed

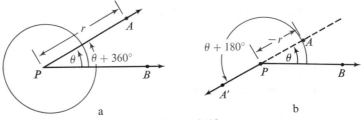

a   b

Figure 8.10

distance from P to A along the negative extension of $\overrightarrow{PA'}$. Thus we see that also $(-r, \theta + 180°)$, and more generally

$$(-r, \theta + 180° + k360°), \quad k \in J,$$

are polar coordinates of A. The pole P itself is represented by $(0, \theta)$ for any θ.

Example   Write four additional sets of polar coordinates for the point having polar co-
                      ordinates (3, 30°).

Solution   With positive values for r, two more pairs of polar coordinates for (3, 30°) are
                      (3, 390°) and (3, −330°). Using negative values for r, we have (−3, 210°) and
                      (−3, −150°). The figures show these cases. Of course there are infinitely many
                      other possible polar coordinates for the same point.

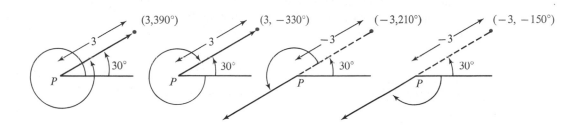

Relationships   Cartesian and polar coordinates of a point can be related by means of the trigono-
between   metric functions. If the pole P in a polar coordinate system is also the origin O in a
polar and   Cartesian coordinate system, and if the polar axis coincides with the positive x-axis
rectangular   of the Cartesian system, as shown in Figure 8.11, then the coordinates (x, y)
coordinates   can be expressed in terms of the polar coordinates (r, θ) by the following equat-
                      ions:

$$x = r \cos \theta \quad \text{and} \quad y = r \sin \theta. \qquad (1)$$

Conversely, we have

$$r = \pm\sqrt{x^2 + y^2}, \qquad \cos \theta = \frac{x}{\pm\sqrt{x^2 + y^2}},$$

$$\sin \theta = \frac{y}{\pm\sqrt{x^2 + y^2}} \quad [(x, y) \neq (0, 0)]. \qquad (2)$$

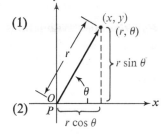

Figure 8.11

The sets of equations (1) and (2) enable us to find
rectangular coordinates for a point with a given pair of polar coordinates, and vice
versa. Often, in determining θ, it is simplest first to determine the quadrant from
the signs of x and y, and then to use the relation

$$\tan \theta = \frac{y}{x}.$$

Example   Find the rectangular coordinates of the point with polar coordinates (4, 30°).
                      Show the graph of the point on a combined polar and rectangular coordinate
                      system.

                                                                      (Solution overleaf)

Solution   Using (1), we obtain

$$x = 4 \cos 30° = 4 \cdot \frac{\sqrt{3}}{2} = 2\sqrt{3},$$

$$y = 4 \sin 30° = 4 \cdot \frac{1}{2} = 2.$$

The rectangular coordinates are $(2\sqrt{3}, 2)$.

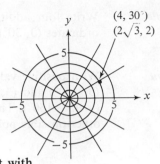

Example   Find a pair of polar coordinates for the point with Cartesian coordinates $(7, -2)$. Show the graph of the point on a combined polar and rectangular coordinate system.

Solution   By (2), $r = \pm\sqrt{x^2 + y^2} = \pm\sqrt{49 + 4} = \pm\sqrt{53}$. Choosing the positive sign, and noting that the point $(7, -2)$ is in the fourth quadrant, we have

$$\tan \theta = \frac{-2}{7},$$

from which

$$\theta \approx -16° \ 00'.$$

A pair of polar coordinates is therefore

$$(\sqrt{53}, -16° \ 00'),$$

where the given angle measure is an approximation.

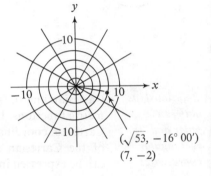

Equations in polar form and in rectangular form   Equations (1) on page 219 can be used to transform Cartesian equations to polar form, and Equations (2) can be used to transform polar equations to Cartesian form. Several examples are shown in the exercise set.

Exercise 8.7

Find four additional sets of polar coordinates $(-360° < \theta \leq 360°)$ for the point with polar coordinates as given.

1. $(6, 485°)$ **2.** $(-3, 518°)$ **3.** $(-2, -450°)$

4. $(5, 720°)$ **5.** $(6, -600°)$ **6.** $(3, -395°)$

Find the Cartesian coordinates of a point with polar coordinates as given.

7. $(5, 45°)$ **8.** $(-3, 30°)$ **9.** $\left(\frac{1}{2}, 330°\right)$

10. $\left(\frac{3}{4}, 225°\right)$ **11.** $(10, -135°)$ **12.** $(-6, -240°)$

Find two sets of polar coordinates, one involving an angle of positive measure and one an angle of negative measure, for the point with Cartesian coordinates as given.

13. $(3\sqrt{2}, 3\sqrt{2})$ **14.** $\left(-\dfrac{\sqrt{3}}{2}, \dfrac{1}{2}\right)$ **15.** $(-1, -\sqrt{3})$

16. $(0, -4)$ **17.** $(0, 0)$ **18.** $(-6, 0)$

Transform the given equation to an equation in polar form.

Example $x^2 + y^2 - 2x + 3 = 0$

Solution From Equations (1) on page 219, we find that $x^2 = r^2 \cos^2 \theta$ and $y^2 = r^2 \sin^2 \theta$. Thus we have
$$r^2 \cos^2 \theta + r^2 \sin^2 \theta - 2r \cos \theta + 3 = 0,$$
from which
$$r^2 (\cos^2 \theta + \sin^2 \theta) - 2r \cos \theta + 3 = 0,$$
$$r^2 - 2r \cos \theta + 3 = 0.$$

19. $x^2 + y^2 = 25$ **20.** $x = 3$ **21.** $y = -4$
22. $x^2 + y^2 - 4y = 0$ **23.** $x^2 + 9y^2 = 9$ **24.** $x^2 - 4y^2 = 4$

Transform the given equation to an equation in Cartesian form free from radicals.

Example $r(1 - 2 \cos \theta) = 3$

Solution Using Equations (2) on page 219, we have
$$\pm\sqrt{x^2 + y^2}\left[1 - 2\left(\frac{x}{\pm\sqrt{x^2 + y^2}}\right)\right] = 3,$$
from which
$$\pm\sqrt{x^2 + y^2} - 2x = 3,$$
$$\pm\sqrt{x^2 + y^2} = 2x + 3.$$
Upon squaring each member, we have the equation
$$x^2 + y^2 = 4x^2 + 12x + 9,$$
from which
$$3x^2 - y^2 + 12x + 9 = 0.$$

25. $r = 5$ **26.** $r = 4 \sin \theta$ **27.** $r = 9 \cos \theta$
28. $r \cos \theta = 3$ **29.** $r(1 - \cos \theta) = 2$ **30.** $r(1 + \sin \theta) = 2$

31. Show by transformation of coordinates that the graph of $r = \sec^2 (\theta/2)$ is a parabola.
32. Show that the graph of $r = \csc^2 (\theta/2)$ is a parabola.

Graph the given equation.

Example $r = 1 + \sin \theta$

Solution   Using selected values for θ ($0° \leq \theta \leq 360°$) we first obtain the following ordered pairs (r, θ), where values for $r = 1 + \sin \theta$ are approximated. The graph is then sketched as shown. Since the sine function is periodic, with period 2π, any other values of θ less than $0°$ or greater than $360°$ would simply yield ordered pairs whose graphs would also be on the curve.

θ	$\sin \theta$	$r = 1 + \sin \theta$
0°	0	1
30°	0.5	1.5
60°	0.9	1.9
90°	1	2
120°	0.9	1.9
150°	0.5	1.5
180°	0	1
210°	−0.5	0.5
240°	−0.9	0.1
270°	−1	0
300°	−0.9	0.1
330°	−0.5	0.5
360°	0	1

33. $r = 4 \sin \theta$

34. $r = 9 \cos \theta$

35. $r = 1 + \cos \theta$

36. $r = 1 - \sin \theta$

37. $r \cos \theta = 3$

38. $r \sin \theta = -3$

39. $r = 4 \sin 2\theta$

40. $r = 4 \cos 2\theta$

41. $r = 3$ *Hint:* All solutions are of the form $(3, \theta)$.

42. $\theta = 30°$ *Hint:* All solutions are of the form $(r, 30°)$.

Chapter Review

[8.1]   *Verify that the given equation is an identity.*

1. $\tan x = \sin x \cdot \sec x$

2. $\cos x - \sin x = (1 - \tan x) \cos x$

3. $\dfrac{1 - \tan^2 \alpha}{\tan \alpha} = \cot \alpha - \tan \alpha$

4. $\dfrac{\cos^2 \alpha}{1 - \sin \alpha} = \dfrac{\cos \alpha}{\sec \alpha - \tan \alpha}$

5. $\dfrac{\sin 2\theta}{1 + \cos 2\theta} = \tan \theta$

6. $(\cos^2 \theta - \sin^2 \theta)^2 + \sin^2 2\theta = 1$

[8.2]   *Specify the given set in terms of rational multiples of π.*

7. $\left\{ x \mid \cos x = \dfrac{\sqrt{3}}{2} \right\}$

8. $\left\{ y \mid \cot y = -\dfrac{1}{\sqrt{3}} \right\}$

Solve the given equation over the specified replacement set.

9. $\sqrt{3} \tan x - 1 = 0; \quad x \in R$

10. $\cos^2 \theta - 1 = 0; \quad 0° \leq \theta \leq 360°$

11. $2 \cos \theta \sin \theta - \cos \theta = 0; \quad 0° \leq \theta \leq 360°$

12. $\sin^2 \alpha - 3 \sin \alpha + 2 = 0; \quad 0^R \leq \alpha \leq 2\pi^R$

[8.3] *Solve the given equation over the interval* $0° \leq \theta \leq 360°$.

13. $\sin 3\theta = \dfrac{\sqrt{3}}{2}$

14. $\cos^2 2\theta + 2 \cos 2\theta + 1 = 0$

Solve the given equation over R.

15. $\sin 2x - \cos 2x = 0$

16. $\tan^2 2x - 3 \tan 2x + 2 = 0$

[8.4] *Find the value of the given expression.*

17. $\text{Arccos} \dfrac{\sqrt{3}}{2}$

18. $\text{Cot}^{-1} \sqrt{3}$

19. $\text{Sec}^{-1} (1.743)$

20. $\text{Arcsin} (-0.8572)$

Find the value of the given expression.

21. $\text{Cos}^{-1} \left(\sin \dfrac{\pi}{3} \right)$

22. $\text{Tan}^{-1} (\cos 0)$

23. $\sin (\text{Cos}^{-1} (-1))$

24. $\cos \left(2 \text{Cos}^{-1} \dfrac{1}{2} \right)$

Express the given expression without using inverse notation.

25. $\cos (\text{Arctan } x)$

26. $\tan (\text{Cos}^{-1} x)$

Solve for x in terms of y.

27. $y = \text{Cos}^{-1} 2x$

28. $y = 2 \text{Tan}^{-1} (x - \pi)$

[8.5] *Write the complex number in the form r* cis *θ.*

29. $2 + 2i$

30. $5 - 5\sqrt{3}\, i$

Write the complex number in the form a + bi.

31. $3 \text{ cis } 150°$

32. $4 \text{ cis } (-60°)$

For the pair of complex numbers, find (a) $z_1 \cdot z_2$ *and (b)* $\dfrac{z_1}{z_2}$. *Express each result in the form* $a + bi$.

33. $z_1 = 2 \text{ cis } 30$ and $z_2 = 9 \text{ cis } 120$
34. $z_1 = 3 - 2i$ and $z_2 = -2 + 5i$

[8.6]   *Use Theorem 8.2 or 8.3 to write the given expression as a complex number of the form* $a + bi$. *Use Table VI as necessary.*

35. $(3 \text{ cis } 45°)^5$

36. $\dfrac{2(1 - i\sqrt{3})^2}{(1 + i)^4}$

Find the nth roots of z by applying Theorem 8.4. Leave the results in trigonometric form and list all n of the nth roots.

37. $z = 1 + i, \quad n = 3$

38. $z = 3\sqrt{3} - 3i, \quad n = 4$

Solve the given equation over C.

39. $x^4 + 1 = 0$

40. $x^5 - 1 = 0$

[8.7] **41.** Find four additional sets of polar coordinates $(-360° < \theta \leq 360°)$ for the point with polar coordinates $(4, -420°)$.

42. Find the Cartesian coordinates of the point with polar coordinates $(6, 135°)$.

43. Transform $x^2 + y^2 - 9x = 0$ to an equation in polar form.

44. Transform $r = 4 \cos \theta$ to an equation in Cartesian form.

45. Graph $r = 4 \cos \theta$.

46. Graph $r = 4$.

9 *Natural-Number Functions*

9.1 *Mathematical Induction*

The material in the present section depends on a special property of the set of natural numbers, or positive integers: $N = \{1, 2, 3, \ldots\}$.

The set N has the following properties:

    a. $1 \in N$.

    b. If $k \in N$, then $k + 1 \in N$.

In addition, *N contains no elements not implied by properties a and b.* These properties underlie the following theorem, called the **principle of mathematical induction,** which we state without proof.

Theorem 9.1   *If a given sentence involving natural numbers n is true for n = 1, and if its truth for n = k implies its truth for n = k + 1, then it is true for every natural number n.*

Requirements of a proof by mathematical induction   We can exploit Theorem 9.1 to prove a number of assertions. Although the technique we shall use is called **proof by mathematical induction,** the argument we shall employ is deductive, as have been all of the other arguments in this book. Proofs by mathematical induction require two things:

    a. A demonstration that the assertion to be proved is true for the natural number 1.

    b. A demonstration that the truth of the assertion for a natural number k implies its truth for $k + 1$.

When these two demonstrations have been made, the principle of mathematical induction assures us that the assertion is true for every natural number.

Example   Prove that the sum of the first n natural numbers is $\dfrac{n(n+1)}{2}$.

Solution   In symbols, we wish to show that

$$1 + 2 + 3 + \cdots + n = \frac{n(n+1)}{2}.$$

We must do two things:

1. We must first show that the assertion is true for $n = 1$, that is, that

$$1 = \frac{1(1+1)}{2},$$

which is true.

2. We must next show that the truth of

$$1 + 2 + 3 + \cdots + k = \frac{k(k+1)}{2}$$

implies the truth of

$$1 + 2 + 3 + \cdots + k + (k+1) = \frac{(k+1)[(k+1)+1]}{2}.$$

That is, we must show that the truth of the assertion for $n = k$ implies its truth for $n = k + 1$. Now, assume the truth of

$$1 + 2 + 3 + \cdots + k = \frac{k(k+1)}{2}.$$

Then, by adding $k + 1$ to each member of this equation, we obtain

$$1 + 2 + 3 + \cdots + k + (k+1) = \frac{k(k+1)}{2} + (k+1)$$

$$= (k+1)\left(\frac{k}{2} + 1\right)$$

$$= (k+1)\left(\frac{k+2}{2}\right)$$

$$= \frac{(k+1)[(k+1)+1]}{2}.$$

Thus the second fact necessary for our proof is established. By the principle of mathematical induction, the assertion is true for every natural number n.

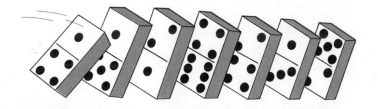

Figure 9.1

This method of proof is often compared to lining up a row of dominoes, with the assumption that whenever one domino is toppled, the one following will topple. One then needs only to topple the first domino ($n = 1$) and the whole row behind it will topple, as indicated in Figure 9.1.

Exercise 9.1

By mathematical induction, prove the validity of the given formula for all positive integral values of n.

1. $\dfrac{1}{2} + \dfrac{2}{2} + \dfrac{3}{2} + \cdots + \dfrac{n}{2} = \dfrac{n(n+1)}{4}$

2. $1 + 3 + 5 + \cdots + (2n - 1) = n^2$

3. $2 + 4 + 6 + \cdots + 2n = n(n + 1)$

4. $2 + 6 + 10 + \cdots + (4n - 2) = 2n^2$

5. $1^2 + 2^2 + 3^2 + \cdots + n^2 = \dfrac{n(n+1)(2n+1)}{6}$

6. $2 + 2^2 + 2^3 + \cdots + 2^n = 2^{n+1} - 2$

7. $1^3 + 3^3 + 5^3 + \cdots + (2n-1)^3 = n^2(2n^2 - 1)$

8. $\dfrac{1}{1 \cdot 2} + \dfrac{1}{2 \cdot 3} + \dfrac{1}{3 \cdot 4} + \cdots + \dfrac{1}{n(n+1)} = \dfrac{n}{n+1}$

9. $1 \cdot 2 + 2 \cdot 3 + 3 \cdot 4 + \cdots + n(n+1) = \dfrac{n(n+1)(n+2)}{3}$

10. $1 \cdot 4 + 2 \cdot 9 + 3 \cdot 16 + \cdots + n(n+1)^2 = \dfrac{1}{12} n(n+1)(n+2)(3n+5)$

11. Show that if $2 + 4 + 6 + \cdots + 2n = n(n+1) + 2$ is true for $n = k$, then it is true for $n = k + 1$. Is it true for every $n \in N$?

12. Show that $n^3 + 11n = 6(n^2 + 1)$ is true for $n = 1, 2,$ and 3. Is it true for every $n \in N$?

9.2   Sequences

Let us consider a class of functions in which each function has as its domain either the set N of positive integers or a subset of successive members of N.

Definition 9.1   *A sequence function is a function having as its domain the set N of positive integers 1, 2, 3, ... A finite-sequence function has as its domain the set of positive integers 1, 2, 3, . . . , n, for some fixed n.*

For example, the function defined by

$$s(n) = n + 3, \quad n \in \{1, 2, 3, \ldots\}, \tag{1}$$

is a sequence function. The elements in the range of such a function, considered in the order

$$s(1), s(2), s(3), s(4), \ldots,$$

are said to form a **sequence**. Similarly, the elements of a finite-sequence function, considered in order, constitute a **finite sequence**.

For example, the sequence associated with (1) is found by successively substituting the numbers $1, 2, 3, \ldots$, for n:

$$s(1) = 1 + 3 = 4,$$

$$s(2) = 2 + 3 = 5,$$

$$s(3) = 3 + 3 = 6,$$

$$s(4) = 4 + 3 = 7,$$

and so on. Thus the first four terms of (1) are 4, 5, 6, and 7. The nth term, or general term, is $n + 3$. As another example, the first five terms of the sequence defined by the equation

$$s(n) = \frac{3}{2n - 1}, \quad n \in \{1, 2, 3, \ldots\},$$

are 3/1, 3/3, 3/5, 3/7, and 3/9; and the twenty-fifth term is

$$s(25) = \frac{3}{2(25) - 1} = \frac{3}{49}.$$

Given several terms in a sequence, we are often able to construct an expression for a general term of a sequence to which they belong. Such a general term is not unique. Thus, if the first three terms in a sequence are $2, 4, 6, \ldots$, we may *surmise* that the general term is $s(n) = 2n$. Note, however, that the sequences for both

$$s(n) = 2n$$

and

$$s(n) = 2n + (n - 1)(n - 2)(n - 3), \quad n \in N,$$

start with 2, 4, 6, but that the two sequences differ for terms following the third.

Sequence notation

The notation ordinarily used for the terms in a sequence is not function notation as such. It is customary to denote a term in a sequence by means of a subscript. Thus, the sequence $s(1), s(2), s(3), s(4), \ldots$ would appear as $s_1, s_2, s_3, s_4, \ldots$.

Let us next consider two special kinds of sequences that have many applications. The first kind can be defined as follows:

Definition 9.2   *An **arithmetic progression** is a sequence defined by equations of the form*

$$s_1 = a, \qquad s_{n+1} = s_n + d,$$

where $a, d \in R$, *and* $n \in N$.

General term of an arithmetic progression

Since each term in such a sequence is obtained from the preceding term by adding d, d is called the **common difference**. Thus $3, 7, 11, 15, \ldots$ is an arithmetic progression, in which $s_1 = 3$ and $d = 4$.

For each arithmetic progression, the general term is established by the following theorem.

Theorem 9.2   *The nth term in the sequence defined by*

$$s_1 = a, \qquad s_{n+1} = s_n + d,$$

where $a, d \in R$, and $n \in N$, is

$$s_n = a + (n - 1)d. \tag{2}$$

Proof We shall use mathematical induction. That (2) is true for the natural number 1 is evident by direct substitution of 1 in (2):

$$s_1 = a + (1 - 1)d = a.$$

If now we assume that (2) is true for the natural number k, then we have

$$s_k = a + (k - 1)d.$$

By the defining equation, we accordingly have

$$s_{k+1} = s_k + d.$$

Replacing s_k in this expression with $a + (k - 1)d$, we obtain

$$s_{k+1} = a + (k - 1)d + d$$
$$= a + kd$$
$$= a + [(k + 1) - 1]d,$$

and the principle of mathematical induction assures us that the relationship (2) is valid for all natural numbers.

The second kind of sequence we shall consider can be defined as follows:

Definition 9.3   *A geometric progression is a sequence defined by equations of the form*

$$s_1 = a, \qquad s_{n+1} = rs_n,$$

where $a, r \in R$, $a \neq 0$, $r \neq 0$, and $n \in N$.

Thus, $3, 9, 27, 81, \ldots$ is a geometric progression in which each term except the first is obtained by multiplying the preceding term by 3. Since the effect of multiplying the terms in this way is to produce a fixed ratio between any two successive terms, the multiplier, r, is called the **common ratio**.

General term of a geometric progression

The general term for a geometric progression is that established by the following theorem. The proof by induction is left as an exercise.

Theorem 9.3   *The nth term in the sequence defined by*

$$s_1 = a, \quad s_{n+1} = rs_n,$$

where $a, r \in R$, $a \neq 0$, $r \neq 0$, and $n \in N$, is

$$s_n = ar^{n-1}.$$

Exercise 9.2

Find the first four terms in the sequence with general term as given.

Examples   a. $s_n = \dfrac{n(n+1)}{2}$

b. $s_n = (-1)^n 2^n$

Solutions   a. $s_1 = \dfrac{1(1+1)}{2} = 1$

b. $s_1 = (-1)^1 2^1 = -2$

$s_2 = \dfrac{2(2+1)}{2} = 3$

$s_2 = (-1)^2 2^2 = 4$

$s_3 = \dfrac{3(3+1)}{2} = 6$

$s_3 = (-1)^3 2^3 = -8$

$s_4 = \dfrac{4(4+1)}{2} = 10;$

$s_4 = (-1)^4 2^4 = 16;$

$1, 3, 6, 10$

$-2, 4, -8, 16$

1. $s_n = n - 5$ 2. $s_n = 2n - 3$ 3. $s_n = \dfrac{n^2 - 2}{2}$

4. $s_n = \dfrac{3}{n^2 + 1}$ 5. $s_n = 1 + \dfrac{1}{n}$ 6. $s_n = \dfrac{n}{2n - 1}$

7. $s_n = \dfrac{n(n-1)}{2}$ 8. $s_n = \dfrac{5}{n(n+1)}$ 9. $s_n = (-1)^n$

10. $s_n = (-1)^{n+1}$ 11. $s_n = \dfrac{(-1)^n(n-2)}{n}$ 12. $s_n = (-1)^{n-1} 3^{n+1}$

Write the next three terms in the given arithmetic progression.

Examples   a. $5, 9, \ldots$

b. $x, x - a, \ldots$

Solutions   Find the common difference and then continue the sequence.

a. $d = 9 - 5 = 4;$

b. $d = (x - a) - x = -a;$

$13, 17, 21$

$x - 2a, x - 3a, x - 4a$

13. $3, 7, \ldots$ **14.** $-6, -1, \ldots$ **15.** $x, x + 1, \ldots$

16. $a, a + 5, \ldots$ **17.** $2x + 1, 2x + 4, \ldots$ **18.** $3a, 5a, \ldots$

Write the next four terms in the given geometric progression.

Examples   a. $3, 6, \ldots$ b. $x, 2, \ldots$

Solutions   Find the common ratio, and then continue the sequence.

a. $r = \dfrac{6}{3} = 2;$ b. $r = \dfrac{2}{x} \quad (x \neq 0);$

     $12, 24, 48, 96$ $\dfrac{4}{x}, \dfrac{8}{x^2}, \dfrac{16}{x^3}, \dfrac{32}{x^4}$

19. $2, 8, \ldots$ **20.** $4, 8, \ldots$ **21.** $\dfrac{2}{3}, \dfrac{4}{3}, \ldots$

22. $\dfrac{1}{2}, -\dfrac{3}{2}, \ldots$ **23.** $\dfrac{a}{x}, -1, \ldots$ **24.** $\dfrac{a}{b}, \dfrac{a}{bc}, \ldots$

Example   Find the general term and the fourteenth term of the *arithmetic* progression $-6, -1, \ldots$.

Solution   Find the common difference.

$$d = -1 - (-6) = 5$$

Use $s_n = a + (n - 1)d$.

$$s_n = -6 + (n - 1)5 = 5n - 11$$

$$s_{14} = 5(14) - 11 = 59$$

25. Find the general term and the seventh term in the arithmetic progression $7, 11, \ldots$.

26. Find the general term and the twelfth term in the arithmetic progression $2, \dfrac{5}{2}, \ldots$.

27. Find the general term and the twentieth term in the arithmetic progression $3, -2, \ldots$.

28. Find the general term and the ninth term in the arithmetic progression $\dfrac{3}{4}, 2, \ldots$.

Example   Find the general term and also the ninth term of the *geometric* progression $-24, 12, \ldots$.

                                                      (Solution overleaf)

Solution   Find the common ratio.

$$r = \frac{12}{-24} = -\frac{1}{2}$$

Use $s_n = ar^{n-1}$.

$$s_n = -24\left(-\frac{1}{2}\right)^{n-1}$$

$$s_9 = -24\left(-\frac{1}{2}\right)^8 = -\frac{3}{32}$$

29. Find the general term and the sixth term in the geometric progression $48, 96, \ldots$.

30. Find the general term and the eighth term in the geometric progression -3, $\frac{3}{2}, \ldots$.

31. Find the general term and the seventh term in the geometric progression $-\frac{1}{3}, 1, \ldots$.

32. Find the general term and the ninth term in the geometric progression -81, $-27, \ldots$.

33. If the third term in an arithmetic progression is 7 and the eighth term is 17, find the common difference. What are the first and the twentieth terms?

34. If the fifth term of an arithmetic progression is -16 and the twentieth term is -46, what is the twelfth term?

35. Which term in the arithmetic progression $4, 1, \ldots$ is -77?

36. What is the twelfth term in an arithmetic progression in which the second term is x and the third term is y?

37. Find the first term of a geometric progression with fifth term 48 and ratio 2.

38. Find two different values for x so that $-\frac{3}{2}, x, -\frac{8}{27}$ will be in geometric progression.

39. Prove Theorem 9.3 by mathematical induction.

9.3   Series

Associated with any sequence is a *series*.

Definition 9.4   *A series is the indicated sum of the terms in a sequence.*

For example, with the finite sequence

$$4, 7, 10, \ldots, 3n + 1,$$

for a given counting number n, there is associated the finite series

$$S_n = 4 + 7 + 10 + \cdots + (3n + 1);$$

similarly, with the finite sequence

$$x, x^2, x^3, x^4, \ldots, x^n,$$

there is associated the finite series

$$S_n = x + x^2 + x^3 + x^4 + \cdots + x^n.$$

Since the terms in the series are the same as those in the sequence, we can refer to the first term or the second term or the general term of a series in the same manner as we do for a sequence.

Sum of the first n terms of an arithmetic progression

Consider the series S_n of the first n terms of the general arithmetic progression

$$S_n = a + (a + d) + (a + 2d) + \cdots + [a + (n - 1)d], \tag{1}$$

and then consider the same series written as

$$S_n = s_n + (s_n - d) + (s_n - 2d) + \cdots + [s_n - (n - 1)d], \tag{2}$$

where the terms are displayed in reverse order. Adding (1) and (2) term by term, we have

$$S_n + S_n = (a + s_n) + (a + s_n) + (a + s_n) + \cdots + (a + s_n),$$

where the term $(a + s_n)$ occurs n times. Then

$$2S_n = n(a + s_n),$$

$$S_n = \frac{n}{2}(a + s_n). \tag{3}$$

If (3) is rewritten as

$$S_n = n\left(\frac{a + s_n}{2}\right),$$

we observe that the sum is given by the product of the number of terms in the series and the average of the first and last terms. The validity of (3) can be established by mathematical induction and is left as an exercise.

An alternate form for (3) is obtained by substituting $a + (n - 1)d$ for s_n in (3) to obtain

$$2S_n = n(a + s_n),$$

$$S_n = \frac{n}{2}(a + [a + (n - 1)d]),$$

$$S_n = \frac{n}{2}[2a + (n - 1)d], \tag{3a}$$

where the sum is now expressed in terms of a, n, and d.

Sum of the first n terms of a geometric progression

To find an explicit representation for the sum of a given number of terms in a geometric progression in terms of a, r, and n, we employ a device somewhat similar to the one used in finding the sum in an arithmetic progression. Consider the geometric series (4) containing n terms, and the series (5) obtained by multiplying both members of (4) by r:

$$S_n = a + ar + ar^2 + ar^3 + \cdots + ar^{n-2} + ar^{n-1}, \tag{4}$$

$$rS_n = ar + ar^2 + ar^3 + ar^4 + \cdots + ar^{n-1} + ar^n. \tag{5}$$

When we subtract (5) from (4), all terms in the right-hand members except the first term in (4) and the last term in (5) vanish, yielding

$$S_n - rS_n = a - ar^n.$$

Factoring S_n from the left-hand member gives

$$(1 - r)S_n = a - ar^n,$$

$$S_n = \frac{a - ar^n}{1 - r}, \tag{6}$$

if $r \neq 1$, and we have a formula for the sum of the first n terms of a geometric progression. Establishing the validity of Equation (6), which can be accomplished by mathematical induction, is left as an exercise.

An alternative expression for (6) can be obtained by first writing

$$S_n = \frac{a - r(ar^{n-1})}{1 - r},$$

and then, since $s_n = ar^{n-1}$, expressing this as

$$S_n = \frac{a - rs_n}{1 - r}, \tag{7}$$

where the sum is now given in terms of a, s_n, and r.

Sigma notation

A series for which the general term is known can be represented in a very convenient, compact way by means of what is called **sigma,** or **summation, notation.** The Greek letter $\sum$ (sigma) is used to denote a sum. For example,

$$S_n = 4 + 7 + 10 + \cdots + (3n + 1)$$

can be written

$$S_n = \sum_{j=1}^{n} (3j + 1),$$

where we understand that S_n is the series having terms obtained by replacing j in the expression $3j + 1$ with the numbers $1, 2, 3, \ldots, n$, successively. Similarly,

$$S_4 = \sum_{j=3}^{6} j^2$$

appears in expanded form as

$$S_4 = 3^2 + 4^2 + 5^2 + 6^2,$$

where the first value for j is 3 and the last is 6.

The variable used in conjunction with summation notation is called the **index of summation,** and the set of integers over which we sum (in this case, $\{3, 4, 5, 6\}$) is called the **range of summation**.

Notation for an infinite sum

To indicate that a series has an infinite number of terms, we cannot use the notation S_n for the sum, because there is no value to substitute for n. We therefore adopt notation such as

$$S_\infty = \sum_{j=1}^{\infty} \frac{1}{2^j} \tag{8}$$

to indicate that there is no last term in a series. In expanded form, the infinite series (8) is given by

$$S_\infty = \frac{1}{2} + \frac{1}{4} + \frac{1}{8} + \cdots.$$

Exercise 9.3

Write the given series in expanded form.

Examples a. $\displaystyle\sum_{j=2}^{5}(j^2 + 1)$ b. $\displaystyle\sum_{k=1}^{\infty}(-1)^k 2^{k+1}$

Solutions a. $j = 2,\quad 2^2 + 1 = 5;$ b. $k = 1,\quad (-1)^1 2^{1+1} = (-1)(4) = -4;$

$\qquad\quad j = 3,\quad 3^2 + 1 = 10;$ $\qquad k = 2,\quad (-1)^2 2^{2+1} = (1)(8) = 8;$

$\qquad\quad j = 4,\quad 4^2 + 1 = 17;$ $\qquad k = 3,\quad (-1)^3 2^{3+1} = (-1)(16) = -16.$

$\qquad\quad j = 5,\quad 5^2 + 1 = 26.$ $\qquad \displaystyle\sum_{k=1}^{\infty}(-1)^k 2^{k+1} = -4 + 8 - 16 + \cdots$

$\qquad\displaystyle\sum_{j=2}^{5}(j^2 + 1) = 5 + 10 + 17 + 26$

1. $\displaystyle\sum_{j=1}^{4} j^2$ 2. $\displaystyle\sum_{j=1}^{4}(3j - 2)$ 3. $\displaystyle\sum_{j=1}^{3}\frac{(-1)^j}{2^j}$

4. $\displaystyle\sum_{i=3}^{5}\frac{(-1)^{i+1}}{i - 2}$ 5. $\displaystyle\sum_{k=0}^{\infty}\frac{1}{5^k}$ 6. $\displaystyle\sum_{k=0}^{\infty}\frac{k}{1 + k}$

Write the given series in sigma notation.

Examples a. $5 + 8 + 11 + 14$ b. $x^2 + x^4 + x^6$ c. $\dfrac{3}{5} + \dfrac{5}{7} + \dfrac{7}{9} + \cdots$

Solutions Find an expression for the general term and write in sigma notation.

a. $3j + 2$ b. x^{2j} c. $\dfrac{2j + 1}{2j + 3}$

$\quad\displaystyle\sum_{j=1}^{4}(3j + 2)$ $\quad\displaystyle\sum_{j=1}^{3}x^{2j}$ $\quad\displaystyle\sum_{j=1}^{\infty}\frac{2j + 1}{2j + 3}$

7. $x + x^3 + x^5 + x^7$ 8. $x^3 + x^5 + x^7 + x^9 + x^{11}$

9. $1 + 4 + 9 + 16 + 25$ 10. $\dfrac{1}{3} + \dfrac{1}{9} + \dfrac{1}{27} + \dfrac{1}{81}$

11. $1 \cdot 2 + 2 \cdot 3 + 3 \cdot 4 + 4 \cdot 5 + \cdots$ **12.** $\dfrac{1}{2} + \dfrac{2}{3} + \dfrac{3}{4} + \dfrac{4}{5} + \cdots$

13. $\dfrac{2}{1} + \dfrac{3}{2} + \dfrac{4}{3} + \dfrac{5}{4} + \cdots$ **14.** $\dfrac{1}{1} + \dfrac{2}{3} + \dfrac{3}{5} + \dfrac{4}{7} + \cdots$

Find the sum.

Example $\displaystyle\sum_{j=1}^{12} (4j + 1)$

Solution   Write the first two or three terms in expanded form:

$$5 + 9 + 13 + \cdots.$$

This is an arithmetic series. The first term is 5 and the common difference is 4. Therefore we can use

$$S_n = \frac{n}{2}[2a + (n-1)d]$$

to obtain

$$S_{12} = \frac{12}{2}[2(5) + (12 - 1)4] = 324.$$

15. $\displaystyle\sum_{j=1}^{7} (2j + 1)$ **16.** $\displaystyle\sum_{j=1}^{21} (3j - 2)$ **17.** $\displaystyle\sum_{j=3}^{15} (7j - 1)$

18. $\displaystyle\sum_{j=10}^{20} (2j - 3)$ **19.** $\displaystyle\sum_{k=1}^{8} \left(\frac{1}{2}k - 3\right)$ **20.** $\displaystyle\sum_{k=1}^{100} k$

Example $\displaystyle\sum_{j=2}^{5} \left(\frac{1}{3}\right)^j$

Solution   Write the first two terms in expanded form:

$$\left(\frac{1}{3}\right)^2 + \left(\frac{1}{3}\right)^3 + \cdots.$$

This is a geometric series in which the first term is $\dfrac{1}{9}$, the ratio is $\dfrac{1}{3}$, and $n = 4$.
Therefore we can use $S_n = \dfrac{a - ar^n}{1 - r}$ to obtain

$$S_4 = \frac{\dfrac{1}{9} - \dfrac{1}{9}\left(\dfrac{1}{3}\right)^4}{1 - \dfrac{1}{3}} = \frac{40}{243}.$$

21. $\displaystyle\sum_{j=1}^{6} 3^j$ **22.** $\displaystyle\sum_{j=1}^{4} (-2)^j$ **23.** $\displaystyle\sum_{k=3}^{7} \left(\frac{1}{2}\right)^{k-2}$

24. $\displaystyle\sum_{j=3}^{12} 2^{j-5}$ **25.** $\displaystyle\sum_{j=1}^{6} \left(\frac{1}{3}\right)^j$ **26.** $\displaystyle\sum_{j=1}^{4} (3 + 2^j)$

27. Find the sum of all even integers n, for $13 < n < 29$.

28. Find the sum of all integral multiples of 7 between 8 and 110.

29. How many bricks will there be in a pile one brick in thickness if there are 27 bricks in the bottom row, 25 in the second row, and so forth, to the top row, which has one brick?

30. If there are a total of 256 bricks in a pile arranged in the manner of those in Exercise 29, how many bricks are there in the third row from the bottom of the pile?

31. By mathematical induction, prove that for an arithmetic progression, $S_n = \dfrac{n}{2}(a + s_n)$ for all positive integral values of n.

32. Show that the sequence formed by adding the corresponding terms in two arithmetic progressions is an arithmetic progression.

33. Show that the sum of the terms in two series with terms in arithmetic progression can be written as a series with terms in arithmetic progression.

34. By mathematical induction, prove that for a geometric progression, $S_n = \dfrac{a - ar^n}{1 - r}$ for all positive integral values of n, provided $r \neq 1$.

9.4   The Binomial Theorem

There are situations, as in the binomial expansion below, in which it is necessary to write the product of several consecutive positive integers. To facilitate writing products of this type, we use a special symbol $n!$ (read "n factorial" or "factorial n"), which is defined by

$$n! = n(n - 1)(n - 2) \cdots (3)(2)(1).$$

Thus

$$5! = 5 \cdot 4 \cdot 3 \cdot 2 \cdot 1 \quad \text{(read "five factorial"),}$$

and

$$8! = 8 \cdot 7 \cdot 6 \cdot 5 \cdot 4 \cdot 3 \cdot 2 \cdot 1 \quad \text{(read "eight factorial").}$$

Factorial notation can also be used to represent products of consecutive positive integers, beginning with integers different from 1. For example,

$$8 \cdot 7 \cdot 6 \cdot 5 = \frac{8!}{4!},$$

because

$$\frac{8!}{4!} = \frac{8 \cdot 7 \cdot 6 \cdot 5 \cdot 4 \cdot 3 \cdot 2 \cdot 1}{4 \cdot 3 \cdot 2 \cdot 1} = 8 \cdot 7 \cdot 6 \cdot 5.$$

Since

$$n! = n(n - 1)(n - 2)(n - 3) \cdots 5 \cdot 4 \cdot 3 \cdot 2 \cdot 1$$

and

$$(n - 1)! = (n - 1)(n - 2)(n - 3) \cdots 5 \cdot 4 \cdot 3 \cdot 2 \cdot 1,$$

for $n > 1$, we can write the recursive relationship

$$n! = n(n - 1)!.$$ (1)

For example,

$$7! = 7 \cdot 6!,$$

$$27! = 27 \cdot 26!,$$

$$(n + 2)! = (n + 2)(n + 1)!.$$

If (1) is to hold also for $n = 1$, then we must have

$$1! = 1 \cdot (1 - 1)!$$

or

$$1! = 1 \cdot 0!.$$

Therefore, for consistency, we define $0!$ by

$$0! = 1.$$

A special case of the use of factorial notation occurs in the formula

$$\binom{n}{r} = \frac{n!}{r!\,(n - r)!}.$$

As you will see in Section 9.6, the symbol $\binom{n}{r}$ is read "the number of combinations of n things taken r at a time"; it denotes the number of r-element subsets of a set containing n members. As examples, we have

$$\binom{5}{3} = \frac{5!}{3!\,(5 - 3)!} = \frac{5!}{3!\,2!} = \frac{5 \cdot 4 \cdot 3!}{3!\,(2 \cdot 1)} = 10,$$

$$\binom{5}{1} = \frac{5!}{1!\,(5 - 1)!} = \frac{5!}{4!} = \frac{5 \cdot 4!}{4!} = 5,$$

and

$$\binom{5}{0} = \frac{5!}{0!\,(5 - 0)!} = \frac{5!}{5!} = 1.$$

The series obtained by expanding a binomial of the form

$$(a + b)^n$$

is particularly useful in certain branches of mathematics. Starting with familiar examples, where n takes the values 1, 2, 3, 4, and 5 in turn, we can show by direct multiplication that

$$(a + b)^1 = a + b,$$

$$(a + b)^2 = a^2 + 2ab + b^2,$$

$$(a + b)^3 = a^3 + 3a^2b + 3ab^2 + b^3,$$

$$(a + b)^4 = a^4 + 4a^3b + 6a^2b^2 + 4ab^3 + b^4,$$

$$(a + b)^5 = a^5 + 5a^4b + 10a^3b^2 + 10a^2b^3 + 5ab^4 + b^5.$$

We observe that in each case:

1. The first term is a^n, and the last term is b^n.

2. The variable factors of the second term are $a^{n-1}b^1$, and the coefficient is n, which can be written in the form
$$\frac{n}{1!}.$$

3. The variable factors of the third term are $a^{n-2}b^2$, and the coefficient can be written in the form
$$\frac{n(n-1)}{2!}.$$

4. The variable factors of the fourth term are $a^{n-3}b^3$, and the coefficient can be written in the form
$$\frac{n(n-1)(n-2)}{3!}.$$

The foregoing expansions suggest the following result, known as the **binomial theorem.** Since its proof is quite lengthy, it is omitted.

Theorem 9.4   *For each natural number* n,

$$(a+b)^n = a^n + \frac{n}{1!}a^{n-1}b + \frac{n(n-1)}{2!}a^{n-2}b^2 + \frac{n(n-1)(n-2)}{3!}a^{n-3}b^3$$

$$+ \cdots + \frac{n(n-1)(n-2)\cdots(n-r+2)}{(r-1)!}a^{n-r+1}b^{r-1} + \cdots + b^n, \quad (2)$$

where r *is the number of the term,* $r > 1$.

For example,

$$(x-2)^4 = x^4 + \frac{4}{1!}x^3(-2)^1 + \frac{4\cdot3}{2!}x^2(-2)^2 + \frac{4\cdot3\cdot2}{3!}x(-2)^3 + \frac{4\cdot3\cdot2\cdot1}{4!}(-2)^4$$

$$= x^4 - 8x^3 + 24x^2 - 32x + 16.$$

In this case, $a = x$ and $b = -2$ in the binomial expansion.

Observe that the coefficients of the terms in the binomial expansion (2) can be represented as follows.

1st term: $\qquad \binom{n}{0} = \frac{n!}{0!\,n!} = 1,$

2nd term: $\qquad \binom{n}{1} = \frac{n!}{1!\,(n-1)!} = \frac{n\cdot(n-1)!}{1!\cdot(n-1)!} = \frac{n}{1!},$

3rd term: $\qquad \binom{n}{2} = \frac{n!}{2!\,(n-2)!} = \frac{n(n-1)(n-2)!}{2!\,(n-2)!} = \frac{n(n-1)}{2!},$

rth term: $\qquad \binom{n}{r-1} = \frac{n!}{(r-1)!\,(n-r+1)!}$

$$= \frac{n(n-1)(n-2)\cdots(n-r+2)(n-r+1)!}{(r-1)!\,(n-r+1)!}$$

$$= \frac{n(n-1)(n-2)\cdots(n-r+2)}{(r-1)!}.$$

Hence, the binomial expansion (2) can be written as

$$(a + b)^n = \binom{n}{0}a^n + \binom{n}{1}a^{n-1}b + \binom{n}{2}a^{n-2}b^2 + \binom{n}{3}a^{n-3}b^3 + \cdots$$

$$+ \binom{n}{r-1}a^{n-r+1}b^{r-1} + \cdots + \binom{n}{n}b^n. \quad (3)$$

For example,

$$(x - 2)^4 = \binom{4}{0}x^4 + \binom{4}{1}x^3(-2)^1 + \binom{4}{2}x^2(-2)^2 + \binom{4}{3}x(-2)^3 + \binom{4}{4}(-2)^4.$$

Simplifying the coefficients, we obtain the same result as above,

$$(x - 2)^4 = x^4 - 8x^3 + 24x^2 - 32x + 16.$$

Note that for $r > 1$ the rth term in a binomial expansion is given by

$$\binom{n}{r-1}a^{n-r+1}b^{r-1} = \frac{n(n-1)(n-2)(n-r+2)}{(r-1)!}a^{n-r+1}b^{r-1}. \quad (4)$$

For example, by the left-hand member of (4), the seventh term of $(x - 2)^{10}$ is

$$\binom{10}{6}x^4(-2)^6 = \frac{10!}{6!\,4!}x^4(-2)^6 = \frac{10 \cdot 9 \cdot 8 \cdot 7 \cdot 6!}{6! \cdot 4 \cdot 3 \cdot 2 \cdot 1}x^4(64)$$

$$= 13{,}440x^4,$$

while, by the right-hand member of (4), we have

$$\frac{10 \cdot 9 \cdot 8 \cdot 7 \cdot 6 \cdot 5}{6 \cdot 5 \cdot 4 \cdot 3 \cdot 2 \cdot 1}x^4(64) = 13{,}440x^4.$$

Exercise 9.4

1. Write $(2n)!$ in expanded form for $n = 4$.
2. Write $2n!$ in expanded form for $n = 4$.
3. Write $n(n - 1)!$ in expanded form for $n = 6$.
4. Write $2n(2n - 1)!$ in expanded form for $n = 2$.

Write in expanded form and simplify.

Examples   a. $\dfrac{7!}{4!}$ b. $\dfrac{4!\,6!}{8!}$

Solutions   a. $\dfrac{7 \cdot 6 \cdot 5 \cdot 4!}{4!} = 210$ b. $\dfrac{4 \cdot 3 \cdot 2 \cdot 1 \cdot 6!}{8 \cdot 7 \cdot 6!} = \dfrac{3}{7}$

5. $5!$ **6.** $7!$ **7.** $\dfrac{9!}{7!}$ **8.** $\dfrac{12!}{11!}$

9. $\dfrac{5!\,7!}{8!}$ **10.** $\dfrac{12!\,8!}{16!}$ **11.** $\dfrac{8!}{2!\,(8-2)!}$ **12.** $\dfrac{10!}{4!\,(10-4)!}$

Write the product in factorial notation.

Examples   a. $1 \cdot 2 \cdot 3 \cdot 4 \cdot 5 \cdot 6$ b. $11 \cdot 12 \cdot 13 \cdot 14$ c. 150

Solutions   a. $6!$ b. $\dfrac{14!}{10!}$ c. $\dfrac{150!}{149!}$

13. $1 \cdot 2 \cdot 3$ **14.** $1 \cdot 2 \cdot 3 \cdot 4 \cdot 5$ **15.** $3 \cdot 4 \cdot 5 \cdot 6$

16. 7 **17.** $8 \cdot 7 \cdot 6$ **18.** $28 \cdot 27 \cdot 26 \cdot 25 \cdot 24$

Write the given expression in factorial notation and simplify.

Examples   a. $\dbinom{6}{2}$ b. $\dbinom{4}{4}$

Solutions   a. $\dbinom{6}{2} = \dfrac{6!}{2!\,(6-2)!} = \dfrac{6!}{2!\,4!}$ b. $\dbinom{4}{4} = \dfrac{4!}{4!\,(4-4)!}$

$\qquad\qquad\qquad\quad = \dfrac{6 \cdot 5 \cdot 4!}{2 \cdot 1 \cdot 4!} = 15$ $\qquad = \dfrac{4!}{4!\,0!} = 1$

19. $\dbinom{6}{5}$ **20.** $\dbinom{4}{2}$ **21.** $\dbinom{3}{3}$ **22.** $\dbinom{5}{5}$

23. $\dbinom{7}{0}$ **24.** $\dbinom{2}{0}$ **25.** $\dbinom{5}{2}$ **26.** $\dbinom{5}{3}$

Write the given expression in factored form and show the first three factors and the last three factors.

Example $(2n + 1)!$

Solution $(2n + 1)(2n)(2n - 1) \cdot \cdots \cdot 3 \cdot 2 \cdot 1$

27. $n!$ **28.** $(n + 4)!$ **29.** $(3n)!$

30. $3n!$ **31.** $(n - 2)!$ **32.** $(3n - 2)!$

Simplify the given expression.

Examples   a. $\dfrac{(n-1)!}{(n-3)!}$ b. $\dfrac{(n-1)!\,(2n)!}{2n!\,(2n-2)!}$

Solutions a. $\dfrac{(n-1)(n-2)(n-3)!}{(n-3)!}$ b. $\dfrac{(n-1)!\,(2n)(2n-1)(2n-2)!}{2(n)(n-1)!\,(2n-2)!}$

$\qquad\qquad (n-1)(n-2)$ $2n-1$

33. $\dfrac{(n+2)!}{n!}$ **34.** $\dfrac{(n+2)!}{(n-1)!}$ **35.** $\dfrac{(n+1)(n+2)!}{(n+3)!}$

36. $\dfrac{(2n+4)!}{(2n+2)!}$ **37.** $\dfrac{(2n)!\,(n-2)!}{4(2n-2)!\,(n)!}$ **38.** $\dfrac{(2n+1)!\,(2n-1)!}{[(2n)!]^2}$

Expand.

Example $(a-3b)^4$

Solution From the binomial expansion (2),

$$(a-3b)^4 = a^4 + \frac{4}{1!}a^3(-3b) + \frac{4\cdot3}{2!}a^2(-3b)^2 + \frac{4\cdot3\cdot2}{3!}a(-3b)^3$$

$$+ \frac{4\cdot3\cdot2\cdot1}{4!}(-3b)^4$$

$$= a^4 - 12a^3b + 54a^2b^2 - 108ab^3 + 81b^4.$$

Alternatively, from the binomial expansion (3),

$$(a-3b)^4 = \binom{4}{0}a^4 + \binom{4}{1}a^3(-3b) + \binom{4}{2}a^2(-3b)^2 + \binom{4}{3}a(-3b)^3 + \binom{4}{4}(-3b)^4,$$

which also simplifies to the expression obtained above.

39. $(x+3)^5$ **40.** $(2x+y)^4$ **41.** $(x-3)^4$ **42.** $(2x-1)^5$

43. $\left(2x-\dfrac{y}{2}\right)^3$ **44.** $\left(\dfrac{x}{3}+3\right)^5$ **45.** $\left(\dfrac{x}{2}+2\right)^6$ **46.** $\left(\dfrac{2}{3}-a^2\right)^4$

Write the first four terms in the given expansion. Do not simplify the terms.

Example $(x+2y)^{15}$

Solution You may write either

$$(x+2y)^{15} = x^{15} + \frac{15}{1!}x^{14}(2y) + \frac{15\cdot14}{2!}x^{13}(2y)^2 + \frac{15\cdot14\cdot13}{3!}x^{12}(2y)^3$$

or

$$(x + 2y)^{15} = \binom{15}{0}x^{15} + \binom{15}{1}x^{14}(2y) + \binom{15}{2}x^{13}(2y)^2 + \binom{15}{3}x^{12}(2y)^3.$$

47. $(x + y)^{20}$ **48.** $(x - y)^{15}$ **49.** $(a - 2b)^{12}$

50. $(2a - b)^{12}$ **51.** $(x - \sqrt{2})^{10}$ **52.** $\left(\dfrac{x}{2} + 2\right)^8$

Find the indicated power to the nearest hundredth.

Example $(0.97)^7$

Solution   Either form of the binomial expansion (2) or (3) can be used. We shall use (2).

$$(0.97)^7 = (1 - 0.03)^7$$

$$= 1^7 + \frac{7}{1!}(1)^6(-0.03)^1 + \frac{7 \cdot 6}{2!}(1)^5(-0.03)^2$$

$$+ \frac{7 \cdot 6 \cdot 5}{3!}(1)^4(-0.03)^3 + \cdots$$

$$= 1 - 0.21 + 0.0189 - 0.000945 + \cdots$$

$$\approx 0.807955$$

Hence, to the nearest hundredth, $(0.97)^7 = 0.81$.

53. $(1.02)^{10}$ *Hint:* $1.02 = (1 + 0.02)$.
54. $(1.01)^{15}$

55. If an amount of money P is invested at 4% compounded annually, the amount A present at the end of n years is given by $A = P(1 + 0.04)^n$. Find the amount A (to the nearest dollar) if \$1000 was invested for 10 years.

56. In Problem 55, find the amount present at the end of 20 years.

Find the specified term.

Example $(x - 2y)^{12}$, the seventh term

Solution   In Formula (4) use $n = 12$ and $r = 6$.

$$\frac{12!}{6!\,(12 - 6)!}\,x^6(-2y)^6 = 59{,}136x^6y^6$$

57. $(a - b)^{15}$, the sixth term   **58.** $(x + 2)^{12}$, the fifth term
59. $(x - 2y)^{10}$, the fifth term   **60.** $(a^3 - b)^9$, the seventh term

61. Given that the binomial formula holds for $(1 + x)^n$ where n is a negative integer:

   a. Write the first four terms of $(1 + x)^{-1}$.

   b. Find the first four terms of the quotient $1/(1 + x)$ by dividing $(1 + x)$ into 1.

   Compare the results of (a) and (b).

62. Given that the binomial formula (2) holds as an infinite "sum" for $(1 + x)^n$, where n is a rational number and $|x| < 1$, find to two decimal places:

   a. $\sqrt{1.02}$;   **b.** $\sqrt{0.99}$.

9.5   *Basic Counting Principles; Permutations*

Associated with each finite set A is a nonnegative integer n, namely the number of elements in A. Hence, we have a function from the set of all finite sets to the set of nonnegative integers. The symbolism $n(A)$ is used to denote the number of elements in the range of this set function n. For example, if

$$A = \{5, 7, 9\}, \quad B = \{1/2, 0, 3, -5, 7\}, \quad C = \emptyset,$$

then

$$n(A) = 3, \quad n(B) = 5, \quad \text{and} \quad n(C) = 0.$$

When discussing sets in this and the following section, it will be convenient to recall three fundamental operations on sets which we first considered in Sections 1.1 and 2.1.

1. The **union** of two sets A and B, denoted by $A \cup B$, is the set of all elements that belong either to A *or* to B *or* to both.

2. The **intersection** of two sets A and B, denoted by $A \cap B$, is the set of all elements that belong to *both* A and B.

3. The **Cartesian product** of two sets A and B, denoted by $A \times B$, is the set of all ordered pairs (x, y) such that $x \in A$ and $y \in B$.

Counting properties

All the sets with which we shall hereafter be concerned are assumed to be finite sets. We then have the following properties, called **counting properties**, for the function n.

   I $n(A \cup B) = n(A) + n(B),$ if $A \cap B = \emptyset$.

Thus, if A and B are disjoint sets then the number of elements in their union is the sum of the number of elements in A and the number of elements in B.

   For example, suppose there are five roads from town R to town S, and two railroads from town R to town S. If A is the set of roads and B the set of railroads from R to S, then $n(A) = 5$, $n(B) = 2$, and $n(A \cup B) = 5 + 2 = 7$; thus there are seven ways one can go from town R to town S by driving or riding on a train.

II $n(A \cup B) = n(A) + n(B) - n(A \cap B), \quad \text{if } A \cap B \neq \emptyset.$

That is, if A and B overlap, then to count the number of elements in $A \cup B$, we might add the number of elements in A to the number of elements in B. But since any elements in the intersection of A and B are counted twice in this process (once in A and once in B), we must subtract the number of such elements from the sum $n(A) + n(B)$ to obtain the number of elements in $A \cup B$, as suggested in Figure 9.2.

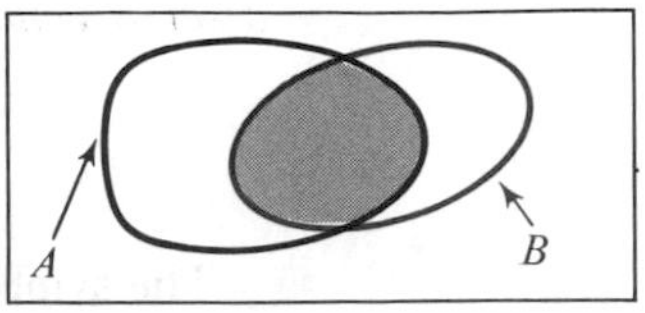

Figure 9.2

For example, suppose there are fifteen unrelated girls and seventeen unrelated boys in a mathematics class, and suppose that there are precisely two brother-sister pairs in the class. If A denotes the set of different families represented by the girls and B denotes the set of different families represented by the boys, then the number of different families represented by all of the members of the class is

$$n(A) + n(B) - n(A \cap B) = 15 + 17 - 2 = 30.$$

Actually, Property II is a consequence of Property I.

III $n(A \times B) = n(A) \cdot n(B).$

This asserts that the number of elements in the Cartesian product of sets A and B is the product of the number of elements in A and the number of elements in B.

For example, suppose again that there are five roads from town R to town S (set A), and further suppose that there are three roads from town S to town T (set B). Then for each element of A there are three elements of B, and the total possible ways one can drive from R to T via S is

$$n(A \times B) = n(A) \cdot n(B) = 5 \cdot 3 = 15.$$

Permutations   Given the set of digits $A = \{1, 2, 3\}$, how many different three-digit numerals can be constructed from the members of A if no member is used more than once? The answer to this question can be obtained by simply listing the different three-digit numerals, 1 2 3, 1 3 2, 2 1 3, 2 3 1, 3 1 2, 3 2 1, and counting them. Such a procedure would be quite impracticable, however, if the number of members of the given set of digits were very large. Another way to arrive at the same conclusion is by applying the third counting property. If we let A denote the set of possible first digits in the foregoing numerals, then $n(A) = 3$. Since no numeral may be used more than once, and since one numeral has already been used for a first digit, there remain but two possibilities for the second digit. If B denotes the set of possible second digits after the first digit has been chosen, then $n(B) = 2$. By similar reasoning, if C is the set of possible third digits after the first two have been chosen, then $n(C) = 1$. By the third counting property (applied twice), we find that

$$n(A \times B \times C) = [n(A) \cdot n(B)] \cdot n(C) = 3 \cdot 2 \cdot 1 = 6.$$

Each of the three-digit numerals discussed above is called a **permutation** of the elements of the set of numerals $\{1, 2, 3\}$.

Definition 9.5   *A **permutation** of a set A is an ordering (first, second, etc.) of the members of A.*

With Definition 9.5, we can state the following result.

Theorem 9.5 *Let $P_{n,n}$ denote the number of distinct permutations of the members of a set A containing n members. Then*

$$P_{n,n} = n!. \tag{1}$$

The symbol $P_{n,n}$ (or sometimes $_nP_n$, or P_n^n) is read "the number of permutations of *n* things taken *n* at a time."

Proof Let A_1 denote the set of possible selections for the first member. Then $A_1 = A$, and $n(A_1) = n(A) = n$. Having made a first selection, let A_2 denote the set of possible second selections. Then A_2 is a subset of A, and

$$n(A_2) = n(A_1) - 1 = n - 1.$$

A continuation of this procedure, together with successive application of the third counting principle, leads to

$$P_{n,n} = n(A_1) \cdot n(A_2) \cdot \cdots \cdot n(A_n) = n \cdot (n - 1) \cdot (n - 2) \cdot \cdots \cdot 1 = n!,$$

as was to be proved.

Example

In how many ways can nine men be assigned positions to form distinct baseball teams?

Solution

Let A denote the set of men, so that $n(A) = 9$. The total number of ways in which 9 men can be assigned 9 positions on a team, or, in other words, the number of possible permutations of the members of a 9-element set is, by Equation (1),

$$P_{9,9} = 9! = 9 \cdot 8 \cdot 7 \cdots 1 = 362,880.$$

Theorem 9.6   *Let $P_{n,r}$ denote the number of permutations of the members, taken r at a time, of a set A containing n members; that is, let $P_{n,r}$ be the number of distinct orderings of r elements when there is a set A of n elements from which to choose. Then*

$$P_{n,r} = n(n - 1)(n - 2) \cdots (n - r + 1). \tag{2}$$

The proof follows the proof of Theorem 9.5, except that the last subset considered is such that $n(A_r) = n - r + 1$.

Example

In how many ways can a basketball team be formed by choosing players for the five positions from a set of ten players?

Solution

Let A denote the set of players, so that $n(A) = 10$. Then from (2) and the fact that a basketball team consists of 5 players, we have

$$P_{10,5} = 10 \cdot 9 \cdot 8 \cdots (10 - 5 + 1) = 10 \cdot 9 \cdot 8 \cdot 7 \cdot 6 = 30,240.$$

An alternative expression for $P_{n,r}$ can be obtained by observing that

$$P_{n,r} = n(n-1)(n-2)\cdots(n-r+1)$$

$$= \frac{n(n-1)(n-2)\cdots(n-r+1)(n-r)!}{(n-r)!},$$

so that

$$P_{n,r} = \frac{n!}{(n-r)!}. \qquad (3)$$

Although Theorems 9.5 and 9.6 are often convenient to use, the same results can be obtained by using counting Property III. In fact, the use of Property III is the most direct method of finding the number of permutations if an element in a set can be selected for more than one position.

Example   In how many ways can three students be assigned a grade of A, B, C, or D?

Solution   Sometimes a simple diagram, such as ___, ___, ___, designating a sequence, is a helpful preliminary device. Since each student may receive any one of four different grades, the sequence would appear as 4, 4, 4. From counting Property III, there are $4 \cdot 4 \cdot 4$, or 64, possible ways the grades may be assigned.

Distinguishable permutations   The problem of finding the number of distinguishable permutations of n objects taken n at a time, if some of the objects are identical, requires a little more careful analysis. As an example, consider the number of permutations of the letters of the word *DIVISIBLE*. We can make a distinction between the three I's by assigning subscripts to each so that we have nine distinct letters,

$$D, I_1, V, I_2, S, I_3, B, L, E.$$

The number of permutations of these nine letters is of course 9!. If the letters other than I_1, I_2, and I_3 are retained in the position they occupy in a permutation of the above nine letters, I_1, I_2, and I_3 can be permuted among themselves 3! ways. Thus, if P is the number of *distinguishable* permutations of the letters

$$D, I, V, I, S, I, B, L, E,$$

then, since for each of these there are 3! ways in which the I's can be permuted without otherwise changing the order of the other letters, it follows that

$$3! \cdot P = 9!,$$

from which

$$P = \frac{9!}{3!}.$$

As another example, consider the letters of the word *MISSISSIPPI*. There would exist 11! distinguishable permutations of the letters in this word if each letter were distinct. Note, however, that the letters S and I each appear four times and the letter P appears twice. Reasoning as we did in the previous example, we see that the number P of distinguishable permutations of the letters in *MISSISSIPPI* is given by

$$4!\,4!\,2! \cdot P = 11!,$$

from which

$$P = \frac{11!}{4!\,4!\,2!}.$$

Exercise 9.5

For the given sets, find $n(A \cap B)$, $n(A \cup B)$,   and $n(A \times B)$.

Example $A = \{a, b, c\}, \quad B = \{c, d\}$

Solution $A \cap B = \{c\}$. Therefore $n(A \cap B) = 1$.

$n(A \cup B) = n(A) + n(B) - n(A \cap B) = 3 + 2 - 1 = 4$.

$n(A \times B) = n(A) \cdot n(B) = 3 \cdot 2 = 6$.

1. $A = \{d, e\}, \quad B = \{e, f, g, h\}$ 2. $A = \{e\}, \quad B = \{a, b, c, d\}$
3. $A = \{1, 2, 3, 4\}, \quad B = \{3, 4, 5, 6\}$ 4. $A = \{1, 2\}, \quad B = \{3, 4, 5\}$
5. $A = \{1, 2\}, \quad B = \{1, 2\}$ 6. $A = \emptyset, \quad B = \{2, 3, 4\}$

Example   In how many different ways can three members of a class be assigned a grade of A, B, C, or D so that no two members receive the same grade?

Solution   Since the first student may receive any one of four different grades, the second student may then receive any one of three different grades, and the third student may then receive any one of two different grades, the sequence would appear as

$$\underline{4}, \underline{3}, \underline{2}.$$

From counting Property III, there are $4 \cdot 3 \cdot 2$, or 24, possible ways the grades may be assigned. In this case, we could have obtained the same result directly from Theorem 9.6, since $P_{4,3} = 4 \cdot 3 \cdot 2 = 24$.

In each exercise, a digit or letter may be used more than once unless stated otherwise.

7. How many different two-digit numerals can be formed from the digits 5 and 6?
8. How many different two-digit numerals can be formed from the digits 7, 8, and 9?
9. In how many different ways can four students be seated in a row?
10. In how many different ways can five students be seated in a row?
11. In how many different ways can four questions on a true-false test be answered?
12. In how many different ways can five questions on a true-false test be answered?
13. In how many ways can you write different three-digit numerals from $\{2, 3, 4, 5\}$?
14. How many different seven-digit telephone numbers can be formed from the set of digits $\{1, 2, 3, 4, 5, 6, 7, 8, 9, 0\}$?
15. In how many ways can you write different three-digit numerals, using $\{2, 3, 4, 5\}$, if no digit is to be used more than once in each numeral?

16. How many different seven-digit telephone numbers can be formed from the set of digits {1, 2, 3, 4, 5, 6, 7, 8, 9, 0} if no digit is to be used more than once in any number?

17. How many three-letter arrangements can be formed from $\{A, N, S, W, E, R\}$?

18. How many different three-letter arrangements can be formed from $\{A, N, S, W, E, R\}$ if no letter is to be used more than once in any arrangement?

19. How many four-digit numerals for positive odd integers can be formed from {1, 2, 3, 4, 5}?

20. How many four-digit numerals for positive even integers can be formed from the set {1, 2, 3, 4, 5}?

21. How many numerals for positive integers less than 500 can be formed from {3, 4, 5}?

22. How many numerals for positive odd integers less than 500 can be formed from {3, 4, 5}?

23. How many numerals for positive even integers less than 500 can be formed from {3, 4, 5}?

24. How many numerals for positive even integers between 400 and 500, inclusive, can be formed from {3, 4, 5}?

25. How many permutations of the elements of $\{P, R, I, M, E\}$ end in a vowel?

26. How many permutations of the elements of $\{P, R, O, D, U, C, T\}$ end in a vowel?

27. Find the number of distinguishable permutations of the letters in the word *LIMIT*.

28. Find the number of distinguishable permutations of the letters in the word

    COMBINATION.

29. Find the number of distinguishable permutations of the letters in the word

    COLORADO.

30. Find the number of distinguishable permutations of the letters in the word

    TALLAHASSEE.

31. Show that $P_{5,3} = 5(P_{4,2})$.   32. Show that $P_{5,r} = 5(P_{4,r-1})$.

33. Show that $P_{n,3} = n(P_{n-1,2})$.

34. Show that $P_{n,3} - P_{n,2} = (n-3)(P_{n,2})$.

35. Solve for n: $P_{n,5} = 5(P_{n,4})$.   36. Solve for n: $P_{n,5} = 9(P_{n-1,4})$.

Example   In how many ways can four students be seated around a circular table?

Solution   In any such arrangement (which is called a **circular permutation**), there is no first position. Each person can take four different initial positions without affecting the arrangement. Thus, there are

$$\frac{4!}{4} = 6 \quad \text{arrangements.}$$

(In general, there are $n!/n$, or $(n-1)!$, circular permutations of n things taken n at a time).

37. In how many ways can five students be seated around a circular table?

38. In how many ways can six students be seated around a circular table?

39. In how many ways can six students be seated around a circular table if a certain two must be seated together?

40. In how many ways can three different keys be arranged on a key ring? *Hint:* Arrangements should be considered identical if one can be obtained from the other by turning the ring over. In general, there are only $(1/2)(n-1)!$ distinct arrangements of n keys on a ring $(n \geq 3)$.

9.6   *Combinations*

An additional counting concept of importance in many areas of mathematics is that of finding the number of distinct r-element subsets of an n-element set with no reference to relative order of the elements in the subset. For example, five different playing cards can be arranged in 5! permutations, but to a poker player they represent the same hand. The set of five cards (with no reference to the arrangement of the cards) is called a *combination*.

Definition 9.6   *A subset of an n-element set A is called a **combination**.*

The counting of combinations is related to the counting of permutations. From Theorem 9.6, we know that the number of distinct permutations of n elements of a set A taken r at a time is given by

$$P_{n,r} = \frac{n!}{(n-r)!}.$$

With this in mind, consider the following result concerning the number $\binom{n}{r}^*$ that was introduced in Section 9.4.

Theorem 9.7   *The number $\binom{n}{r}$ of distinct combinations of the members, taken r at a time, of a set A containing n members is given by*

$$\binom{n}{r} = \frac{P_{n,r}}{r!}. \tag{1}$$

Proof   There are, by definition, $\binom{n}{r}$ r-element subsets of the set A, where $n(A) = n$. Also, from Theorem 9.5, each of these subsets has $r!$ permutations of its members. There are therefore $\binom{n}{r} r!$ permutations of n elements of A taken r at a time. Thus

---

* The symbol $C_{n,r}$ is sometimes used to represent this number.

$$P_{n,r} = \binom{n}{r} r!,$$

from which we obtain

$$\binom{n}{r} = \frac{P_{n,r}}{r!},$$

as was to be shown

Thus, to find the number of r-element subsets of an n-element set A, we count the number of permutations of the elements of A taken r at a time, and then divide by the number of possible permutations of an r-element set. This seems very much like counting a set of people by counting the number of arms and legs and dividing the result by 4, but this approach gives us a very useful expression for the number we seek, $\binom{n}{r}$. Since

$$P_{n,r} = n(n-1)(n-2)\cdots(n-r+1),$$

it follows that

$$\binom{n}{r} = \frac{P_{n,r}}{r!} = \frac{n(n-1)(n-2)\cdots(n-r+1)}{r!}. \tag{2}$$

In how many ways can a committee of five be selected from a set of twelve persons?

Example

What we wish here is the number of 5-element subsets of a 12-element set. From (2), we have

Solution

$$\binom{12}{5} = \frac{12 \cdot 11 \cdot 10 \cdot 9 \cdot 8}{5 \cdot 4 \cdot 3 \cdot 2 \cdot 1} = 792.$$

By Equation (3) on page 247, we have the alternative expression

$$\binom{n}{r} = \frac{P_{n,r}}{r!} = \frac{n!}{r!\,(n-r)!}. \tag{3}$$

Since the numbers $\binom{n}{r}$ are the coefficients in the binomial expansion, and since these coefficients are symmetric, we have the following plausible assertion.

Theorem 9.8 $\binom{n}{r} = \binom{n}{n-r}.$

Proof   From (3), we have

$$\binom{n}{r} = \frac{n!}{r!\,(n-r)!}$$

and

$$\binom{n}{n-r} = \frac{n!}{(n-r)!\,[n-(n-r)]!} = \frac{n!}{(n-r)!\,r!},$$

and the theorem is proved.

Theorem 9.8 is plausible also since each time a distinct set of r objects is chosen, a distinct set of $n - r$ objects remains unchosen.

Exercise 9.6

Example

How many different amounts of money can be formed from a penny, a nickel, a dime, and a quarter?

Solution

We want to find the total number of combinations that can be formed by taking the coins 1, 2, 3, and 4 at a time. By (3) we have,

$$\binom{4}{1} = \frac{4!}{1!\,3!} = 4, \quad \binom{4}{2} = \frac{4!}{2!\,2!} = 6, \quad \binom{4}{3} = \frac{4!}{3!\,1!} = 4, \quad \binom{4}{4} = \frac{4!}{4!\,0!} = 1,$$

and the total number of combinations is 15. Clearly each combination gives a different amount.

1. How many different amounts of money can be formed from a penny, a nickel, and a dime?

2. How many different amounts of money can be formed from a penny, a nickel, a dime, a quarter, and a half-dollar?

3. How many different committees of four persons each can be chosen from a group of six persons?

4. How many different committees of four persons each can be chosen from a group of ten persons?

5. In how many different ways can a set of five cards be selected from a standard bridge deck containing 52 cards?

6. In how many different ways can a set of 13 cards be selected from a standard bridge deck of 52 cards?

7. In how many different ways can a hand consisting of five spades, five hearts, and three diamonds be selected from a standard bridge deck of 52 cards?

8. In how many different ways can a hand consisting of ten spades, one heart, one diamond, and one club be selected from a standard bridge deck of 52 cards?

9. In how many different ways can a hand consisting of either five spades, five hearts, five diamonds, or five clubs be selected from a standard bridge deck?

10. In how many different ways can a hand consisting of three aces and two cards that are not aces be selected from a standard bridge deck?

11. A combination of three balls is picked at random from a box containing five red, four white, and three blue balls. In how many ways can the set chosen contain at least one white ball?

12. In Exercise 11, in how many ways can the set chosen contain at least one white and one blue ball?

13. A set of five distinct points lies on a circle. How many inscribed triangles can be drawn having all their vertices in this set?

14. A set of ten distinct points lies on a circle. How many inscribed quadrilaterals can be drawn having all their vertices in this set?

15. A set of ten distinct points lies on a circle. How many inscribed hexagons can be drawn having all their vertices in this set?

16. Given $\binom{n}{3} = \binom{50}{47}$, find n.

Chapter Review

[9.1] *By mathematical induction, prove the given formula for all positive integral values of n.*

1. $3 + 6 + 9 + \cdots + 3n = \dfrac{3n(n+1)}{2}$

2. $\dfrac{1}{2} + \dfrac{1}{4} + \dfrac{1}{8} + \cdots + \dfrac{1}{2^n} = 1 - \dfrac{1}{2^n}$

[9.2] *Write the next three terms of the given arithmetic progression.*

3. $7, 10, \ldots$ **4.** $a, a - 2, \ldots$

Write the next three terms in the given geometric progression.

5. $-2, 6, \ldots$ **6.** $\dfrac{2}{3}, 1, \ldots$

7. Find the general term and the seventh term of the arithmetic progression $-3, 2, \ldots$.

8. Find the general term and the fifth term of the geometric progression $-2, \dfrac{2}{3}, \ldots$.

9. If the fourth term of an arithmetic progression is 13 and the ninth term is 33, find the seventh term.

10. Which term in a geometric progression $-\dfrac{2}{9}, \dfrac{2}{3}, \ldots$ is 54?

[9.3] **11.** Write $\displaystyle\sum_{k=2}^{5} k(k-1)$ in expanded form.

12. Write $x^2 + x^3 + x^4 + \cdots$ in sigma notation.

13. Find the value for $\displaystyle\sum_{j=3}^{9} (3j - 1)$.

14. Find the value for $\displaystyle\sum_{j=1}^{5} \left(\dfrac{1}{3}\right)^j$.

[9.4] **15.** Write $n(n-3)!$ in expanded form for $n = 5$.

Write the given expression in expanded form and simplify.

16. $\dfrac{8!\,3!}{7!}$ **17.** $\dbinom{7}{2}$ **18.** $\dfrac{(n-1)!}{n!\,(n+1)!}$

19. Write the first four terms of the binomial expansion of $(x - 2y)^{10}$.

20. Find the eighth term in the expansion of $(x - 2y)^{10}$.

[9.5] **21.** How many different two-digit numerals can be formed from $\{6, 7, 8, 9\}$?

22. How many different ways can six questions on a true-false test be answered?

23. How many four-digit numerals for positive odd integers can be formed from $\{3, 4, 5, 6,\}$?

24. How many distinguishable permutations can be formed using the letters in the word *TENNIS*?

[9.6] **25.** How many different committees of 5 persons can be formed from a group of 12 persons?

26. In how many different ways can a hand consisting of 3 spades, 5 hearts, 4 diamonds, and 1 club be selected from a standard bridge deck of 52 cards?

27. A box contains 4 red, 6 white, and 2 blue marbles. In how many ways can you select 3 marbles from the box if at least one of the marbles chosen is white?

28. How many hexagons can be drawn whose vertices are members of a set of 9 fixed points on the circle?

10

Analytic Geometry in Three-Space

Equations in Three Variables

If each ordered pair of real numbers (x, y) is itself paired with a real number z, the pairing can be represented by the ordered triple (x, y, z). The set of all such ordered triples is sometimes represented by the symbol R^3. This denotes the Cartesian product of $R \times R$, or R^2, with R, that is, $(R \times R) \times R$.

Solutions of
equations

An equation in three variables, such as

$$x^2 + 2y^2 - 3z - 4 = 0, \tag{1}$$

has **ordered triples** of real numbers as solutions, just as an equation in two variables has ordered pairs of real numbers as solutions. For example, $(1, 0, -1)$ is a solution of (1) because if x, y, and z are replaced with 1, 0, and -1, respectively, the result is

$$1^2 + 2(0)^2 - 3(-1) - 4 = 0,$$
$$1 + 0 + 3 - 4 = 0,$$
$$0 = 0,$$

which is a true statement. On the other hand, $(1, 1, 1)$ is not a solution of (1) because

$$1^2 + 2(1)^2 - 3(1) - 4 \neq 0.$$

Example

Find a second solution in R^3 of (1) above.

Solution

Select any arbitrary real-number replacements for *any two* variables, say 2 for x and 1 for y, and determine the value of the third variable, in this case z. Thus,

$$2^2 + 2(1)^2 - 3z - 4 = 0,$$

$$z = \frac{2}{3},$$

and another solution of (1) is $\left(2, 1, \dfrac{2}{3}\right)$.

Example   Find the solution of (1) on page 255 with first and second components zero.

Solution   Substituting 0 for x and 0 for y, we have

$$0 + 0 - 3z - 4 = 0,$$

$$z = -\frac{4}{3},$$

and the solution is $\left(0, 0, -\frac{4}{3}\right)$.

A solution in R^3 of an equation such as $x = 3$ is any ordered triple of the form $(3, y, z)$, where y and z may be any real numbers, just as a solution in R^2 is any ordered pair of the form $(3, y)$, where y may be any real number. Similarly, solutions in R^3 of an equation such as $y = 4$ are of the form $(x, 4, z)$, and solutions in R^3 of an equation such as $z = 5$ are of the form $(x, y, 5)$. Furthermore, solutions in R^3 of an equation such as $x^2 + y^2 = 25$ are ordered triples (x, y, z) such that $x^2 + y^2 = 25$ and z is any real number.

R^3 and geometric space

Each member of R^3 can be paired with a point in space by using an extension of a standard Cartesian coordinate system of the plane. If each of three mutually perpendicular lines in space intersecting at a point is scaled (or coordinatized) with origin at the point of intersection, each line then becomes an **axis** of a three-dimensional coordinate system, as suggested by Figure 10.1. The planes determined by the axes taken in pairs are called **coordinate planes,** and we identify them by using the letters associated with the two axes they contain; that is, we refer to them as the *xy*-plane, the *xz*-plane, and the *yz*-plane.

The orientation of the axes in Figure 10.1 constitutes a **right-hand system.** Interchanging the *x*- and *y*-axes produces a **left-hand system.** In this text we shall use only a right-hand system.

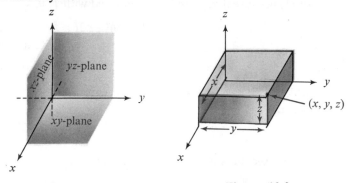

Figure 10.1   *Figure 10.2*

By associating each component of an ordered triple (x, y, z) with the directed (perpendicular) distance from a coordinate plane to a point in space, a one-to-one correspondence can be established between R^3 and the set of all points in space. One way to visualize the pairing of an ordered triple (x, y, z) with its graph in space is suggested by Figure 10.2. The rectangular prism with one vertex at the origin, as shown, will have the point paired with (x, y, z) as vertex opposite the origin. For example, the graphs of $(2, 3, 5)$, $(3, 4, -2)$, and $(1, -5, 2)$ are shown in Figure 10.3. Of course it is not necessary to sketch the entire rectangular prism in order to locate a point.

The coordinate planes separate space into eight regions called **octants.** The region in which all the coordinates of each point are positive numbers is called the **first octant.** The remaining octants are not ordinarily assigned numbers.

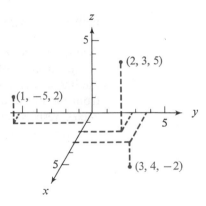

Figure 10.3

Distance between points in space

The formula for the distance between two points in a plane (see page 36) has a natural extension to three dimensions. Figure 10.4 shows points $P(x_1, y_1, z_1)$ and $Q(x_2, y_2, z_2)$ as opposite vertices of a rectangular prism with faces *parallel* to the coordinate planes. Vertices R and S in the bottom face are also shown, together with their coordinates. Since P, R, and S are coplanar points in a plane parallel to the xy-plane, the distance from P to S is given by

$$d_1 = \sqrt{(x_2 - x_1)^2 + (y_2 - y_1)^2}.$$

Then, because P, Q, and S are coplanar points in a plane parallel to the z-axis, we have

$$d_2 = |z_2 - z_1|.$$

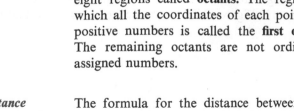

From the Pythagorean theorem, we obtain

$$d^2 = d_1^2 + d_2^2,$$

$$d^2 = (x_2 - x_1)^2 + (y_2 - y_1)^2 + (z_2 - z_1)^2,$$

Figure 10.4

from which

$$d = \sqrt{(x_2 - x_1)^2 + (y_2 - y_1)^2 + (z_2 - z_1)^2}.$$

This is known as the **distance formula** in R^3.

Example

Find the distance from $(6, 3, -6)$ to $(10, 0, 6)$.

Solution

Setting $(x_2, y_2, z_2) = (6, 3, -6)$ and $(x_1, y_1, z_1) = (10, 0, 6)$, we have

$$d = \sqrt{(6 - 10)^2 + (3 - 0)^2 + (-6 - 6)^2}$$

$$= \sqrt{(-4)^2 + 3^2 + (-12)^2}$$

$$= \sqrt{16 + 9 + 144} = \sqrt{169} = 13.$$

Direction angles

Recall from Section 2.2 that we defined the direction of a line in 2-space in terms of the quotient $(y_2 - y_1)/(x_2 - x_1)$, where x_1, y_1, x_2, and y_2 are the coordinates of two points $P_1(x_1, y_1)$ and $P_2(x_2, y_2)$ in the line. This quotient was called the *slope* of the line. Note that the slope of the line is simply the tangent of the angle that the line through P_1 and P_2 forms with the positive x-axis.

The direction of a line in 3-space can also be specified in terms of the angles that the line makes with the coordinate axes. The **direction angles** α, β, and γ of a ray from the origin are those angles formed by the ray and the positive x-, y-, and z-axes, respectively (as shown in Figure 10.5), where the measures of the angles are

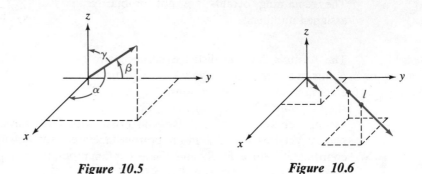

Figure 10.5   Figure 10.6

between $0°$ and $180°$, inclusive. The direction angles of any line in 3-space are the direction angles of a ray from the origin parallel to the line (Figure 10.6).

Direction
cosines

The cosines of the direction angles α, β, and γ of a ray are called the **direction cosines** of the ray.

Example

Find the direction cosines of the ray directed along the positive y-axis.

Solution

The direction angles are

$$\alpha = 90°, \quad \beta = 0°, \quad \text{and} \quad \gamma = 90°.$$

Hence,

$$\cos \alpha = 0, \quad \cos \beta = 1, \quad \text{and} \quad \cos \gamma = 0,$$

and the direction cosines are $0, 1, 0$.

Example

Find two sets of direction cosines of rays parallel to the x-axis.

Solution

One set of direction angles is

$$\alpha = 0°, \quad \beta = 90°, \quad \text{and} \quad \gamma = 90°,$$

with direction cosines $1, 0, 0$. A second set of direction angles is

$$\alpha = 180°, \quad \beta = 90°, \quad \text{and} \quad \gamma = 90°,$$

with direction cosines $-1, 0, 0$.

Exercise 10.1

Find the solution(s) with the given components of the given equation in R³.

Example $x^2 + 2y^2 - z^2 = 4$;   (0, 3, ?)

Solution   Substituting 0 for x and 3 for y yields

$$(0)^2 + 2(3)^2 - z^2 = 4,$$
$$-z^2 = -14,$$
$$z = \sqrt{14} \quad \text{or} \quad z = -\sqrt{14}.$$

Hence the solutions are $(0, 3, \sqrt{14})$ and $(0, 3, -\sqrt{14})$.

1. $x - y + z = 2$;   **a.** (2, 1, ?),   **b.** (-1, ?, 3),   **c.** (?, 2, -2)
2. $x^2 - 3y - z = 4$;   **a.** (1, -1, ?),   **b.** (4, ?, 2),   **c.** (?, -2, 0)
3. $x^2 + 2y^2 + z = 0$;   **a.** (4, 0, ?),   **b.** (0, ?, -4),   **c.** (?, 0, 5)
4. $3x^2 - y^2 + 2z^2 = 2$;   **a.** (0, -1, ?),   **b.** (2, ?, 0),   **c.** (?, 3, 0)

In Exercises 5 and 6, find the value of the constant k if the given ordered triple is a solution of the given equation in R³.

Example $x + 4y - kz = 3$;   (1, 2, 3)

Solution $1 + 4(2) - 3k = 3$; hence $k = 2$.

5. $x - ky + z = 4$;   **a.** (2, -1, 3)   **b.** (0, 1, 0)   **c.** (2, 2, -2)
6. $z = x^2 - ky^2$;   **a.** (4, 1, 7)   **b.** (-1, 2, 3)   **c.** (0, 4, 0)

Find the solutions of the given equation in R³ where

a.   *the first and second components are zero,*
b.   *the first and third components are zero, and*
c.   *the second and third components are zero.*

Example $x^2 + 3y - z = 4$

Solution   a. $(0)^2 + 3(0) - z = 4$; hence $z = -4$, and $(0, 0, -4)$ is a solution.

b. $(0)^2 + 3y - (0) = 4$; hence $y = \dfrac{4}{3}$, and $\left(0, \dfrac{4}{3}, 0\right)$ is a solution.

c. $x^2 + 3(0) - (0) = 4$; hence $x = 2$ or $x = -2$, and $(2, 0, 0)$ and $(-2, 0, 0)$ are solutions.

7. $x + 2y + 3z = 6$ **8.** $x - 3y + 4z = 12$

9. $2x^2 + y^2 + z^2 = 4$ **10.** $x^2 + 3y^2 - z = 12$

11. $x^2 - y - z^2 = 9$ **12.** $4x^2 - 2y^2 - z = 8$

13. $z = 4x^2 - y^2$ **14.** $z = x^2 + 2y^2$

Specify the form of a solution for the given equation in R^3 with the given component.

Example $x^2 + y = 4$,   where $x = 3$

Solution   Substituting 3 for x, we have
$$(3)^2 + y = 4,$$
$$y = -5.$$

Hence a solution in R^3 is of the form $(3, -5, z)$, where z can be any real number.

15. $x^2 - y = 6$, where $x = 2$ **16.** $y^2 + 3z = 12$, where $y = 3$

17. $x^2 - z^2 = 9$, where $z = 2$ **18.** $x^2 + y^2 = 4$, where $x = 0$

19. $y = 3$ **20.** $z = -4$ **21.** $x = -2$ **22.** $y = 0$

Graph the given set of ordered triples.

Example $\{(4, 5, 2), (2, -6, 3), (0, 0, 4)\}$

Solution   The graph is shown at the right.

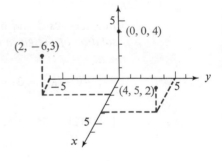

23. $\{(0, 0, 2), (3, 0, 0), (0, 4, 0)\}$

24. $\{(0, 0, -2), (-3, 0, 0), (0, -5, 0)\}$

25. $\{(2, 3, 0), (3, 0, 4), (0, 2, 5)\}$

26. $\{(-3, 1, 0), (4, 0, -2), (0, -3, 1)\}$

27. $\{(4, 3, 4), (7, 6, 1), (1, 5, 3)\}$

28. $\{(3, 3, 6), (2, -3, 3), (2, 4, -3)\}$

29. $\{(-5, 3, 1), (-5, -4, 2), (1, 2, -6)\}$

30. $\{(5, 5, 5), (5, 5, -5), (5, -5, -5)\}$

Find the distance between the points with given coordinates.

Example $(2, -1, 4)$ and $(-3, 4, 5)$

Solution   From the distance formula in R^3, we have
$$d = \sqrt{(x_2 - x_1)^2 + (y_2 - y_1)^2 + (z_2 - z_1)^2}$$
$$= \sqrt{(-3 - 2)^2 + [4 - (-1)]^2 + (5 - 4)^2}$$
$$= \sqrt{25 + 25 + 1} = \sqrt{51}.$$

31. $(4, 2, -1)$ and $(5, -1, 2)$ **32.** $(3, -3, 0)$ and $(0, 2, -1)$

33. $(3, 3, -5)$ and $(1, 4, -2)$ **34.** $(-6, 1, 3)$ and $(-4, 4, 2)$

35. $(0, -2, 4)$ and $(-3, 1, 2)$ **36.** $(-5, 4, 6)$ and $(2, -7, -2)$

37. Find the direction cosines of a ray directed along the negative x-axis.

38. Find the direct cosines of a ray directed along the positive z-axis.

39. Find two sets of direction cosines of rays parallel to the y-axis.

40. Find two sets of direction cosines of rays parallel to the z-axis.

10.2   *Graphs of First-Degree Equations in R³*

Just as the graph of a first-degree equation in two variables

$$Ax + By + C = 0 \quad (A, B, C \in R)$$

is a straight line in R^2, the graph of a first-degree equation in three variables

$$Ax + By + Cz + D = 0 \quad (A, B, C, D \in R)$$

is a plane in R^3.

If the equation $2x + y + z - 6 = 0$ is solved for z in terms of x and y, we obtain

$$z = -2x - y + 6.$$

In this form, the equation apparently serves to pair every ordered pair (x, y) of real numbers with exactly one real number z. This pairing constitutes a function, with domain the set R^2. In set notation, we can represent the function as

$$\{((x, y), z) \mid z = -2x - y + 6\}$$

or

$$\{(x, y, z) \mid z = -2x - y + 6\}. \qquad (1)$$

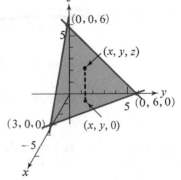

Figure 10.7

Such a function is said to be a function of the two variables x and y.

Figure 10.7 shows the part of the plane that is the graph of the first-degree function of two variables specified in (1) above.

By a method similar to the intercept method used to graph first-degree equations in two variables in R^2, we can obtain the intercepts of a plane in a three-dimensional system by assigning zero to two of the variables and then finding the associated value for the third variable. In the example above, the ordered triples $(3, 0, 0)$, $(0, 6, 0)$, and $(0, 0, 6)$ are solutions of the equation; the respective intercepts of its graph in the x-axis, y-axis, and z-axis are 3, 6, and 6.

Because pictures of surfaces in three dimensions are difficult to draw, we shall focus our attention mainly on **plane sections of surfaces,** which are the curves formed where a plane intersects a surface. In particular, we shall examine sections in planes parallel to the coordinate planes and in the coordinate planes themselves.

Planes parallel to coordinate planes   In R^3, equations of the form

$$x = k, \quad y = k, \quad \text{and} \quad z = k,$$

where k is a constant, determine planes parallel to coordinate planes, as shown in a, b, and c of Figure 10.8. This is simply a reflection of the fact that ordered triples

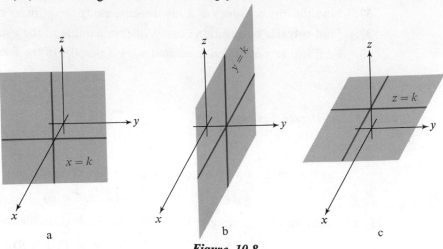

a   b   c

Figure 10.8

of the form

$$(k, y, z), \quad (x, k, z), \quad \text{and} \quad (x, y, k)$$

have graphs located at a fixed distance $|k|$ on one side or the other of one of the coordinate planes, with the side depending on the sign of k.

Exercise 10.2

As suggested by the example on page 261, the graph in R^3 of a first-degree equation, of the form $ax + by + cz = d$, is a plane. Find the intercepts of the graph of the given first-degree equation on the coordinate axes and show the graph in one octant.

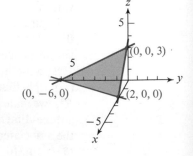

Example $3x - y + 2z = 6$

Solution   The ordered triples corresponding to the points of intersection of the plane and the coordinate axes are shown in the figure.

1. $x + 3y + 2z = 6$
3. $2x + 4y + 3z = 12$
5. $3x - y + 3z = 6$
7. $2x + y - 2z = 8$

2. $x + 4y + 2z = 8$
4. $3x + y + 2z = 6$
6. $2x - 2y + z = 4$
8. $-3x + 2y + z = 6$

Example $2x + y = 6$

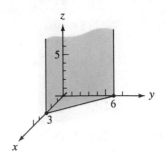

Solution $2x + y = 6$ is equivalent to $2x + y + 0z = 6$.
The intercepts in the x-axis and y-axis are 3 and
6, respectively. Their graphs and the graph of the
equation in R^3 (a plane parallel to the z-axis) are
shown in the figure.

     9. $x + 3y = 6$ **10.** $3x + 4y = 12$
    11. $2x + z = 8$ **12.** $3x + 2z = 12$
    13. $2y + z = 4$ **14.** $2y + 3z = 6$

Graph the given set in R^3.

Example $\{(x, y, z) \mid z = 3\}$

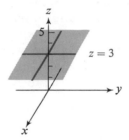

Solution   The graph is shown at the right.

    15. $\{(x, y, z) \mid x = 3\}$
    16. $\{(x, y, z) \mid y = 4\}$
    17. $\{(x, y, z) \mid z = 5\}$
    18. $\{(x, y, z) \mid z = -2\}$

Graph the given equation in R^3.

    19. $y = -6$ **20.** $z = -3$
    21. $x = 5$ **22.** $y = 0$

    23.   Graph $x = 4$
       a.   in R.   **b.**   in R^2.   **c.**   in R^3.
    24.   Consider the equation, $ax + by + cz + d = 0$, of a plane in R^3. State the
       conditions on a, b, and c such that:
       a.   The plane is not parallel to any coordinate plane or axis.
       b.   The plane is parallel to the x-axis; the y-axis; the z-axis.
       c.   The plane is parallel to the xy-plane; the xz-plane; the yz-plane.

It can be shown that the distance, d, of a point $P(x_1, y_1, z_1)$ from a plane
$$ax + by + cz + d = 0, \quad \text{where} \quad a, b, c, d \in R,$$
is given by

$$d = \frac{|ax_1 + by_1 + cz_1 + d|}{\sqrt{a^2 + b^2 + c^2}}.$$

*Use this formula to find the distance from the point with given coordinates to the
corresponding plane with the given equation in Exercises 25–28 on page 264.*

25. $(5, 6, 3)$; $9x + 4y + 5z - 1 = 0$

26. $(2, 1, 8)$; $4x + 6y + 7z + 2 = 0$

27. $(-2, 7, -5)$; $x - 5y - 2z + 6 = 0$

28. $(3, -2, -4)$; $x - 8y + 9z + 3 = 0$

10.3   *Graphs in R^3*

Traces

We can use planes parallel to coordinate planes (see Section 10.2) to help sketch surfaces in space by sketching sections of the surface determined by the planes. In many cases, sections parallel to a particular one of the coordinate planes offer a better picture than sections parallel to either of the other coordinate planes. This suggests that some care should be given to selecting the most appropriate planes to use. Of course, in sketching any surface the first thing to do is draw the sections in the *coordinate planes*. These particular sections are called **traces.** To identify the traces of a surface, we look at the equation of the surface when *each of the variables has, in turn, the value* 0.

For example, to sketch the graph of

$$z = -2x - y + 6$$

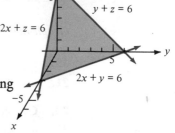

shown in Figure 10.7 on page 261, we can find the traces in the coordinate planes as follows:

In *xy*-plane, set $z = 0$ to obtain $2x + y = 6.$

In *yz*-plane, set $x = 0$ to obtain $y + z = 6.$

In *xz*-plane, set $y = 0$ to obtain $2x + z = 6.$

Then each of these traces can easily be sketched using their intercepts with the coordinate axes.

For example, consider the equation

$$2x^2 + 2y^2 + z^2 = 72. \tag{1}$$

To find an equation for the trace in the *xy*-plane, we set z equal to 0. Thus this trace has equation $2x^2 + 2y^2 = 72$, with graph a circle, as shown in Figure 10.9-a.

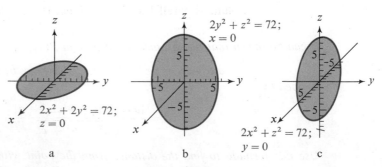

Figure 10.9

To find the trace in the *yz*-plane, we set *x* equal to 0. This trace has equation $2y^2 + z^2 = 72$, with graph an ellipse, as shown in Figure 10.9-b.

To find the trace in the *xz*-plane, we set *y* equal to 0. This trace has equation $2x^2 + z^2 = 72$, with graph an ellipse, as shown in Figure 10.9-c.

Now several additional sections for the surface associated with (1) will suggest the appearance of the entire surface. Rewriting (1) equivalently as

$$2x^2 + 2y^2 = 72 - z^2,$$

we observe that for values $|z| < \sqrt{72}$, the equation has the form

$$x^2 + y^2 = k, \quad k > 0,$$

the graph of which is a circle for all such k. This means that sections which are graphs of such equations can aid us in completing the sketch of the entire surface. First let us take *z* equal to 2 and 6, for example, in turn and obtain the resulting equations and their respective sections as shown in Figure 10.10-a. Similar sections can be obtained below the *xy*-plane for *z* equal to −2 and −6.

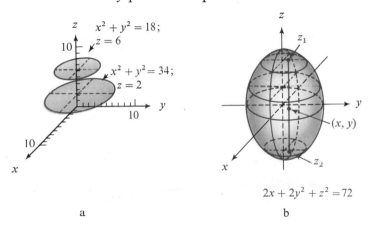

a   b

Figure 10.10

Using all the information available from the traces in the coordinate planes and the sections obtained for selected values of *z*, we can sketch the entire surface, as shown in Figure 10.10-b.

Because the equation $2x^2 + 2y^2 + z^2 = 72$ pairs each ordered pair in $\{(x, y) \mid 2x^2 + 2y^2 < 72\}$ with *more than one value z*, the set of ordered triples determined by the equation is not a function. By solving for *z*, however, we have

$$z = \sqrt{72 - 2x^2 - 2y^2} \quad \text{or} \quad z = -\sqrt{72 - 2x^2 - 2y^2},$$

and either of these equations defines a function with domain

$$\{(x, y) \mid 2x^2 + 2y^2 \le 72\}$$

and with range either

$$\{z \mid 0 \le z \le \sqrt{72}\} \quad \text{or} \quad \{z \mid -\sqrt{72} \le z \le 0\}.$$

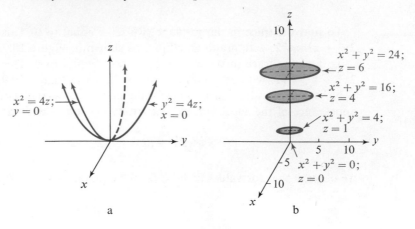

Figure 10.11

Now consider, as a second example, the graph of

$$x^2 + y^2 = 4z.$$

We can first sketch the traces of the surface in the coordinate planes, as shown in Figure 10.11-a. Notice that the trace in the xy-plane, with equation $x^2 + y^2 = 0$, is the point at the origin. By inspection, we observe that the sections that will best delineate the surface are those parallel to (and above) the xy-plane. For $z < 0$, we have $x^2 + y^2 < 0$, which has no solution with real-number components. We can therefore let $z = 1$, $z = 4$, and $z = 6$, for example, in turn and sketch the associated sections.

The equation of the section in the plane $z = 1$ is

$$x^2 + y^2 = 4,$$

the equation for the section in the plane $z = 4$ is

$$x^2 + y^2 = 16,$$

and the equation for the section in the plane $z = 6$ is

$$x^2 + y^2 = 24.$$

The sections are shown in Figure 10.11-b.

Using all the information obtained from the traces in the coordinate planes and sections in the planes corresponding to $z = 1$, $z = 4$, and $z = 6$, we complete the graph of the entire surface, as shown in Figure 10.12.

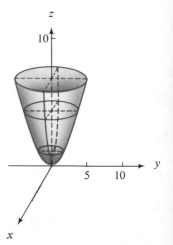

Figure 10.12

Quadric surfaces   Information concerning intercepts, sections of surfaces, and traces can be used to sketch a wide variety of surfaces. Some of these surfaces, their equations, and relationships between coefficients of six types of quadric surface (surfaces given by quadratic equations), are shown below. The ability to recognize the general form of a surface from a particular equation helps to determine which sections would be most helpful in sketching its graph.

In each equation, $a, b, c \in R$ and $a, b, c > 0$:

Ellipsoid

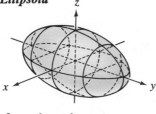

$$\frac{x^2}{a^2} + \frac{y^2}{b^2} + \frac{z^2}{c^2} = 1$$

If $a = b = c$, a sphere.

Elliptic paraboloid

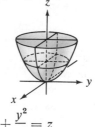

$$\frac{x^2}{a^2} + \frac{y^2}{b^2} = z$$

If $a = b$, a circular paraboloid.

Elliptic hyperboloid of one sheet

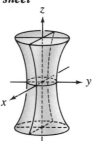

$$\frac{x^2}{a^2} + \frac{y^2}{b^2} - \frac{z^2}{c^2} = 1$$

Elliptic hyperboloid of two sheets

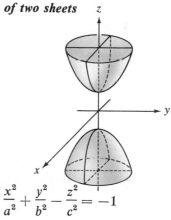

$$\frac{x^2}{a^2} + \frac{y^2}{b^2} - \frac{z^2}{c^2} = -1$$

If $a = b$, circular hyperboloids.

Hyperbolic paraboloid

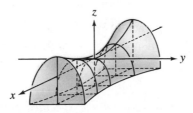

$$\frac{x^2}{a^2} - \frac{y^2}{b^2} = z$$

Elliptic cone

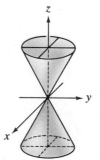

$$\frac{x^2}{a^2} + \frac{y^2}{b^2} = \frac{z^2}{c^2}$$

If $a = b$, a circular cone.

Exercise 10.3

Graph the given set.

Example $\{(x, y, z) \mid x^2 + 4z^2 = 16\}$

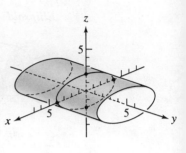

Solution   The defining equation $x^2 + 4z^2 = 16$ can be written equivalently as $x^2 + 0y^2 + 4z^2 = 16$. We first graph the trace in the xz-plane. Since y can take any value, each section parallel to the xz-plane has the same graph as the trace in that plane. The surface is shown in the figure and is called an **elliptic cylinder**.

1. $\{(x, y, z) \mid x^2 - z = 0\}$ 2. $\{(x, y, z) \mid x^2 + y^2 = 25\}$
3. $\{(x, y, z) \mid 4y^2 + z^2 = 16\}$ 4. $\{(x, y, z) \mid 4x^2 - y = 0\}$

Graph the solution set in R^3 of the given equation.

5. $4x^2 + y^2 = 16$ 6. $x^2 + z^2 = 16$
7. $y^2 + z = 4$ 8. $x^2 - 4y^2 = 16$

Graph the trace on the specified coordinate plane for the given equation if such trace exists.

Example $4x^2 - y^2 - z^2 = -16$; xy-plane

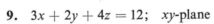

Solution   On the xy-plane, $z = 0$.

For $z = 0$,

$4x^2 - y^2 = -16$,

$y^2 - 4x^2 = 16$.

9. $3x + 2y + 4z = 12$; xy-plane   10. $x - 4y + 2z = 8$; yz-plane
11. $x^2 + 2y^2 + 4z^2 = 16$; xz-plane   12. $x^2 + y + 2z^2 = 4$; yz-plane
13. $4x^2 - y^2 + z = 36$; xy-plane   14. $x^2 + 3y^2 - z^2 = 16$; xz-plane
15. $4x^2 - y^2 - 2z^2 = 4$; yz-plane   16. $-x + y - 6z^2 = 12$; xz-plane

Graph the section of the given surface on the specified plane if such section exists.

Example $y = x^2 + z^2$; $y = 4$

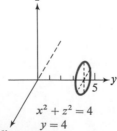

Solution   For $y = 4$,

$x^2 + z^2 = 4$.

17. $4x + 2y + 3z = 13$; $z = 3$
18. $x + 3y + 2z = -6$; $y = -4$
19. $x^2 + y^2 + z^2 = 25$; $x = 3$
20. $x^2 + 4y^2 - 2z^2 = 16$; $z = 3$
21. $x - 4y^2 + z^2 = 0$; $y = 1$
22. $3x^2 + y^2 - z = 4$; $x = 2$
23. $4x^2 - 9y^2 - 2z^2 = 28$; $z = 2$
24. $2x^2 + y - z^2 = -12$; $y = -4$

Name the type of quadric surface represented by the given equation.

Example $x^2 + 4y^2 - 8z = 0$

Solution   Rewriting the equation equivalently as

$$z = \frac{x^2}{8} + \frac{y^2}{2},$$

and comparing this form with the equation $\frac{x^2}{a^2} + \frac{y^2}{b^2} = z$, we observe that its graph is an elliptic paraboloid.

25. $z = 4y^2 + 4x^2$
26. $4z = 3x^2 + 12y^2$
27. $x^2 + y^2 + z^2 = 1$
28. $2x^2 + 3y^2 + 4z^2 = 24$
29. $x^2 + 4y^2 - 3z^2 = -2$
30. $x^2 - 4y^2 + z^2 = 3$

Using the appropriate traces and sections, graph the given equation in R^3.

31. $x^2 + y^2 + z^2 = 9$
32. $2x^2 + 4y^2 + z^2 = 16$
33. $x^2 + 9y^2 + 9z^2 = 36$
34. $x^2 - 2y^2 + z^2 = 16$
35. $x^2 \quad y \mid z^2 = 0$
36. $1x^2 \mid 2y^2 \quad z = 8$
37. $4x^2 + y^2 - z^2 = 0$
38. $4x^2 + y^2 - z^2 = 4$

Chapter Review

[10.1] *Find the solutions with the given components for $2x^2 - 3y + z^2 = 18$.*

1. $(0, 0, ?)$
2. $(0, ?, 0)$
3. $(?, 0, 0)$
4. $(1, 1, ?)$

Find the value of the constant k if the given ordered triple is a solution of the given equation.

5. $3x + ky - z = 2$; $(1, 1, 5)$
6. $2x^2 + y^2 - kz = 3$; $(0, 3, -1)$
7. Specify the form of a solution in R^3 for $4x^2 - y = 16$, where $x = 1$.
8. Specify the form of a solution in R^3 for $y^2 + 3z^2 = 12$, where $y = 0$.
9. Graph the set of ordered triples $\{(0, 0, 4), (0, -2, 0), (5, 0, 0)\}$.
10. Graph the set of ordered triples $\{(0, 2, 5), (3, 0, 4), (2, 3, 0)\}$.

Find the distance between the points with the given coordinates.

11. $(5, 1, -2)$ and $(6, -3, 1)$ **12.** $(2, -4, 0)$ and $(-1, 6, 2)$

[10.2] *Find the intercepts of the graph of the given first-degree equation on the coordinate axes and show the graph of the plane in one octant.*

13. $2x - y + z = 4$ **14.** $x + 3y + 3z = 6$

Graph the solution set of the first-degree equation in R^3.

15. $y = 3$ **16.** $z = 4$
17. $x + 2y = 6$ **18.** $6x - 3y = 12$

[10.3] *Graph the solution set of the second-degree equation in R^3.*

19. $y^2 - z = 4$ **20.** $x^2 + y^2 = 9$

21. Graph the trace of $4x^2 + y^2 - z^2 = 16$ on the yz-plane.
22. Graph the section of the surface $x^2 + 2y^2 - z^2 = 9$ on the plane $y = 2$.

Name the type of quadric surface represented by the given equation.

23. $2y = 5x^2 + 6z^2$ **24.** $x^2 + y^2 + z^2 = 10$
25. $3x^2 + y^2 - 2z^2 = 1$ **26.** $x^2 - y^2 + 3z^2 = -2$

Graph the given equation.

27. $x^2 + 4y^2 + 4z^2 = 16$ **28.** $2x^2 + 4y^2 - z = 4$

Appendix—Additional Topics of Analytic Geometry

A.1   *Parametric Representations of Relations*

Relations are sometimes defined by systems of equations of the form

$$x = g(t),$$
$$y = h(t),$$

where the variables representing elements in the domain and range, in this case x and y, respectively, are each related to a third variable, in this case t. Such equations are called **parametric equations** of the relation, and the variable t is called a **parameter.** For any arbitrary value of t in the common domain of g and h, the values for x and y are determined. For example, we can obtain some ordered pairs (x, y) and the graph of the relation defined by

$$x = t \quad \text{and} \quad y = 1 - t$$

(Figure A.1.1) by assigning some values to t, as shown in the table below, and obtaining the corresponding values for x and y.

t	$x = t$	$y = 1 - t$	(x, y)
-3	-3	4	$(-3, 4)$
-2	-2	3	$(-2, 3)$
-1	-1	2	$(-1, 2)$
0	0	1	$(0, 1)$
1	1	0	$(1, 0)$
2	2	-1	$(2, -1)$
3	3	-2	$(3, -2)$

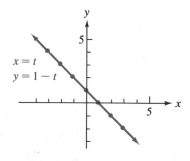

Figure A.1.1

Parametric to Cartesian form   Sometimes a Cartesian equation for a locus can be obtained if we are given a set of parametric equations. Thus, if $x = t$ and $y = 1 - t$, as in the foregoing example, we can eliminate the parameter by substituting x for t in $y = 1 - t$, to obtain

$$y = 1 - x \quad \text{or} \quad x + y = 1,$$

the Cartesian form of a linear equation.

Eliminating a parameter does not always yield an equivalent solution set for parametric equations. For example, if $x = \cos t$ and $y = \cos 2t$, we can eliminate the parameter by using the trigonometric identity

$$\cos 2t = 2\cos^2 t - 1$$

and writing

$$y = \cos 2t = 2\cos^2 t - 1,$$

from which, because $x = \cos t$, we have

$$y = 2x^2 - 1.$$

Note that this is the equation for a parabola with graph as shown in Figure A.1.2. However, because in the original parametric equations $x = \cos t$, we know that $|x| \leq 1$, and since $y = \cos 2t$, we know that $|y| \leq 1$. Hence, without additional restrictions to the domain and range, the Cartesian equation does not have the same graph as the parametric equations. The graph of the parametric equations is that portion shown in color on the figure.

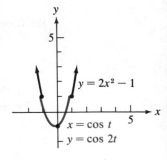

Figure A.1.2

Cartesian to parametric form

You can always obtain one set of parametric equations for a curve (or parts of a curve) having a given Cartesian equation by using $y = tx$, where t is a parameter.

Example

Find a set of parametric equations for the curve with equation $x^2 + y^2 - 4x = 0$.

Solution

Let $y = tx$, and replace y in the Cartesian equation with tx.

$$x^2 + (tx)^2 - 4x = 0$$
$$x^2 + t^2x^2 - 4x = 0$$
$$(1 + t^2)x^2 - 4x = 0$$
$$x[(1 + t^2)x - 4] = 0$$

We thus have either $x = 0$ or $(1 + t^2)x - 4 = 0$.

If $x = 0$, then $y = tx = 0$.

If $(1 + t^2)x - 4 = 0$, then

$$x = \frac{4}{1 + t^2}.$$

Since

$$y = tx = t\left(\frac{4}{1 + t^2}\right) = \frac{4t}{1 + t^2},$$

we have

$$x = \frac{4}{1+t^2} \quad \text{and} \quad y = \frac{4t}{1+t^2}$$

as parametric equations for the Cartesian equation $x^2 + y^2 - 4x = 0$.

Notice in the foregoing equation, that the origin $(0, 0)$, which is on the graph of $x^2 + y^2 - 4x = 0$, is not on the graph of the parametric equations—that is, there is no value of t for which $x = 0$.

Exercise A.1

  a.   *Sketch the graph of the given parametric equations.*

  b.   *Eliminate the parameter and give a Cartesian equation for the graph.*

Example $x = t^2 - 3, \quad y = t + 1$

Solution   a.   Since we need ordered pairs (x, y) in order to determine the graph, we can begin by making a table of values using appropriate replacements for t. In this case we shall construct the table in a horizontal form.

t	-4	-3	-2	-1	0	1	2	3	4
x	13	6	1	-2	-3	-2	1	6	13
y	-3	-2	-1	0	1	2	3	4	5

We can thus locate the points associated with the ordered pairs (x, y), connect them in order of increasing t, and sketch the graph. (It appears to be a parabola.)

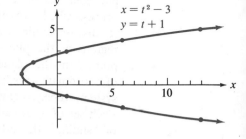

  b.   We can eliminate the parameter t from the original equations by first writing the equation $y = t + 1$ equivalently as

$$t = y - 1$$

and then substituting $y - 1$ for t in the equation $x = t^2 - 3$.

$$x = (y - 1)^2 - 3$$
$$x = y^2 - 2y + 1 - 3$$
$$x = y^2 - 2y - 2$$

From this Cartesian form and our earlier work, we can now assert that the graph is indeed a parabola.

1. $x = 2 - 3t$; $y = 3t - 1$
2. $x = 4t + 6$; $y = 2t - 3$
3. $x = 3t - 2$; $y = 3 - 2t$
4. $x = t + 2$; $y = 2t - 3$
5. $x = t^2$; $y = 1 - t$
6. $x = t + 1$; $y = t^2 - 1$
7. $x = 2 + \dfrac{1}{t}$; $y = \dfrac{1}{t} + 1$
8. $x = \dfrac{2}{t} + 1$; $y = \dfrac{3}{t} - 1$
9. $x = \sin t$; $y = \cos^2 t$
10. $x = \cos 2t - 1$; $y = \sin t$
11. $x = \cos \alpha$; $y = \sin \alpha$
12. $x = 2 + \cos t$; $y = \cos 2t$
13. $x = 1 - \cos t$; $y = 1 + \sin t$
14. $x = 2 \cos \alpha$; $y = 2 \sin \alpha$
15. $x = 4 \cos \alpha$; $y = 3 \sin \alpha$
16. $x = 2 \cos \alpha$; $y = 5 \sin \alpha$

Use the relationship $y = tx$ to obtain parametric equations for the graph of the given Cartesian equation.

17. $3x - y = 4$
18. $x^2 - y^2 = 9$
19. $2x + 5y = -1$
20. $x^2 + 4y^2 = 100$
21. $x^2 + y^2 = 16$
22. $y = 2x^2 - 3x + 1$

A.2   Translation of Axes

The equations of the circles, parabolas, ellipses, and hyperbolas discussed in Chapter 3 were developed for the special case (except for the parabola) where the central point, or point about which the curve is symmetric, is located at the origin. (The parabola has no central point.) By using a process called **translation of axes,** it is possible to discuss conic sections located in other regions of the plane. Figure A.2.1 shows a point P in the plane together with two sets of coordinate axes, an xy-system and an $x'y'$-system, whose corresponding axes are parallel. Each set of axes can be viewed as the result of "sliding" the other set across the plane while the axes remain parallel to their original position, suggesting the terminology "translation of axes."

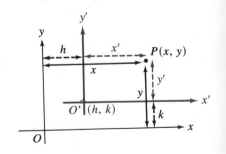

Figure A.2.1

If the origin O' of the $x'y'$-system shown in Figure A.2.1 corresponds to the point (h, k) of the xy-system, then the $x'y'$-coordinates of the point P in the plane are related to the xy-coordinates of P by the equations

$$x = x' + h,$$
$$y = y' + k. \tag{1}$$

These equations can be solved for x' and y' to obtain the equivalent equations

$$x' = x - h,$$
$$y' = y - k. \tag{2}$$

Example

The origin in an xy-system of coordinates is translated into the point $(4, -3)$, which becomes the origin in an $x'y'$-system. Under this translation,
(a) what are the $x'y'$-coordinates of the point whose xy-coordinates are $(5, 8)$, and
(b) what is the $x'y'$-equation for $x + 5y = 3$?

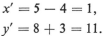

Solution

(a) Using Equations (2) with $h = 4$ and $k = -3$, we have

$$x' = x - 4,$$
$$y' = y + 3,$$

from which, for $(x, y) = (5, 8)$,

$$x' = 5 - 4 = 1,$$
$$y' = 8 + 3 = 11.$$

Therefore, the $x'y'$-coordinates are $(1, 11)$

(b) Using Equations (1), with $h - 4$ and $k - -3$, we have

$$x = x' + 4,$$
$$y = y' - 3.$$

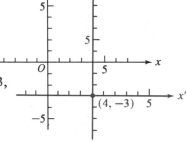

Replacing x and y in $x + 5y = 3$ with the right-hand members of these latter equations we have

$$(x' + 4) + 5(y' - 3) = 3,$$

from which we obtain the desired equation,

$$x' + 5y' = 14.$$

Now consider a parabola with vertex at the point $P(h, k)$ in an xy-coordinate system and with its axis parallel to the x-axis, as shown in Figure A.2.2. If an $x'y'$-coordinate system had origin at P, then the $x'y'$-equation of the parabola would have the form

$$y'^2 = 2px'.$$

As before, the constant $|p|$ represents the distance between the focus and the directrix. By Equation (2), the xy-equation for the parabola would then have the form

$$(y - k)^2 = 2p(x - h). \tag{3}$$

Since in Figure A.2.2, $p > 0$, the coordinates of the focus must then be $(p/2 + h, k)$, while an equation for the directrix is $x = -p/2 + h$.

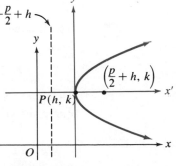

Figure A.2.2

Similar considerations lead to the conclusions given in Table A.1 regarding circles, ellipses, and hyperbolas with center at the point (h, k) in the xy-plane. The data in the table can be used to help find an equation when the requisite information about a graph is given.

Table A.1

Curve	x, y-equation	Major axis parallel to	Foci	Vertices
Circle	$(x - h)^2 + (y - k)^2 = r^2$	—	—	—
Ellipse $c^2 = a^2 - b^2$	$\dfrac{(x - h)^2}{a^2} + \dfrac{(y - k)^2}{b^2} = 1$	x-axis	$(h \pm c, k)$	$(h \pm a, k)$
	$\dfrac{(x - h)^2}{b^2} + \dfrac{(y - k)^2}{a^2} = 1$	y-axis	$(h, k \pm c)$	$(h, k \pm a)$
Hyperbola $c^2 = a^2 + b^2$	$\dfrac{(x - h)^2}{a^2} - \dfrac{(y - k)^2}{b^2} = 1$	x-axis	$(h \pm c, k)$	$(h \pm a, k)$
	$\dfrac{(y - k)^2}{a^2} - \dfrac{(x - h)^2}{b^2} = 1$	y-axis	$(h, k \pm c)$	$(h, k \pm a)$

Example   Find an equation for the ellipse with major axis of length 12 and foci at $(-1, 1)$ and $(-1, 5)$. Sketch the curve.

Solution   The equal x-coordinates of the foci tell us that the major axis of the ellipse is parallel to the y-axis. Since the center is the midpoint of the segment with the foci as endpoints, the coordinates of the center are

$$\left(-1, \frac{5 + 1}{2}\right) = (-1, 3),$$

and we see that c, the distance from the center to each focus, is 2. Because the major axis has length 12, it follows that a, the distance from the center to each vertex, is 6. Hence, $a^2 = 36$. From the fact that $b^2 = a^2 - c^2$, we have

$$b^2 = 36 - 4 = 32$$

The desired equation then is

$$\frac{(x + 1)^2}{32} + \frac{(y - 3)^2}{36} = 1.$$

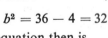

Using $a = 6$ and $b = 4\sqrt{2}$, and starting with the center at $(-1, 3)$, we sketch the curve shown at the right.

We can use the equations of translation (2) together with the process of completing the square to identify and sketch the graph of an equation of the form

$$Ax^2 + By^2 + Dx + Ey + F = 0 \quad (A^2 + B^2 \neq 0). \tag{4}$$

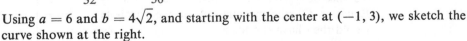

We do this by first completing the square in x and y in Equation (4), to obtain an equation of the form

$$A(x - h)^2 + B(y - k)^2 = C.$$

Then using Equations (2) and substituting x' for $x - h$ and y' for $y - k$, we transform this to

$$Ax'^2 + By'^2 = C.$$

If A, B, and $C \neq 0$, this equation has as its graph a central conic, as detailed in Section 3.2. In standard form, we would have

$$\frac{x'^2}{C/A} + \frac{y'^2}{C/B} = 1.$$

Example

Find an equation in standard form in the $x'y'$-system, for the graph of

$$8x^2 + 4y^2 + 24x - 4y + 1 = 0.$$

Solution

We first rewrite the given equation equivalently as

$$8(x^2 + 3x \qquad) + 4(y^2 - y \qquad) = -1$$

and complete the square in x and y to obtain

$$8\left(x^2 + 3x + \frac{9}{4}\right) + 4\left(y^2 - y + \frac{1}{4}\right) = -1 + 18 + 1,$$

$$8\left(x + \frac{3}{2}\right)^2 + 4\left(y - \frac{1}{2}\right)^2 = 18.$$

If now we take $h = -3/2$ and $k = 1/2$ in Equations (1), we find upon substituting $x' - 3/2$ for x and $y' + 1/2$ for y that this latter equation is equivalent to

$$4x'^2 + 2y'^2 = 9.$$

In standard form, we have

$$\frac{x'^2}{9/4} + \frac{y'^2}{9/2} = 1,$$

the equation of an ellipse, with $a = 3/2$ and $b = 3/\sqrt{2}$.

It is sometimes convenient to use an auxiliary coordinate system as an aid in sketching graphs. For example, to graph

$$8x^2 + 4y^2 + 24x - 4y + 1 = 0$$

in the foregoing example, we can use the $x'y'$-coordinate system with origin at $(-3/2, 1/2)$ in the xy-system, and then graph

$$\frac{x'^2}{9/4} + \frac{y'^2}{9/2} = 1.$$

First, we sketch the $x'y'$-axes, and identify the x' and y' intercepts, $\pm 3/2$ and $\pm 3/\sqrt{2}$, and then we complete the graph as shown in Figure A.2.3.

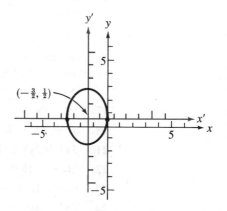

Figure A.2.3

Exercise A.2

For Exercises 1–8, assume that the origin in an xy-system of coordinates is translated into the first point given, which then becomes the origin of an x′y′-system.

1. $(4, 4)$; find the $x'y'$-coordinates of the point whose xy-coordinates are $(-3, 1)$.

2. $(-3, 6)$; find the $x'y'$-coordinates of the point whose xy-coordinates are $(7, 0)$.

3. $(-5, -2)$; find the $x'y'$-coordinates of the point whose xy-coordinates are $(0, -3)$.

4. $(7, 0)$; find the $x'y'$-coordinates of the point whose xy-coordinates are $(0, 7)$.

5. $(3, 4)$; find the $x'y'$-equation for $x - 2y = 8$.

6. $(-1, 6)$; find the $x'y'$-equation for $3x + y = 1$.

7. $(2, -5)$; find the $x'y'$-equation for $x^2 - 2y^2 = 6$.

8. $(-6, -1)$; find the $x'y'$-equation for $x^2 + 2y = -2$.

Find an equation in the xy-system for each curve. Sketch the curve.

9. Circle with center at $(-5, 1)$ and radius 3

10. Ellipse with major axis of length 10 and foci at $(3, 7)$ and $(3, -1)$

11. Parabola with vertex at $(3, 3)$ and directrix with equation $y = 1$

12. Hyperbola with foci at $(-1, 2)$ and $(-7, 2)$ and a vertex at $(-3, 2)$

13. Ellipse with vertices at $(3, 2)$ and $(-7, 2)$ and length of minor axis 6

14. Parabola with focus at $(3, 2)$ and vertex at $(3, 4)$

15. Hyperbola with center at $(3, 2)$, one focus at $(8, 2)$, and one vertex at $(0, 2)$

16. Hyperbola with center at $(-1, 5)$, length of major axis 8, and length of minor axis 6

Find an equation in the x′y′-system and in standard form for the graph of the given equation in x and y.

17. $y^2 - 4y = 12x - 52$ 18. $x^2 + 4x + 8y = 4$

19. $x^2 - 6x + 3 = -y^2 - y$ 20. $x^2 + y^2 = 6x - y + 3$

21. $3x^2 + 2y^2 + 12x - 4y = -2$

22. $16x^2 + 25y^2 - 32x + 100y - 284 = 0$

23. $9x^2 - 16y^2 - 90x + 64y + 17 = 0$

24. $9x^2 - 4y^2 = -61 - 54x - 16y$

25–32. Use an auxiliary $x'y'$-system to graph the equation in Exercises 17–24.

A.3   ***Rotation of Axes***

Another useful transformation of axes is that of rotation. If we visualize all points on the plane as remaining fixed while we rigidly rotate the x- and y-coordinate axes through an angle α into a new set of x'- and y'-coordinate axes, then we have the situation depicted in Figure A.3.1. Thus, a point $P(x, y)$ with polar coordinates

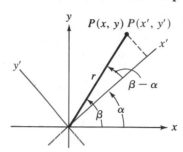

Figure A.3.1

$$x = r \cos \beta \quad \text{and} \quad y = r \sin \beta$$

relative to the xy-coordinate axes, will have polar coordinates

$$x' = r \cos (\beta - \alpha)$$

and   (1)

$$y' = r \sin (\beta - \alpha)$$

relative to its $x'y'$-coordinate axes. Using the identities

$$\cos (\beta - \alpha) = \cos \beta \cos \alpha + \sin \beta \sin \alpha$$

and

$$\sin (\beta - \alpha) = \sin \beta \cos \alpha - \cos \beta \sin \alpha,$$

we can rewrite (1) equivalently as

$$x' = r \cos \beta \cos \alpha + r \sin \beta \sin \alpha$$

and

$$y' = r \sin \beta \cos \alpha - r \cos \beta \sin \alpha.$$

Since $x = r \cos \beta$ and $y = r \sin \beta$, the latter equations become

$$x' = x \cos \alpha + y \sin \alpha \quad \text{and} \quad y' = -x \sin \alpha + y \cos \alpha. \quad (2)$$

If Equations (2) are solved for x and y in terms of x' and y', we then have

$$x = x' \cos \alpha - y' \sin \alpha \quad \text{and} \quad y = x' \sin \alpha + y' \cos \alpha \quad (3)$$

Equations (2) and (3) relate the xy-coordinates and the $x'y'$-coordinates of a point in the plane under a rotation of axes through an angle α.

Example   Find the $x'y'$-coordinates of the point P with xy-coordinates $(5, -2)$ under a rotation of axes through an angle measuring $60°$.

Solution   We have $m°(\alpha) = 60°$, so that, from (2)

$$x' = 5 \cos 60° + (-2) \sin 60° \quad \text{and} \quad y' = -5 \sin 60° + (-2) \cos 60°.$$

Then, since $\cos 60° = 1/2$ and $\sin 60° = \sqrt{3}/2$, we have

$$x' = 5\left(\frac{1}{2}\right) + (-2)\frac{\sqrt{3}}{2} = \frac{5}{2} - \sqrt{3}$$

(continued overleaf)

and

$$y' = -5\left(\frac{\sqrt{3}}{2}\right) + (-2)\left(\frac{1}{2}\right) = \frac{-5\sqrt{3}}{2} - 1.$$

Thus, the new coordinates of P are $\left(\dfrac{5 - 2\sqrt{3}}{2}, \dfrac{-5\sqrt{3} - 2}{2}\right)$.

Now consider the general second-degree equation in two variables,

$$Ax^2 + Bxy + Cy^2 + Dx + Ey + F = 0. \tag{4}$$

If $B \neq 0$, we can use a rotation of axes to remove the $x'y'$-term from the transformed equation and produce an equivalent one of the form (4) on page 276. Under a rotation through an angle α, we can replace x and y in (4) above with the right-hand members of Equation (3) to obtain

$$A(x' \cos \alpha - y' \sin \alpha)^2 + B(x' \cos \alpha - y' \sin \alpha)(x' \sin \alpha + y' \cos \alpha)$$
$$+ C(x' \sin \alpha + y' \cos \alpha)^2 + D(x' \cos \alpha - y' \sin \alpha)$$
$$+ E(x' \sin \alpha + y' \cos \alpha) + F$$
$$= 0.$$

If this equation is simplified, we arrive at one of the form

$$A'x'^2 + B'x'y' + C'y'^2 + D'x' + E'y' + F' = 0,$$

where

$$A' = A \cos^2 \alpha + B \cos \alpha \sin \alpha + C \sin^2 \alpha,$$
$$B' = 2(C - A) \sin \alpha \cos \alpha + B(\cos^2 \alpha - \sin^2 \alpha),$$
$$C' = A \sin^2 \alpha - B \cos \alpha \sin \alpha + C \cos^2 \alpha,$$
$$D' = D \cos \alpha + E \sin \alpha,$$
$$E' = -D \sin \alpha + E \cos \alpha,$$
$$F' = F.$$

If, now, we recall that $\sin 2\alpha = 2 \sin \alpha \cos \alpha$, and $\cos 2\alpha = \cos^2 \alpha - \sin^2 \alpha$, the equation for B' becomes

$$B' = (C - A) \sin 2\alpha + B \cos 2\alpha.$$

If we wish $B' = 0$, we have two cases to consider. On the one hand, if $C = A$, then

$$B' = B \cos 2\alpha = 0,$$

providing

$$\cos 2\alpha = 0.$$

This will occur when $m°(\alpha) = 45° + k(90°)$, for $k \in J$. On the other hand, if $C \neq A$, then

$$B' = (C - A) \sin 2\alpha + B \cos 2\alpha = 0,$$

providing

$$(A - C) \sin 2\alpha = B \cos 2\alpha,$$

from which

$$\tan 2\alpha = \frac{B}{A - C}.$$

Clearly, there are infinitely many angles of rotation we can select, but, in particular, we can restrict the angle of rotation to a positive acute angle.

Example

Use a rotation of axes to identify the graph of $3x^2 + 4\sqrt{3}\,xy - y^2 = 15$ and sketch the graph.

Solution

We first eliminate the $x'y'$-term, using

$$\tan 2\alpha = \frac{B}{A - C} = \frac{4\sqrt{3}}{3 - (-1)} = \frac{4\sqrt{3}}{4} = \sqrt{3}.$$

By inspection, $m°(2\alpha) = 60°$, from which $m°(\alpha) = 30°$, so that

$$\sin \alpha = \frac{1}{2} \quad \text{and} \quad \cos \alpha = \frac{\sqrt{3}}{2}.$$

From the equations on page 280, we have

$$A' = 3\left(\frac{\sqrt{3}}{2}\right)^2 + 4\sqrt{3}\left(\frac{\sqrt{3}}{2}\right)\left(\frac{1}{2}\right) + (-1)\left(\frac{1}{2}\right)^2 = 5,$$

$$B' = 0, \quad \left(\text{because } \tan 2\alpha = \frac{B}{A - C} = \sqrt{3}\right)$$

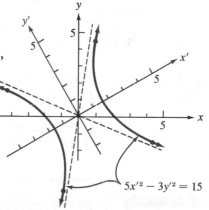

$$C' = 3\left(\frac{1}{2}\right)^2 - 4\sqrt{3}\left(\frac{\sqrt{3}}{2}\right)\left(\frac{1}{2}\right) + (-1)\left(\frac{\sqrt{3}}{2}\right)^2 = -3,$$

$$F = 15.$$

The new equation is then

$$5x'^2 - 3y'^2 = 15,$$

which is the equation of a hyperbola. Its graph is shown at the right.

Exercise A.3

Find the $x'y'$-coordinates of a point whose coordinates are given under a rotation of axes through an angle with the given measure.

1. $(-3, 5)$; $30°$ 2. $(7, -1)$; $45°$ 3. $(8, 6)$; $60°$
4. $(-2, -2)$; $90°$ 5. $(-6, 8)$; $120°$ 6. $(4, -3)$; $135°$
7. $(6, -5)$; $150°$ 8. $(-3, 0)$; $225°$

In Exercises 9–18, use a rotation of axes to eliminate the $x'y'$-term; then, when necessary, use a translation to eliminate the first-degree terms from the resulting equation. Identify the graph of the original equation.

9. $3x^2 - 10xy + 3y^2 = -32$ 10. $2x^2 + 3xy + 2y^2 = 25$
11. $x^2 + \sqrt{3}\,xy + 2y^2 = 18$ 12. $2x^2 - 2\sqrt{3}\,xy + 4y^2 = 9$
13. $x^2 + 3xy + 5y^2 = 22$ 14. $19x^2 + 6xy + 11y^2 = 20$
15. $3x^2 + 10xy + 3y^2 - 2x - 14y = 15$
16. $3x^2 + 2\sqrt{3}\,xy + y^2 - 8x + 8\sqrt{3}\,y = -4$

17. $x^2 + 2xy + y^2 + 2x - 4y + 5 = 0$

18. $x^2 + 4xy + 5y^2 - 8x + 3y + 12 = 0$

19. Prove that $A + C$ is invariant (does not change) under a rotation of axes.

20. Prove that $B^2 - 4AC$ is invariant under a rotation of axes.

Appendix Review

[A.1] a. *Sketch the graph of the given parametric equations.*
       b. *Eliminate the parameter and give a Cartesian equation for the graph.*

1. $x = 2 + t; \quad y = 3t^2$ 2. $x = 2 + \cos t; \quad y = 1 - \sin t$

Use the relationship $y = tx$ to obtain parametric equations for the graph of each Cartesian equation.

3. $x - 4y = 3$ 4. $2x^2 + 3y^2 = 6$

[A.2] *For Exercises 5 and 6, assume that the origin in an xy-system of coordinates is translated into the first point given, which then becomes the origin of an $x'y'$-system.*

5. $(2, -1)$; find the $x'y'$-coordinates of the point whose xy-coordinates are $(-5, 3)$.

6. $(-3, 0)$; find the $x'y'$-equation for $2x^2 - y = 1$.

Find an equation in the xy-system for each curve. Sketch the curve.

7. Circle with center at $(1, -3)$ and radius 4.

8. Parabola with focus at $(5, 0)$ and vertex at $(7, 0)$.

Find an equation in the x' y'-system and in standard form for the graph of the given equation in x and y.

9. $x^2 - 2x = 6y + 14$ 10. $2x^2 + y^2 - x + 4y - 5 = 0$

11. Use an auxiliary system to graph the equation in Exercise 9.

12. Use an auxiliary system to graph the equation in Exercise 10.

[A.3] *Find the $x'y'$-coordinates of a point whose coordinates are given under a rotation of axes through an angle with the given measure.*

13. $(2, -1); \quad 210°$ 14. $(0, 4); \quad 315°$

Use a rotation of axes to eliminate the $x'y'$-term; then, when necessary, use a translation to eliminate the first-degree terms from the resulting equation. Identify the graph of the equation.

15. $x^2 + 4xy + y^2 = 20$ 16. $5x^2 - \sqrt{3}\,xy + 4y^2 - \sqrt{3}\,x + y = 7$

Table I Squares, square roots, and prime factors

No.	Sq.	Sq. Rt.	Factors	No.	Sq.	Sq. Rt.	Factors
1	1	1.000		51	2,601	7.141	$3 \cdot 17$
2	4	1.414	2	52	2,704	7.211	$2^2 \cdot 13$
3	9	1.732	3	53	2,809	7.280	53
4	16	2.000	2^2	54	2,916	7.348	$2 \cdot 3^3$
5	25	2.236	5	55	3,025	7.416	$5 \cdot 11$
6	36	2.449	$2 \cdot 3$	56	3,136	7.483	$2^3 \cdot 7$
7	49	2.646	7	57	3,249	7.550	$3 \cdot 19$
8	64	2.828	2^3	58	3,364	7.616	$2 \cdot 29$
9	81	3.000	3^2	59	3,481	7.681	59
10	100	3.162	$2 \cdot 5$	60	3,600	7.746	$2^2 \cdot 3 \cdot 5$
11	121	3.317	11	61	3,721	7.810	61
12	144	3.464	$2^2 \cdot 3$	62	3,844	7.874	$2 \cdot 31$
13	169	3.606	13	63	3,969	7.937	$3^2 \cdot 7$
14	196	3.742	$2 \cdot 7$	64	4,096	8.000	2^6
15	225	3.873	$3 \cdot 5$	65	4,225	8.062	$5 \cdot 13$
16	256	4.000	2^4	66	4,356	8.124	$2 \cdot 3 \cdot 11$
17	289	4.123	17	67	4,489	8.185	67
18	324	4.243	$2 \cdot 3^2$	68	4,624	8.246	$2^2 \cdot 17$
19	361	4.359	19	69	4,761	8.307	$3 \cdot 23$
20	400	4.472	$2^2 \cdot 5$	70	4,900	8.367	$2 \cdot 5 \cdot 7$
21	441	4.583	$3 \cdot 7$	71	5,041	8.426	71
22	484	4.690	$2 \cdot 11$	72	5,184	8.485	$2^3 \cdot 3^2$
23	529	4.796	23	73	5,329	8.544	73
24	576	4.899	$2^3 \cdot 3$	74	5,476	8.602	$2 \cdot 37$
25	625	5.000	5^2	75	5,625	8.660	$3 \cdot 5^2$
26	676	5.099	$2 \cdot 13$	76	5,776	8.718	$2^2 \cdot 19$
27	729	5.196	3^3	77	5,929	8.775	$7 \cdot 11$
28	784	5.292	$2^2 \cdot 7$	78	6,084	8.832	$2 \cdot 3 \cdot 13$
29	841	5.385	29	79	6,241	8.888	79
30	900	5.477	$2 \cdot 3 \cdot 5$	80	6,400	8.944	$2^4 \cdot 5$
31	961	5.568	31	81	6,561	9.000	3^4
32	1,024	5.657	2^5	82	6,724	9.055	$2 \cdot 41$
33	1,089	5.745	$3 \cdot 11$	83	6,889	9.110	83
34	1,156	5.831	$2 \cdot 17$	84	7,056	9.165	$2^2 \cdot 3 \cdot 7$
35	1,225	5.916	$5 \cdot 7$	85	7,225	9.220	$5 \cdot 17$
36	1,296	6.000	$2^2 \cdot 3^2$	86	7,396	9.274	$2 \cdot 43$
37	1,369	6.083	37	87	7,569	9.327	$3 \cdot 29$
38	1,444	6.164	$2 \cdot 19$	88	7,744	9.381	$2^3 \cdot 11$
39	1,521	6.245	$3 \cdot 13$	89	7,921	9.434	89
40	1,600	6.325	$2^3 \cdot 5$	90	8,100	9.487	$2 \cdot 3^2 \cdot 5$
41	1,681	6.403	41	91	8,281	9.539	$7 \cdot 13$
42	1,764	6.481	$2 \cdot 3 \cdot 7$	92	8,464	9.592	$2^2 \cdot 23$
43	1,849	6.557	43	93	8,649	9.644	$3 \cdot 31$
44	1,936	6.633	$2^2 \cdot 11$	94	8,836	9.695	$2 \cdot 47$
45	2,025	6.708	$3^2 \cdot 5$	95	9,025	9.747	$5 \cdot 19$
46	2,116	6.782	$2 \cdot 23$	96	9,216	9.798	$2^5 \cdot 3$
47	2,209	6.856	47	97	9,409	9.849	97
48	2,304	6.928	$2^4 \cdot 3$	98	9,604	9.899	$2 \cdot 7^2$
49	2,401	7.000	7^2	99	9,801	9.950	$3^2 \cdot 11$
50	2,500	7.071	$2 \cdot 5^2$	100	10,000	10.000	$2^2 \cdot 5^2$

Table II Common logarithms

x	0	1	2	3	4	5	6	7	8	9
1.0	.0000	.0043	.0086	.0128	.0170	.0212	.0253	.0294	.0334	.0374
1.1	.0414	.0453	.0492	.0531	.0569	.0607	.0645	.0682	.0719	.0755
1.2	.0792	.0828	.0864	.0899	.0934	.0969	.1004	.1038	.1072	.1106
1.3	.1139	.1173	.1206	.1239	.1271	.1303	.1335	.1367	.1399	.1430
1.4	.1461	.1492	.1523	.1553	.1584	.1614	.1644	.1673	.1703	.1732
1.5	.1761	.1790	.1818	.1847	.1875	.1903	.1931	.1959	.1987	.2014
1.6	.2041	.2068	.2095	.2122	.2148	.2175	.2201	.2227	.2253	.2279
1.7	.2304	.2330	.2355	.2380	.2405	.2430	.2455	.2480	.2504	.2529
1.8	.2553	.2577	.2601	.2625	.2648	.2672	.2695	.2718	.2742	.2765
1.9	.2788	.2810	.2833	.2856	.2878	.2900	.2923	.2945	.2967	.2989
2.0	.3010	.3032	.3054	.3075	.3096	.3118	.3139	.3160	.3181	.3201
2.1	.3222	.3243	.3263	.3284	.3304	.3324	.3345	.3365	.3385	.3404
2.2	.3424	.3444	.3464	.3483	.3502	.3522	.3541	.3560	.3579	.3598
2.3	.3617	.3636	.3655	.3674	.3692	.3711	.3729	.3747	.3766	.3784
2.4	.3802	.3820	.3838	.3856	.3874	.3892	.3909	.3927	.3945	.3962
2.5	.3979	.3997	.4014	.4031	.4048	.4065	.4082	.4099	.4116	.4133
2.6	.4150	.4166	.4183	.4200	.4216	.4232	.4249	.4265	.4281	.4298
2.7	.4314	.4330	.4346	.4362	.4378	.4393	.4409	.4425	.4440	.4456
2.8	.4472	.4487	.4502	.4518	.4533	.4548	.4564	.4579	.4594	.4609
2.9	.4624	.4639	.4654	.4669	.4683	.4698	.4713	.4728	.4742	.4757
3.0	.4771	.4786	.4800	.4814	.4829	.4843	.4857	.4871	.4886	.4900
3.1	.4914	.4928	.4942	.4955	.4969	.4983	.4997	.5011	.5024	.5038
3.2	.5051	.5065	.5079	.5092	.5105	.5119	.5132	.5145	.5159	.5172
3.3	.5185	.5198	.5211	.5224	.5237	.5250	.5263	.5276	.5289	.5302
3.4	.5315	.5328	.5340	.5353	.5366	.5378	.5391	.5403	.5416	.5428
3.5	.5441	.5453	.5465	.5478	.5490	.5502	.5514	.5527	.5539	.5551
3.6	.5563	.5575	.5587	.5599	.5611	.5623	.5635	.5647	.5658	.5670
3.7	.5682	.5694	.5705	.5717	.5729	.5740	.5752	.5763	.5775	.5786
3.8	.5798	.5809	.5821	.5832	.5843	.5855	.5866	.5877	.5888	.5899
3.9	.5911	.5922	.5933	.5944	.5955	.5966	.5977	.5988	.5999	.6010
4.0	.6021	.6031	.6042	.6053	.6064	.6075	.6085	.6096	.6107	.6117
4.1	.6128	.6138	.6149	.6160	.6170	.6180	.6191	.6201	.6212	.6222
4.2	.6232	.6243	.6253	.6263	.6274	.6284	.6294	.6304	.6314	.6325
4.3	.6335	.6345	.6355	.6365	.6375	.6385	.6395	.6405	.6415	.6425
4.4	.6435	.6444	.6454	.6464	.6474	.6484	.6493	.6503	.6513	.6522
4.5	.6532	.6542	.6551	.6561	.6571	.6580	.6590	.6599	.6609	.6618
4.6	.6628	.6637	.6646	.6656	.6665	.6675	.6684	.6693	.6702	.6712
4.7	.6721	.6730	.6739	.6749	.6758	.6767	.6776	.6785	.6794	.6803
4.8	.6812	.6821	.6830	.6839	.6848	.6857	.6866	.6875	.6884	.6893
4.9	.6902	.6911	.6920	.6928	.6937	.6946	.6955	.6964	.6972	.6981
5.0	.6990	.6998	.7007	.7016	.7024	.7033	.7042	.7050	.7059	.7067
5.1	.7076	.7084	.7093	.7101	.7110	.7118	.7126	.7135	.7143	.7152
5.2	.7160	.7168	.7177	.7185	.7193	.7202	.7210	.7218	.7226	.7235
5.3	.7243	.7251	.7259	.7267	.7275	.7284	.7292	.7300	.7308	.7316
5.4	.7324	.7332	.7340	.7348	.7356	.7364	.7372	.7380	.7388	.7396
x	0	1	2	3	4	5	6	7	8	9

Table II (*continued*)   285

x	0	1	2	3	4	5	6	7	8	9
5.5	.7404	.7412	.7419	.7427	.7435	.7443	.7451	.7459	.7466	.7474
5.6	.7482	.7490	.7497	.7505	.7513	.7520	.7528	.7536	.7543	.7551
5.7	.7559	.7566	.7574	.7582	.7589	.7597	.7604	.7612	.7619	.7627
5.8	.7634	.7642	.7649	.7657	.7664	.7672	.7679	.7686	.7694	.7701
5.9	.7709	.7716	.7723	.7731	.7738	.7745	.7752	.7760	.7767	.7774
6.0	.7782	.7789	.7796	.7803	.7810	.7818	.7825	.7832	.7839	.7846
6.1	.7853	.7860	.7868	.7875	.7882	.7889	.7896	.7903	.7910	.7917
6.2	.7924	.7931	.7938	.7945	.7952	.7959	.7966	.7973	.7980	.7987
6.3	.7993	.8000	.8007	.8014	.8021	.8028	.8035	.8041	.8048	.8055
6.4	.8062	.8069	.8075	.8082	.8089	.8096	.8102	.8109	.8116	.8122
6.5	.8129	.8136	.8142	.8149	.8156	.8162	.8169	.8176	.8182	.8189
6.6	.8195	.8202	.8209	.8215	.8222	.8228	.8235	.8241	.8248	.8254
6.7	.8261	.8267	.8274	.8280	.8287	.8293	.8299	.8306	.8312	.8319
6.8	.8325	.8331	.8338	.8344	.8351	.8357	.8363	.8370	.8376	.8382
6.9	.8388	.8395	.8401	.8407	.8414	.8420	.8426	.8432	.8439	.8445
7.0	.8451	.8457	.8463	.8470	.8476	.8482	.8488	.8494	.8500	.8506
7.1	.8513	.8519	.8525	.8531	.8537	.8543	.8549	.8555	.8561	.8567
7.2	.8573	.8579	.8585	.8591	.8597	.8603	:8609	.8615	.8621	.8627
7.3	.8633	.8639	.8645	.8651	.8657	.8663	.8669	.8675	.8681	.8686
7.4	.8692	.8698	.8704	.8710	.8716	.8722	.8727	.8733	.8739	.8745
7.5	.8751	.8756	.8762	.8768	.8774	.8779	.8785	.8791	.8797	.8802
7.6	.8808	.8814	.8820	.8825	.8831	.8837	.8842	.8848	.8854	.8859
7.7	.8865	.8871	.8876	.8882	.8887	.8893	.8899	.8904	.8910	.8915
7.8	.8921	.8927	.8932	.8938	.8943	.8949	.8954	.8960	.8965	.8971
7.9	.8976	.8982	.8987	.8993	.8998	.9004	.9009	.9015	.9020	.9025
8.0	.9031	.9036	.9042	.9047	.9053	.9058	.9063	.9069	.9074	.9079
8.1	.9085	.9090	.9096	.9101	.9106	.9112	.9117	.9122	.9128	.9133
8.2	.9138	.9143	.9149	.9154	.9159	.9165	.9170	.9175	.9180	.9186
8.3	.9191	.9196	.9201	.9206	.9212	.9217	.9222	.9227	.9232	.9238
8.4	.9243	.9248	.9253	.9258	.9263	.9269	.9274	.9279	.9284	.9289
8.5	.9294	.9299	.9304	.9309	.9315	.9320	.9325	.9330	.9335	.9340
8.6	.9345	.9350	.9355	.9360	.9365	.9370	.9375	.9380	.9385	.9390
8.7	.9395	.9400	.9405	.9410	.9415	.9420	.9425	.9430	.9435	.9440
8.8	.9445	.9450	.9455	.9460	.9465	.9469	.9474	.9479	.9484	.9489
8.9	.9494	.9499	.9504	.9509	.9513	.9518	.9523	.9528	.9533	.9538
9.0	.9542	.9547	.9552	.9557	.9562	.9566	.9571	.9576	.9581	.9586
9.1	.9590	.9595	.9600	.9605	.9609	.9614	.9619	.9624	.9628	.9633
9.2	.9638	.9643	.9647	.9652	.9657	.9661	.9666	.9671	.9675	.9680
9.3	.9685	.9689	.9694	.9699	.9703	.9708	.9713	.9717	.9722	.9727
9.4	.9731	.9736	.9741	.9745	.9750	.9754	.9759	.9763	.9768	.9773
9.5	.9777	.9782	.9786	.9791	.9795	.9800	.9805	.9809	.9814	.9818
9.6	.9823	.9827	.9832	.9836	.9841	.9845	.9850	.9854	.9859	.9863
9.7	.9868	.9872	.9877	.9881	.9886	.9890	.9894	.9899	.9903	.9908
9.8	.9912	.9917	.9921	.9926	.9930	.9934	.9939	.9943	.9948	.9952
9.9	.9956	.9961	.9965	.9969	.9974	.9978	.9983	.9987	.9991	.9996
x	0	1	2	3	4	5	6	7	8	9

Table III *Exponential functions*

x	e^x	e^{-x}	x	e^x	e^{-x}
0.00	1.0000	1.0000	1.5	4.4817	0.2231
0.01	1.0101	0.9901	1.6	4.9530	0.2019
0.02	1.0202	0.9802	1.7	5.4739	0.1827
0.03	1.0305	0.9705	1.8	6.0496	0.1653
0.04	1.0408	0.9608	1.9	6.6859	0.1496
0.05	1.0513	0.9512	2.0	7.3891	0.1353
0.06	1.0618	0.9418	2.1	8.1662	0.1225
0.07	1.0725	0.9324	2.2	9.0250	0.1108
0.08	1.0833	0.9331	2.3	9.9742	0.1003
0.09	1.0942	0.9139	2.4	11.023	0.0907
0.10	1.1052	0.9048	2.5	12.182	0.0821
0.11	1.1163	0.8958	2.6	13.464	0.0743
0.12	1.1275	0.8869	2.7	14.880	0.0672
0.13	1.1388	0.8781	2.8	16.445	0.0608
0.14	1.1503	0.8694	2.9	18.174	0.0550
0.15	1.1618	0.8607	3.0	20.086	0.0498
0.16	1.1735	0.8521	3.1	22.198	0.0450
0.17	1.1853	0.8437	3.2	24.533	0.0408
0.18	1.1972	0.8353	3.3	27.113	0.0369
0.19	1.2092	0.8270	3.4	29.964	0.0334
0.20	1.2214	0.8187	3.5	33.115	0.0302
0.21	7.2337	0.8106	3.6	36.598	0.0273
0.22	1.2461	0.8025	3.7	40.447	0.0247
0.23	1.2586	0.7945	3.8	44.701	0.0224
0.24	1.2712	0.7866	3.9	49.402	0.0202
0.25	1.2840	0.7788	4.0	54.598	0.0183
0.30	1.3499	0.7408	4.1	60.340	0.0166
0.35	1.4191	0.7047	4.2	66.686	0.0150
0.40	1.4918	0.6703	4.3	73.700	0.0136
0.45	1.5683	0.6376	4.4	81.451	0.0123
0.50	1.6487	0.6065	4.5	90.017	0.0111
0.55	1.7333	0.5769	4.6	99.484	0.0101
0.60	1.8221	0.5488	4.7	109.95	0.0091
0.65	1.9155	0.5220	4.8	121.51	0.0082
0.70	2.0138	0.4966	4.9	134.29	0.0074
0.75	2.1170	0.4724	5.0	148.41	0.0067
0.80	2.2255	0.4493	5.5	244.69	0.0041
0.85	2.3396	0.4274	6.0	403.43	0.0025
0.90	2.4596	0.4066	6.5	665.14	0.0015
0.95	2.5857	0.3867	7.0	1096.6	0.0009
1.0	2.7183	0.3679	7.5	1808.0	0.0006
1.1	3.0042	0.3329	8.0	2981.0	0.0003
1.2	3.3201	0.3012	8.5	4914.8	0.0002
1.3	3.6693	0.2725	9.0	8103.1	0.0001
1.4	4.0552	0.2466	10.0	22026	0.00005

Table IV Natural logarithms of numbers

n	$\log_e n$	n	$\log_e n$	n	$\log_e n$
	*	4.5	1.5041	9.0	2.1972
0.1	7.6974	4.6	1.5261	9.1	2.2083
0.2	8.3906	4.7	1.5476	9.2	2.2192
0.3	8.7960	4.8	1.5686	9.3	2.2300
0.4	9.0837	4.9	1.5892	9.4	2.2407
0.5	9.3069	5.0	1.6094	9.5	2.2513
0.6	9.4892	5.1	1.6292	9.6	2.2618
0.7	9.6433	5.2	1.6487	9.7	2.2721
0.8	9.7769	5.3	1.6677	9.8	2.2824
0.9	9.8946	5.4	1.6864	9.9	2.2925
1.0	0.0000	5.5	1.7047	10	2.3026
1.1	0.0953	5.6	1.7228	11	2.3979
1.2	0.1823	5.7	1.7405	12	2.4849
1.3	0.2624	5.8	1.7579	13	2.5649
1.4	0.3365	5.9	1.7750	14	2.6391
1.5	0.4055	6.0	1.7918	15	2.7081
1.6	0.4700	6.1	1.8083	16	2.7726
1.7	0.5306	6.2	1.8245	17	2.8332
1.8	0.5878	6.3	1.8405	18	2.8904
1.9	0.6419	6.4	1.8563	19	2.9444
2.0	0.6931	6.5	1.8718	20	2.9957
2.1	0.7419	6.6	1.8871	25	3.2189
2.2	0.7885	6.7	1.9021	30	3.4012
2.3	0.8329	6.8	1.9169	35	3.5553
2.4	0.8755	6.9	1.9315	40	3.6889
2.5	0.9163	7.0	1.9459	45	3.8067
2.6	0.9555	7.1	1.9601	50	3.9120
2.7	0.9933	7.2	1.9741	55	4.0073
2.8	1.0296	7.3	1.9879	60	4.0943
2.9	1.0647	7.4	2.0015	65	4.1744
3.0	1.0986	7.5	2.0149	70	4.2485
3.1	1.1314	7.6	2.0281	75	4.3175
3.2	1.1632	7.7	2.0412	80	4.3820
3.3	1.1939	7.8	2.0541	85	4.4427
3.4	1.2238	7.9	2.0669	90	4.4998
3.5	1.2528	8.0	2.0794	100	4.6052
3.6	1.2809	8.1	2.0919	110	4.7005
3.7	1.3083	8.2	2.1041	120	4.7875
3.8	1.3350	8.3	2.1163	130	4.8676
3.9	1.3610	8.4	2.1282	140	4.9416
4.0	1.3863	8.5	2.1401	150	5.0106
4.1	1.4110	8.6	2.1518	160	5.0752
4.2	1.4351	8.7	2.1633	170	5.1358
4.3	1.4586	8.8	2.1748	180	5.1930
4.4	1.4816	8.9	2.1861	190	5.2470

* Subtract 10 for $n < 1$. Thus $\log_e 0.1 = 7.6974 - 10 = -2.3026$.

Table V   Values of circular functions

Real Number x or θ radians	θ degrees	sin x or sin θ	csc x or csc θ	tan x or tan θ	cot x or cot θ	sec x or sec θ	cos x or cos θ
0.00	0° 00′	0.0000	No value	0.0000	No value	1.000	1.000
.01	0° 34′	.0100	100.0	.0100	100.0	1.000	1.000
.02	1° 09′	.0200	50.00	.0200	49.99	1.000	0.9998
.03	1° 43′	.0300	33.34	.0300	33.32	1.000	0.9996
.04	2° 18′	.0400	25.01	.0400	24.99	1.001	0.9992
0.05	2° 52′	0.0500	20.01	0.0500	19.98	1.001	0.9988
.06	3° 26′	.0600	16.68	.0601	16.65	1.002	.9982
.07	4° 01′	.0699	14.30	.0701	14.26	1.002	.9976
.08	4° 35′	.0799	12.51	.0802	12.47	1.003	.9968
.09	5° 09′	.0899	11.13	.0902	11.08	1.004	.9960
0.10	5° 44′	0.0998	10.02	0.1003	9.967	1.005	0.9950
.11	6° 18′	.1098	9.109	.1104	9.054	1.006	.9940
.12	6° 53′	.1197	8.353	.1206	8.293	1.007	.9928
.13	7° 27′	.1296	7.714	.1307	7.649	1.009	.9916
.14	8° 01′	.1395	7.166	.1409	7.096	1.010	.9902
0.15	8° 36′	0.1494	6.692	0.1511	6.617	1.011	0.9888
.16	9° 10′	.1593	6.277	.1614	6.197	1.013	.9872
.17	9° 44′	.1692	5.911	.1717	5.826	1.015	.9856
.18	10° 19′	.1790	5.586	.1820	5.495	1.016	.9838
.19	10° 53′	.1889	5.295	.1923	5.200	1.018	.9826
0.20	11° 28′	0.1987	5.033	0.2027	4.933	1.020	0.9801
.21	12° 02′	.2085	4.797	.2131	4.692	1.022	.9780
.22	12° 36′	.2182	4.582	.2236	4.472	1.025	.9759
.23	13° 11′	.2280	4.386	.2341	4.271	1.027	.9737
.24	13° 45′	.2377	4.207	.2447	4.086	1.030	.9713
0.25	14° 19′	0.2474	4.042	0.2553	3.916	1.032	0.9689
.26	14° 54′	.2571	3.890	.2660	3.759	1.035	.9664
.27	15° 28′	.2667	3.749	.2768	3.613	1.038	.9638
.28	16° 03′	.2764	3.619	.2876	3.478	1.041	.9611
.29	16° 37′	.2860	3.497	.2984	3.351	1.044	.9582
0.30	17° 11′	0.2955	3.384	0.3093	3.233	1.047	0.9553
.31	17° 46′	.3051	3.278	.3203	3.122	1.050	.9523
.32	18° 20′	.3146	3.179	.3314	3.018	1.053	.9492
.33	18° 54′	.3240	3.086	.3425	2.920	1.057	.9460
.34	19° 29′	.3335	2.999	.3537	2.827	1.061	.9428
0.35	20° 03′	0.3429	2.916	0.3650	2.740	1.065	0.9394
.36	20° 38′	.3523	2.839	.3764	2.657	1.068	.9359
.37	21° 12′	.3616	2.765	.3879	2.578	1.073	.9323
.38	21° 46′	.3709	2.696	.3994	2.504	1.077	.9287
.39	22° 21′	.3802	2.630	.4111	2.433	1.081	.9249
0.40	22° 55′	0.3894	2.568	0.4228	2.365	1.086	0.9211
.41	23° 29′	.3986	2.509	.4346	2.301	1.090	.9171
.42	24° 04′	.4078	2.452	.4466	2.239	1.095	.9131
.43	24° 38′	.4169	2.399	.4586	2.180	1.100	.9090
.44	25° 13′	.4259	2.348	.4708	2.124	1.105	.9048
0.45	25° 47′	0.4350	2.299	0.4831	2.070	1.111	0.9004
.46	26° 21′	.4439	2.253	.4954	2.018	1.116	.8961
.47	26° 56′	.4529	2.208	.5080	1.969	1.122	.8916
.48	27° 30′	.4618	2.166	.5206	1.921	1.127	.8870
.49	28° 04′	.4706	2.125	.5334	1.875	1.133	.8823
0.50	28° 39′	0.4794	2.086	0.5463	1.830	1.139	0.8776

Table V (continued) 289

Real Number x or θ radians	θ degrees	sin x or sin θ	csc x or csc θ	tan x or tan θ	cot x or cot θ	sec x or sec θ	cos x or cos θ
0.50	28° 39′	0.4794	2.086	0.5463	1.830	1.139	0.8776
.51	29° 13′	.4882	2.048	.5594	1.788	1.146	.8727
.52	29° 48′	.4969	2.013	.5726	1.747	1.152	.8678
.53	30° 22′	.5055	1.978	.5859	1.707	1.159	.8628
.54	30° 56′	.5141	1.945	.5994	1.668	1.166	.8577
0.55	31° 31′	0.5227	1.913	0.6131	1.631	1.173	0.8525
.56	32° 05′	.5312	1.883	.6269	1.595	1.180	.8473
.57	32° 40′	.5396	1.853	.6410	1.560	1.188	.8419
.58	33° 14′	.5480	1.825	.6552	1.526	1.196	.8365
.59	33° 48′	.5564	1.797	.6696	1.494	1.203	.8309
0.60	34° 23′	0.5646	1.771	0.6841	1.462	1.212	0.8253
.61	34° 57′	.5729	1.746	.6989	1.431	1.220	.8196
.62	35° 31′	.5810	1.721	.7139	1.401	1.229	.8139
.63	36° 06′	.5891	1.697	.7291	1.372	1.238	.8080
.64	36° 40′	.5972	1.674	.7445	1.343	1.247	.8021
0.65	37° 15′	0.6052	1.652	0.7602	1.315	1.256	0.7961
.66	37° 49′	.6131	1.631	.7761	1.288	1.266	.7900
.67	38° 23′	.6210	1.610	.7923	1.262	1.276	.7838
.68	38° 58′	.6288	1.590	.8087	1.237	1.286	.7776
.69	39° 32′	.6365	1.571	.8253	1.212	1.297	.7712
0.70	40° 06′	0.6442	1.552	0.8423	1.187	1.307	0.7648
.71	40° 41′	.6518	1.534	.8595	1.163	1.319	.7584
.72	41° 15′	.6594	1.517	.8771	1.140	1.330	.7518
.73	41° 50′	.6669	1.500	.8949	1.117	1.342	.7452
.74	42° 24′	.6743	1.483	.9131	1.095	1.354	.7385
0.75	42° 58′	0.6816	1.467	0.9316	1.073	1.367	0.7317
.76	43° 33′	.6889	1.452	.9505	1.052	1.380	.7248
.77	44° 07′	.6961	1.436	.9697	1.031	1.393	.7179
.78	44° 41′	.7033	1.422	.9893	1.011	1.407	.7109
.79	45° 16′	.7104	1.408	1.009	.9908	1.421	.7038
0.80	45° 50′	0.7174	1.394	1.030	0.9712	1.435	0.6967
.81	46° 25′	.7243	1.381	1.050	.9520	1.450	.6895
.82	46° 59′	.7311	1.368	1.072	.9331	1.466	.6822
.83	47° 33′	.7379	1.355	1.093	.9146	1.482	.6749
.84	48° 08′	.7446	1.343	1.116	.8964	1.498	.6675
0.85	48° 42′	0.7513	1.331	1.138	0.8785	1.515	0.6600
.86	49° 16′	.7578	1.320	1.162	.8609	1.533	.6524
.87	49° 51′	.7643	1.308	1.185	.8437	1.551	.6448
.88	50° 25′	.7707	1.297	1.210	.8267	1.569	.6372
.89	51° 00′	.7771	1.287	1.235	.8100	1.589	.6294
0.90	51° 34′	0.7833	1.277	1.260	0.7936	1.609	0.6216
.91	52° 08′	.7895	1.267	1.286	.7774	1.629	.6137
.92	52° 43′	.7956	1.257	1.313	.7615	1.651	.6058
.93	53° 17′	.8016	1.247	1.341	.7458	1.673	.5978
.94	53° 51′	.8076	1.238	1.369	.7303	1.696	.5898
0.95	54° 26′	0.8134	1.229	1.398	0.7151	1.719	0.5817
.96	55° 00′	.8192	1.221	1.428	.7001	1.744	.5735
.97	55° 35′	.8249	1.212	1.459	.6853	1.769	.5653
.98	56° 09′	.8305	1.204	1.491	.6707	1.795	.5570
.99	56° 43′	.8360	1.196	1.524	.6563	1.823	.5487
1.00	57° 18′	0.8415	1.188	1.557	0.6421	1.851	0.5403
1.01	57° 52′	.8468	1.181	1.592	.6281	1.880	.5319
1.02	58° 27′	.8521	1.174	1.628	.6142	1.911	.5234
1.03	59° 01′	.8573	1.166	1.665	.6005	1.942	.5148
1.04	59° 35′	.8624	1.160	1.704	.5870	1.975	.5062
1.05	60° 10′	0.8674	1.153	1.743	0.5736	2.010	0.4976

Real Number x or θ radians	θ degrees	sin x or sin θ	csc x or csc θ	tan x or tan θ	cot x or cot θ	sec x or sec θ	cos x or cos θ
1.05	60° 10′	0.8674	1.153	1.743	0.5736	2.010	0.4976
1.06	60° 44′	.8724	1.146	1.784	.5604	2.046	.4889
1.07	61° 18′	.8772	1.140	1.827	.5473	2.083	.4801
1.08	61° 53′	.8820	1.134	1.871	.5344	2.122	.4713
1.09	62° 27′	.8866	1.128	1.917	.5216	2.162	.4625
1.10	63° 02′	0.8912	1.122	1.965	0.5090	2.205	0.4536
1.11	63° 36′	.8957	1.116	2.014	.4964	2.249	.4447
1.12	64° 10′	.9001	1.111	2.066	.4840	2.295	.4357
1.13	64° 45′	.9044	1.106	2.120	.4718	2.344	.4267
1.14	65° 19′	.9086	1.101	2.176	.4596	2.395	.4176
1.15	65° 53′	0.9128	1.096	2.234	0.4475	2.448	0.4085
1.16	66° 28′	.9168	1.091	2.296	.4356	2.504	.3993
1.17	67° 02′	.9208	1.086	2.360	.4237	2.563	.3902
1.18	67° 37′	.9246	1.082	2.247	.4120	2.625	.3809
1.19	68° 11′	.9284	1.077	2.498	.4003	2.691	.3717
1.20	68° 45′	0.9320	1.073	2.572	0.3888	2.760	0.3624
1.21	69° 20′	.9356	1.069	2.650	.3773	2.833	.3530
1.22	69° 54′	.9391	1.065	2.733	.3659	2.910	.3436
1.23	70° 28′	.9425	1.061	2.820	.3546	2.992	.3342
1.24	71° 03′	.9458	1.057	2.912	.3434	3.079	.3248
1.25	71° 37′	0.9490	1.054	3.010	0.3323	3.171	0.3153
1.26	72° 12′	.9521	1.050	3.113	.3212	3.270	.3058
1.27	72° 46′	.9551	1.047	3.224	.3102	3.375	.2963
1.28	72° 20′	.9580	1.044	3.341	.2993	3.488	.2867
1.29	73° 55′	.9608	1.041	3.467	.2884	3.609	.2771
1.30	74° 29′	0.9636	1.038	3.602	0.2776	3.738	0.2675
1.31	75° 03′	.9662	1.035	3.747	.2669	3.878	.2579
1.32	75° 38′	.9687	1.032	3.903	.2562	4.029	.2482
1.33	76° 12′	.9711	1.030	4.072	.2456	4.193	.2385
1.34	76° 47′	.9735	1.027	4.256	.2350	4.372	.2288
1.35	77° 21′	0.9757	1.025	4.455	0.2245	4.566	0.2190
1.36	77° 55′	.9779	1.023	4.673	.2140	4.779	.2092
1.37	78° 30′	.9799	1.021	4.913	.2035	5.014	.1994
1.38	79° 04′	.9819	1.018	5.177	.1931	5.273	.1896
1.39	79° 38′	.9837	1.017	5.471	.1828	5.561	.1798
1.40	80° 13′	0.9854	1.015	5.798	0.1725	5.883	0.1700
1.41	80° 47′	.9871	1.013	6.165	.1622	6.246	.1601
1.42	81° 22′	.9887	1.011	6.581	.1519	6.657	.1502
1.43	81° 56′	.9901	1.010	7.055	.1417	7.126	.1403
1.44	82° 30′	.9915	1.009	7.602	.1315	7.667	.1304
1.45	83° 05′	0.9927	1.007	8.238	0.1214	8.299	0.1205
1.46	83° 39′	.9939	1.006	8.989	.1113	9.044	.1106
1.47	84° 13′	.9949	1.005	9.887	.1011	9.938	.1006
1.48	84° 48′	.9959	1.004	10.98	.0910	11.03	.0907
1.49	85° 22′	.9967	1.003	12.35	.0810	12.39	.0807
1.50	85° 57′	0.9975	1.003	14.10	0.0709	14.14	0.0707
1.51	86° 31′	.9982	1.002	16.43	.0609	16.46	.0608
1.52	87° 05′	.9987	1.001	19.67	.0508	19.69	.0508
1.53	87° 40′	.9992	1.001	24.50	.0408	24.52	.0408
1.54	88° 14′	.9995	1.000	32.46	.0308	32.48	.0308
1.55	88° 49′	0.9998	1.000	48.08	0.0208	48.09	0.0208
1.56	89° 23′	.9999	1.000	92.62	.0108	92.63	.0108
1.57	89° 57′	1.000	1.000	1256	.0008	1256	.0008

Table VI   Values of trigonometric functions

Angle θ									
Degrees	Radians	sin θ	csc θ	tan θ	cot θ	sec θ	cos θ		
0° 00′	.0000	.0000	No value	.0000	No value	1.000	1.0000	1.5708	90° 00′
10	029	029	343.8	029	343.8	000	000	679	50
20	058	058	171.9	058	171.9	000	000	650	40
30	087	087	114.6	087	114.6	000	1.0000	621	30
40	116	116	85.95	116	85.94	000	.9999	592	20
50	145	145	68.76	145	68.75	000	999	563	10
1° 00′	.0175	.0175	57.30	.0175	57.29	1.000	.9998	1.5533	89° 00′
10	204	204	49.11	204	49.10	000	998	504	50
20	233	233	42.98	233	42.96	000	997	475	40
30	262	262	38.20	262	38.19	000	997	446	30
40	291	291	34.38	291	34.37	000	996	417	20
50	320	320	31.26	320	31.24	001	995	388	10
2° 00′	.0349	.0349	28.65	.0349	28.64	1.001	.9994	1.5359	88° 00′
10	378	378	26.45	378	26.43	001	993	330	50
20	407	407	24.56	407	24.54	001	992	301	40
30	436	436	22.93	437	22.90	001	990	272	30
40	465	465	21.49	466	21.47	001	989	243	20
50	495	494	20.23	495	20.21	001	988	213	10
3° 00′	.0524	.0523	19.11	.0524	19.08	1.001	.9986	1.5184	87° 00′
10	553	552	18.10	553	18.07	002	985	155	50
20	582	581	17.20	582	17.17	002	983	126	40
30	611	610	16.38	612	16.35	002	981	097	30
40	640	640	15.64	641	15.60	002	980	068	20
50	669	669	14.96	670	14.92	002	978	039	10
4° 00′	.0698	.0698	14.34	.0699	14.30	1.002	.9976	1.5010	86° 00′
10	727	727	13.76	729	13.73	003	974	981	50
20	756	756	13.23	758	13.20	003	971	952	40
30	785	785	12.75	787	12.71	003	969	923	30
40	814	814	12.29	816	12.25	003	967	893	20
50	844	843	11.87	846	11.83	004	964	864	10
5° 00′	.0873	.0872	11.47	.0875	11.43	1.004	.9962	1.4835	85° 00′
10	902	901	11.10	904	11.06	004	959	806	50
20	931	929	10.76	934	10.71	004	957	777	40
30	960	958	10.43	963	10.39	005	954	748	30
40	.0989	.0987	10.13	.0992	10.08	005	951	719	20
50	.1018	.1016	9.839	.1022	9.788	005	948	690	10
6° 00′	.1047	.1045	9.567	.1051	9.514	1.006	.9945	1.4661	84° 00′
10	076	074	9.309	080	9.255	006	942	632	50
20	105	103	9.065	110	9.010	006	939	603	40
30	134	132	8.834	139	8.777	006	936	573	30
40	164	161	8.614	169	8.556	007	932	544	20
50	193	190	8.405	198	8.345	007	929	515	10
7° 00′	.1222	.1219	8.206	.1228	8.144	1.008	.9925	1.4486	83° 00′
10	251	248	8.016	257	7.953	008	922	457	50
20	280	276	7.834	287	7.770	008	918	428	40
30	309	305	7.661	317	7.596	009	914	399	30
40	338	334	7.496	346	7.429	009	911	370	20
50	367	363	7.337	376	7.269	009	907	341	10
8° 00′	.1396	.1392	7.185	.1405	7.115	1.010	.9903	1.4312	82° 00′
10	425	421	7.040	435	6.968	010	899	283	50
20	454	449	6.900	465	827	011	894	254	40
30	484	478	765	495	691	011	890	224	30
40	513	507	636	524	561	012	886	195	20
50	542	536	512	554	435	012	881	166	10
9° 00′	.1571	.1564	6.392	.1584	6.314	1.012	.9877	1.4137	81° 00′
		cos θ	sec θ	cot θ	tan θ	csc θ	sin θ	Radians	Degrees
								Angle θ	

291

Table VI (continued)

Angle θ									
Degrees	Radians	sin θ	csc θ	tan θ	cot θ	sec θ	cos θ		
9° 00'	.1571	.1564	6.392	.1584	6.314	1.012	.9877	1.4137	81° 00'
10	600	593	277	614	197	013	872	108	50
20	629	622	166	644	6.084	013	868	079	40
30	658	650	6.059	673	5.976	014	863	050	30
40	687	679	5.955	703	871	014	858	1.4021	20
50	716	708	855	733	769	015	853	1.3992	10
10° 00'	.1745	.1736	5.759	.1763	5.671	1.015	.9848	1.3963	80° 00
10	774	765	665	793	576	016	843	934	50
20	804	794	575	823	485	016	838	904	40
30	833	822	487	853	396	017	833	875	30
40	862	851	403	883	309	018	827	846	20
50	891	880	320	914	226	018	822	817	10
11° 00'	.1920	.1908	5.241	.1944	5.145	1.019	.9816	1.3788	79° 00'
10	949	937	164	.1974	5.066	019	811	759	50
20	.1978	965	089	.2004	4.989	020	805	730	40
30	.2007	.1994	5.016	035	915	020	799	701	30
40	036	.2022	4.945	065	843	021	793	672	20
50	065	051	876	095	773	022	787	643	10
12° 00'	.2094	.2079	4.810	.2126	4.705	1.022	.9781	1.3614	78° 00'
10	123	108	745	156	638	023	775	584	50
20	153	136	682	186	574	024	769	555	40
30	182	164	620	217	511	024	763	526	30
40	211	193	560	247	449	025	757	497	20
50	240	221	502	278	390	026	750	468	10
13° 00'	.2269	.2250	4.445	.2309	4.331	1.026	.9744	1.3439	77° 00'
10	298	278	390	339	275	027	737	410	50
20	327	306	336	370	219	028	730	381	40
30	356	334	284	401	165	028	724	352	30
40	385	363	232	432	113	029	717	323	20
50	414	391	182	462	061	030	710	294	10
14° 00'	.2443	.2419	4.134	.2493	4.011	1.031	.9703	1.3265	76° 00'
10	473	447	086	524	3.962	031	696	235	50
20	502	476	4.039	555	914	032	689	206	40
30	531	504	3.994	586	867	033	681	177	30
40	560	532	950	617	821	034	674	148	20
50	589	560	906	648	776	034	667	119	10
15° 00'	.2618	.2588	3.864	.2679	3.732	1.035	.9659	1.3090	75° 00'
10	647	616	822	711	689	036	652	061	50
20	676	644	782	742	647	037	644	032	40
30	705	672	742	773	606	038	636	1.3003	30
40	734	700	703	805	566	039	628	1.2974	20
50	763	728	665	836	526	039	621	945	10
16° 00'	.2793	.2756	3.628	.2867	3.487	1.040	.9613	1.2915	74° 00'
10	822	784	592	899	450	041	605	886	50
20	851	812	556	931	412	042	596	857	40
30	880	840	521	962	376	043	588	828	30
40	909	868	487	.2944	340	044	580	799	20
50	938	896	453	.3026	305	045	572	770	10
17° 00'	.2967	.2924	3.420	.3057	3.271	1.046	.9563	1.2741	73° 00'
10	.2996	952	388	089	237	047	555	712	50
20	.3025	.2979	357	121	204	048	546	683	40
30	054	.3007	326	153	172	048	537	654	30
40	083	035	295	185	140	049	528	625	20
50	113	062	265	217	108	050	520	595	10
18° 00'	.3142	.3090	3.236	.3249	3.078	1.051	.9511	1.2566	72° 00
		cos θ	sec θ	cot θ	tan θ	csc θ	sin θ	Radians	Degrees
								Angle θ	

Table VI (*continued*)   293

Angle θ									
Degrees	Radians	sin θ	csc θ	tan θ	cot θ	sec θ	cos θ		
18° 00′	.3142	.3090	3.236	.3249	3.078	1.051	.9511	1.2566	72° 00′
10	171	118	207	281	047	052	502	537	50
20	200	145	179	314	3.018	053	492	508	40
30	229	173	152	346	2.989	054	483	479	30
40	258	201	124	378	960	056	474	450	20
50	287	228	098	411	932	057	465	421	10
19° 00′	.3316	.3256	3.072	.3443	2.904	1.058	.9455	1.2392	71° 00′
10	345	283	046	476	877	059	446	363	50
20	374	311	3.021	508	850	060	436	334	40
30	403	338	.2996	541	824	061	426	305	30
40	432	365	971	574	798	062	417	275	20
50	462	393	947	607	773	063	407	246	10
20° 00′	.3491	.3420	2.924	.3640	2.747	1.064	.9397	1.2217	70° 00′
10	520	448	901	673	723	065	387	188	50
20	549	475	878	706	699	066	377	159	40
30	578	502	855	739	675	068	367	130	30
40	607	529	833	772	651	069	356	101	20
50	636	557	812	805	628	070	346	072	10
21° 00′	.3665	.3584	2.790	.3839	2.605	1.071	.9336	1.2043	69° 00′
10	694	611	769	872	583	072	325	1.2014	50
20	723	638	749	906	560	074	315	1.1985	40
30	752	665	729	939	539	075	304	956	30
40	782	692	709	.3973	517	076	293	926	20
50	811	719	689	.4006	496	077	283	897	10
22° 00′	.3840	.3746	2.669	.4040	2.475	1.079	.9272	1.1868	68° 00
10	869	773	650	074	455	080	261	839	50
20	898	800	632	108	434	081	250	810	40
30	927	827	613	142	414	082	239	781	30
40	956	854	595	176	394	084	228	752	20
50	985	881	577	210	375	085	216	723	10
23° 00′	.4014	.3907	2.559	.4245	2.356	1.086	.9205	1.1694	67° 00′
10	043	934	542	279	337	088	194	665	50
20	072	961	525	314	318	089	182	636	40
30	102	.3987	508	348	300	090	171	606	30
40	131	.4014	491	383	282	092	159	577	20
50	160	041	475	417	264	093	147	548	10
24° 00′	.4189	.4067	2.459	.4452	2.246	1.095	.9135	1.1519	66° 00′
10	218	094	443	487	229	096	124	490	50
20	247	120	427	522	211	097	112	461	40
30	276	147	411	557	194	099	100	432	30
40	305	173	396	592	177	100	088	403	20
50	334	200	381	628	161	102	075	374	10
25° 00′	.4363	.4226	2.366	.4663	2.145	1.103	.9063	1.1345	65° 00′
10	392	253	352	699	128	105	051	316	50
20	422	279	337	734	112	106	038	286	40
30	451	305	323	770	097	108	026	257	30
40	480	331	309	806	081	109	013	228	20
50	509	358	295	841	066	111	.9001	199	10
26° 00′	.4538	.4384	2.281	.4877	2.050	1.113	.8988	1.1170	64° 00′
10	567	410	268	913	035	114	975	141	50
20	596	436	254	950	020	116	962	112	40
30	625	462	241	.4986	2.006	117	949	083	30
40	654	488	228	.5022	1.991	119	936	054	20
50	683	514	215	059	977	121	923	1.1025	10
27° 00′	.4712	.4540	2.203	.5095	1.963	1.122	.8910	1.0996	63° 00′
		cos θ	sec θ	cot θ	tan θ	csc θ	sin θ	Radians	Degrees
								Angle θ	

Table VI (*continued*)

Angle θ									
Degrees	Radians	sin θ	csc θ	tan θ	· cot θ	sec θ	cos θ		
27° 00′	.4712	.4540	2.203	.5095	1.963	1.122	.8910	1.0996	63° 00′
10	741	566	190	132	949	124	897	966	50
20	771	592	178	169	935	126	884	937	40
30	800	617	166	206	921	127	870	908	30
40	829	643	154	243	907	129	857	879	20
50	858	669	142	280	894	131	843	850	10
28° 00′	.4887	.4695	2.130	.5317	1.881	1.133	.8829	1.0821	62° 00′
10	916	720	118	354	868	134	816	792	50
20	945	746	107	392	855	136	802	763	40
30	.4974	772	096	430	842	138	788	734	30
40	.5003	797	085	467	829	140	774	705	20
50	032	823	074	505	816	142	760	676	10
29° 00′	.5061	.4848	2.063	.5543	1.804	1.143	.8746	1.0647	61° 00′
10	091	874	052	581	792	145	732	617	50
20	120	899	041	619	780	147	718	588	40
30	149	924	031	658	767	149	704	559	30
40	178	950	020	696	756	151	689	530	20
50	207	.4975	010	735	744	153	675	501	10
30° 00′	.5236	.5000	2.000	.5774	1.732	1.155	.8660	1.0472	60° 00′
10	265	025	1.990	812	720	157	646	443	50
20	294	050	980	851	709	159	631	414	40
30	323	075	970	890	698	161	616	385	30
40	352	100	961	930	686	163	601	356	20
50	381	125	951	.5969	675	165	587	327	10
31° 00′	.5411	.5150	1.942	.6009	1.664	1.167	.8572	1.0297	59° 00′
10	440	175	932	048	653	169	557	268	50
20	469	200	923	088	643	171	542	239	40
30	498	225	914	128	632	173	526	210	30
40	527	250	905	168	621	175	511	181	20
50	556	275	896	208	611	177	496	152	10
32° 00′	.5585	.5299	1.887	.6249	1.600	1.179	.8480	1.0123	58° 00′
10	614	324	878	289	590	181	465	094	50
20	643	348	870	330	580	184	450	065	40
30	672	373	861	371	570	186	434	036	30
40	701	398	853	412	560	188	418	1.0007	20
50	730	422	844	453	550	190	403	.9977	10
33° 00′	.5760	.5446	1.836	.6494	1.540	1.192	.8387	.9948	57° 00′
10	789	471	828	536	530	195	371	919	50
20	818	495	820	577	520	197	355	890	40
30	847	519	812	619	511	199	339	861	30
40	876	544	804	661	501	202	323	832	20
50	905	568	796	703	492	204	307	803	10
34° 00′	.5934	.5592	1.788	.6745	1.483	1.206	.8290	.9774	56° 00′
10	963	616	781	787	473	209	274	745	50
20	.5992	640	773	830	464	211	258	716	40
30	.6021	664	766	873	455	213	241	687	30
40	050	688	758	916	446	216	225	657	20
50	080	712	751	.6959	437	218	208	628	10
35° 00′	.6109	.5736	1.743	.7002	1.428	1.221	.8192	.9599	55° 00′
10	138	760	736	046	419	223	175	570	50
20	167	783	729	089	411	226	158	541	40
30	196	807	722	133	402	228	141	512	30
40	225	831	715	177	393	231	124	483	20
50	254	854	708	221	385	233	107	454	10
36° 00′	.6283	.5878	1.701	.7265	1.376	1.236	.8090	.9425	54° 00′
		cos θ	sec θ	cot θ	tan θ	csc θ	sin θ	Radians	Degrees
								Angle θ	

Angle θ									
Degrees	Radians	sin θ	csc θ	tan θ	cot θ	sec θ	cos θ		
36° 00′	.6283	.5878	1.701	.7265	1.376	1.236	.8090	.9425	54° 00′
10	312	901	695	310	368	239	073	396	50
20	341	925	688	355	360	241	056	367	40
30	370	948	681	400	351	244	039	338	30
40	400	972	675	445	343	247	021	308	20
50	429	.5995	668	490	335	249	.8004	279	10
37° 00′	.6458	.6018	1.662	.7536	1.327	1.252	.7986	.9250	53° 00′
10	487	041	655	581	319	255	969	221	50
20	516	065	649	627	311	258	951	192	40
30	545	088	643	673	303	260	934	163	30
40	574	111	636	720	295	263	916	134	20
50	603	134	630	766	288	266	898	105	10
38° 00′	.6632	.6157	1.624	.7813	1.280	1.269	.7880	.9076	52° 00′
10	661	180	618	860	272	272	862	047	50
20	690	202	612	907	265	275	844	.9018	40
30	720	225	606	.7954	257	278	826	.8988	30
40	749	248	601	.8002	250	281	808	959	20
50	778	271	595	050	242	284	790	930	10
39° 00′	.6807	.6293	1.589	.8098	1.235	1.287	.7771	.8901	51° 00′
10	836	316	583	146	228	290	753	872	50
20	865	338	578	195	220	293	735	843	40
30	894	361	572	243	213	296	716	814	30
40	923	383	567	292	206	299	698	785	20
50	952	406	561	342	199	302	679	756	10
40° 00′	.6981	.6428	1.556	.8391	1.192	1.305	.7660	.8727	50° 00′
10	.7010	450	550	441	185	309	642	698	50
20	039	472	545	491	178	312	623	668	40
30	069	494	540	541	171	315	604	639	30
40	098	517	535	591	164	318	585	610	20
50	127	539	529	642	157	322	566	581	10
41° 00′	.7156	.6561	1.524	.8693	1.150	1.325	.7547	.8552	49° 00′
10	185	583	519	744	144	328	528	523	50
20	214	604	514	796	137	332	509	494	40
30	243	626	509	847	130	335	490	465	30
40	272	648	504	899	124	339	470	436	20
50	301	670	499	.8952	117	342	451	407	10
42° 00′	.7330	.6691	1.494	.9004	1.111	1.346	.7431	.8378	48° 00′
10	359	713	490	057	104	349	412	348	50
20	389	734	485	110	098	353	392	319	40
30	418	756	480	163	091	356	373	290	30
40	447	777	476	217	085	360	353	261	20
50	476	799	471	271	079	364	333	232	10
43° 00′	.7505	.6820	1.466	.9325	1.072	1.367	.7314	.8203	47° 00′
10	534	841	462	380	066	371	294	174	50
20	563	862	457	435	060	375	274	145	40
30	592	884	453	490	054	379	254	116	30
40	621	905	448	545	048	382	234	087	20
50	650	926	444	601	042	386	214	058	10
44° 00′	.7679	.6947	1.440	.9657	1.036	1.390	.7193	.8029	46° 00′
10	709	967	435	713	030	394	173	.7999	50
20	738	.6988	431	770	024	398	153	970	40
30	767	.7009	427	827	018	402	133	941	30
40	796	030	423	884	012	406	112	912	20
50	825	050	418	.9942	006	410	092	883	10
45° 00′	.7854	.7071	1.414	1.000	1.000	1.414	.7071	.7854	45° 00′
		cos θ	sec θ	cot θ	tan θ	csc θ	sin θ	Radians	Degrees

Angle θ

Table VI (continued)   295

Odd-Numbered Answers

Exercise 1.1 (page 3)   **1.** $\{3, 4, 5, 6\}$ **3.** $\{\ \}$

5. $\{$Sunday, Monday, Tuesday, Wednesday, Thursday, Friday, Saturday$\}$

7. $\{$natural numbers less than $3\} = \{1, 2\}$ **9.** $\{2\} \neq \{-2\}$ **11.** $\emptyset \neq \{0\}$ **13.** $\{5\} \subset \{4, 5, 6\}$

15. $\emptyset \subset \{4, 5, 6\}$ **17.** $3 \in \{2, 3, 4\}$ **19.** $\{2\} \notin \{2, 3, 4\}$

21. a. $\{5, 6, 7\}$ **b.** $\{5, 6\}, \{5, 7\}, \{6, 7\}$ **c.** $\{5\}, \{6\}, \{7\}$ **d.** $\emptyset$

23. a. $\{2, 4\}$ **b.** $\{1, 2, 3, 4, 6, 8\}$ **c.** A or $\{2, 4, 6, 8\}$ **d.** $\emptyset$ **e.** $\{1, 2, 3, 4, 5, 7\}$ **f.** $\{1, 3\}$

25. $\{x \mid x = 2n - 1, n \text{ a natural number}\}$ **27.** $\{x \mid 2^x = 5\}$ **29.** $\{x \mid x = 3n, n \text{ a natural number}\}$.

Exercise 1.2 (page 8)   **1.** False   **3.** True   **5.** True   **7.** False   **9.** True   **11.** True

13. $\{1, 2, 3, 4, 5\}$ **15.** $\{1, 2, 3, 4, 5, 6\}$ **17.** $\{-9, -8, -7, -6\}$

19. $\{x \mid x \in N\}$ **21.** $\{x \mid x \in R\}$ **23.** $\{x \mid x \in R, x \text{ between } -4 \text{ and } 3\}$

25. a. $\{4\}$ **b.** $\{4, -2, 0\}$ **c.** $\left\{4, -2, \dfrac{2}{5}, 0, -\dfrac{3}{4}\right\}$ **d.** $\left\{4, -2, \dfrac{2}{5}, 0, -\dfrac{3}{4}, \sqrt{2}, \sqrt{7}\right\}$

27. Transitive law for equality, E-3   **29.** Symmetric law for equality, E-2

31. Transitive law for equality, E-3   **33.** Closure law for multiplication, F-6

35. Associative law for multiplication, F-7   **37.** Additive-inverse law, F-4

39. Commutative law for multiplication, F-10   **41.** Commutative law for addition, F-5

43. Distributive law, F-8   **45.** Commutative law for addition, F-5

47. Commutative law for multiplication, F-10   **49.** Distributive law, F-8

51. Closed   **53.** Closed   **55.** Closed   **57.** Not closed

59. $2y + (-5)$ **61.** $-2x + [-(y + z)]$ **63.** $2 \cdot \dfrac{1}{3}$ **65.** $x \cdot \dfrac{1}{3y + 2}$

Exercise 1.3 (page 13)   **1.** 1.1   **3.** 1.8-V   **5.** 1.11-III   **7.** 1.7   **9.** 1.9   **11.** 1.11-VII

13. 1.8-III   **15.** 1.11-IV   **17.** 1.2   **19.** 1.5

Exercise 1.4 (page 17)   **1.** I   **3.** III   **5.** II   **7.** IV   **9.** $7 > 3$ **11.** $-4 < -3$

13. $-1 \leq x \leq 1$ **15.** $x > 0$ **17.** $x \geq 0$ **19.** 3   **21.** 7   **23.** -2

25. $3x$ if $x \geq 0$, $-3x$ if $x < 0$ **27.** $x + 1$ if $x \geq -1$, $-(x + 1)$ if $x < -1$

29. $y - 3$ if $y \geq 3$, $-(y - 3)$ if $y < 3$ **31.** $|-2| < |-5|$ **33.** $-7 < |-1|$

35. $|-3| > 0$ **37.** $2 < 5$ **39.** $7 < 8$ **41.** $|x| < 3$

Exercise 1.5 (page 20)   **1.** $\{x \mid x > 1\}$ **3.** $\{x \mid x > 3\}$ **5.** $\left\{x \mid x \le \frac{13}{2}\right\}$

7. $\left\{x \mid x < 0 \text{ or } x \ge \frac{1}{2}\right\}$ **9.** $\left\{x \mid -\frac{8}{3} < x < -2\right\}$ **11.** $\{x \mid x < 0 \text{ or } 2 < x \le 4\}$

13. $\{x \mid -3 < x < 0 \text{ or } x > 2\}$ **15.** $\{6, -6\}$ **17.** $\{-3, 5\}$ **19.** $\left(\frac{1}{3}, 1\right)$

21. $\{x \mid -2 < x < 2\}$ **23.** $\{x \mid x \le -7 \text{ or } x \ge 1\}$ **25.** $\{x \mid x \le 1 \text{ or } x \ge 4\}$

27. $|x - 2| < 1$ **29.** $|x + 8| \le 1$ **31.** $|4x - 5| \le 19$

33. (I) If $a \ge 0$, then $-a \le 0$ and, by Def. 1.12, $|-a| = -(-a) = a$; also $|a| = a$; hence $|-a| = |a|$.
   (II) If $a < 0$, then $-a > 0$ and, by Def. 1.12, $|-a| = -a$; also $|a| = -a$; hence $|-a| = |a|$.

35. Since $a^2 \ge 0$ for all $a \in R$, $|a^2| = a^2$ (Def. 1.12). If $a \ge 0$, then $|a| = a$, so $|a|^2 = a^2$; while if $a < 0$, then $|a| = -a$, so $|a|^2 = (-a)(-a) = a^2$. Hence $|a^2| = |a|^2 = a^2$.

Exercise 1.6 (page 26)   **1.** $2i$ **3.** $4i\sqrt{2}$ **5.** $6i\sqrt{2}$ **7.** $4 + 2i$ **9.** $2 + 15i\sqrt{2}$

11. $2 + 2i$ **13.** $5 + 5i$ **15.** $5 - 3i$ **17.** $-1 - 2i$ **19.** $1 + i$ **21.** $-\frac{1}{10} + \frac{7}{10}i$

23. $-8 - 6i$ **25.** $4 + 2i$ **27.** $15 + 3i$ **29.** $-1 + 4i\sqrt{5}$ **31.** $-\frac{3}{2}i$ **33.** $\frac{6}{13} + \frac{9}{13}i$

35. $\frac{3}{5} - \frac{4}{5}i$ **37. a.** -1 **b.** 1 **c.** $-i$

39. Let $a + bi$ and $c + di$ be complex numbers. Then $(a + bi) + (c + di) = (a + c) + (b + d)i$. Since a, b, c, d are all real numbers, they obey the commutative law of addition; i.e., $a + c = c + a$ and $b + d = d + b$. Hence $(a + c) + (b + d)i = (c + a) + (d + b)i = (c + di) + (a + bi)$, and the assertion is proved.

41. Let $a + bi$ be a complex number; then there is a unique complex number $0 = 0 + 0i$ such that $(a + bi) + (0 + 0i) = (a + 0) + (b + 0)i = a + bi$ and $(0 + 0i) + (a + bi) = (0 + a) + (0 + b)i = a + bi$. This unique complex number is the additive-identity element.

43. Let $a + bi$ and $c + di$ be complex numbers. Then $(a + bi)(c + di) = (ac - bd) + (ad + bc)i$. Also $(c + di)(a + bi) = (ac - bd) + (ad + bc)i$. Hence $(a + bi)(c + di) = (c + di)(a + bi)$, and the assertion is proved.

45. Let $a + bi$ be a complex number. Then, there is a unique complex number $1 = 1 + 0i$ such that $(a + bi)(1 + 0i) = (a \cdot 1 - b \cdot 0) + (b \cdot 1 + a \cdot 0)i = a + bi$. Since multiplication of complex numbers is commutative, we have $(1 + 0i)(a + bi) = a + bi$, and the assertion is proved.

Chapter 1 Review (page 28)   **1.** $\in$ **2.** $\subset$ **3.** $\subset$ **4.** $\in$ **5.** $\{1, 2, 3, 4\}, \{1, 2, 3\}, \{1, 3, 4\}, \{1, 3\}$
   6. $\{x \mid x = 5n, n \text{ a natural number}\}$ **7.** $\{5\}$ **8.** R or $\{1, 5, 9\}$ **9.** $\emptyset$ **10.** T or $\{4, 6\}$ **11.** $\{3, 21\}$
   12. $\{-8, -1, 0, 3, 21\}$ **13.** $\left\{-8, \frac{-15}{7}, -1, 0, 3, \frac{13}{2}, 21\right\}$ **14.** $\{-\sqrt{5}, \sqrt{50}\}$

15. Cummutative law for addition   **16.** Commutative law for multiplication
17. Multiplicative inverse   **18.** Distributive property   **19.** Additive inverse
20. Distributive property   **21.** Th. 1.1
22. Th. 1.8-IV   **23.** Th. 1.11-II   **24.** Th. 1.10   **25.** Th. 1.8-VI   **26.** Th. 1.11-IV

27. Th. 1.11-Vl   **28.** Th. 1.11-VII   **29.** IV   **30.** II   **31.** $-6 > -9$ **32.** $4 \le y$
33. $y \le 0$ **34.** $y + 2 \ge 0$ **35.** 7   **36.** $x - 5$ if $x - 5 \ge 0$, $5 - x$ if $x - 5 < 0$
37. $x < 4$ **38.** $|x| \ge 6$ **39.** $x \le 27$ **40.** $x > \dfrac{3}{2}$ **41.** $x > \dfrac{7}{2}$ or $x < 3$.   **42** $\left\{1, -\dfrac{1}{3}\right\}$

43.   **44.** $|x - 4| \le 7$ **45.** $5 + i$ **46.** $1 - 5i$

47. $8 - i$ **48.** $-5 + 3i$ **49.** $21 - 20i$ **50.** $-\sqrt{14}$ **51.** $\dfrac{4}{5} - \dfrac{3}{5}i$ **52.** $1 + i$

Exercise 2.1 (page 33)   **1.** $\left\{(0, 6), (1, 4), (2, 2), (-3, 12), \left(\dfrac{2}{3}, \dfrac{14}{3}\right)\right\}$

3. $\left\{(0, 0), (1, -3), (2, 3), \left(-3, -\dfrac{9}{7}\right), \left(\dfrac{2}{3}, -\dfrac{9}{7}\right)\right\}$

5. $\left\{(0, \sqrt{11}), (1, \sqrt{14}), (2, \sqrt{17}), (-3, \sqrt{2}), \left(\dfrac{2}{3}, \sqrt{13}\right)\right\}$

7. **a.** Domain: $\{2, 5, 7\}$, range: $\{3, 7, 8\}$ **b.** A function
9. **a.** Domain: $\{2, 3\}$, range: $\{-1, 4, 6\}$ **b.** Not a function
11. **a.** Domain: $\{5, 6, 7\}$, range: $\{5, 6, 7\}$ **b.** A function   **13.** Domain: $\{x \mid x \in R\}$
15. Domain: $\{x \mid x \in R\}$ **17.** Domain: $\{x \mid x \in R, x \ne 2\}$ **19.** Domain: $\{x \mid x \in R, x \ge 0\}$
21. Domain: $\{x \mid x \in R, -2 \le x \le 2\}$ **23.** Domain: $\{x \mid x \in R, x \ne 0, 1\}$ **25.** Yes   **27.** Yes
29. No.   **31.** No   **33.** 2   **35.** -1 **37.** 9   **39.** a^2 **41.** 1   **43.** $a - 2$ **45.** ± 1
47. ± 3 **49.** **a.** 2 **b.** 0 **c.** 2 **d.** x **51.** **a.** even **b.** odd

Exercise 2.2 (page 38)
1.

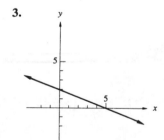

3.

5.

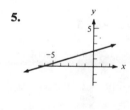

7.

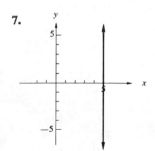

9.

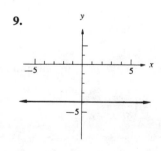

11. Distance, 5; slope, $\dfrac{4}{3}$

13. Distance, 13; slope $\dfrac{12}{5}$ **15.** Distance, $\sqrt{2}$; slope, 1   **17.** $7, 2\sqrt{17}, \sqrt{89}$ **19.** $10, 21, 17$

21. 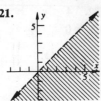   **23.** 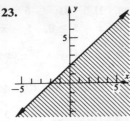   **25.**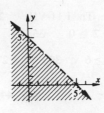

27. 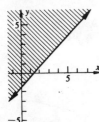   **29.** 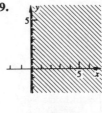   **31.** 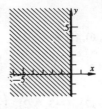   **33.**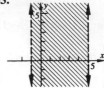

35. 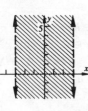   **37.**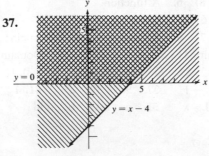

Exercise 2.3 (page 42)   **1.** $x - 2y - 1 = 0$ **3.** $x - y + 1 = 0$ **5.** $y - 2 = 0$

7. $y = -\frac{3}{2} x + \frac{1}{2}$; slope, $-\frac{3}{2}$; intercept, $\frac{1}{2}$ **9.** $y = \frac{1}{3} x - \frac{2}{3}$; slope, $\frac{1}{3}$; intercept, $-\frac{2}{3}$

11. $y = \frac{8}{3} x$; slope, $\frac{8}{3}$; intercept, 0

13. $x - 2y = 0$ **15.** $3x + 2y - 6 = 0$ **17.** $5x + 2y + 10 = 0$ **19.** $6x - 2y + 3 = 0$

21. Since $m = \dfrac{y_2 - y_1}{x_2 - x_1}$ and $y - y_1 = m(x - x_1)$, we have by substitution

$$y - y_1 = \left(\frac{y_2 - y_1}{x_2 - x_1}\right)(x - x_1).$$

Exercise 2.4 (page 45)   **1. a.** $3x + y = 16$ **b.** $x - 3y = 2$ **3. a.** $3x - 2y = 0$ **b.** $2x + 3y = 0$

5. a. $2x + 3y = -7$ **b.** $3x - 2y = -4$ **7. a.** $x + 2y = 5$ **b.** $2x - y = -10$

9. $x + y = 12$ **11.** $7x - 3y = -8$ **13.** $ax - by = \frac{1}{2}(a^2 - b^2)$

15. Slope of $\overline{AB}$; $m_1 = -1$; slope of $\overline{AC}$: $m_2 = 1$; $m_1 m_2 = -1$;   thus, by Th. 1.2, $\overline{AB} \perp \overline{AC}$.
Hence, $\triangle ABC$ is a right triangle.

17. Slope of $\overline{PQ}$: $m_4 = \dfrac{1}{3}$; slope of $\overline{QR}$: $m_2 = \dfrac{-2}{3}$; slope of $\overline{RS}$: $m_3 = \dfrac{1}{3}$;

slope of $\overline{SP}$: $m_4 = \dfrac{-2}{3}$; $m_1 = m_3$ and $m_2 = m_4$, thus, by Th. 2.1, $\overline{PQ}$ is parallel to $\overline{RS}$, and

$\overline{QR}$ is parallel to $\overline{SP}$. Hence, *PQRS* is a parallelogram.

19. From geometry, if L_1 is parallel to L_2, then $\angle P_3P_1Q_1$ is congruent to $\angle P_4P_2Q_2$. Also, $\angle P_1Q_1P_3$ is congruent to $\angle P_2Q_2P_4$, since each is a right angle. Hence, $\triangle P_1Q_1P_3$ is similar to $\triangle P_2Q_2P_4$. Because the lengths of corresponding sides of similar triangles are proportional, it follows that $\dfrac{Q_1P_3}{P_1Q_1} = \dfrac{Q_2P_4}{P_2Q_2}$.

But $\dfrac{Q_1P_3}{P_1Q_1} = \dfrac{y_3}{x_3 - x_1} = m_1$ and $\dfrac{Q_2P_4}{P_2Q_2} = \dfrac{y_4}{x_4 - x_2} = m_2$. Hence, $m_1 = m_2$.

21. $\dfrac{y_1 - y_2}{x_2 - x_1} = -\dfrac{y_2 - y_1}{x_2 - x_1} = -m_2$; $\dfrac{x_2 - x_1}{y_3 - y_1} = \dfrac{1}{m_1}$. Hence, $-m_2 = \dfrac{1}{m_1}$, or $m_1m_2 = -1$.

23. From geometry, if $\overline{PQ}$ is parallel to $\overline{P_2Q_2}$ and if P is the midpoint of $\overline{P_1P_2}$, then Q is the midpoint of $\overline{P_1Q_2}$, or $x_2 - x_1 = 2(x - x_1)$. Then, $x_2 - x_1 = 2x - 2x_1$

$$2x = x_1 + x_2$$
$$x = \frac{x_2 + x_1}{2}.$$

Also, from geometry, if $\overline{PQ}$ is parallel to $\overline{P_2Q_2}$ and if P is the midpoint of $\overline{P_1P_2}$, then $\overline{PQ}$ is one-half of $\overline{P_2Q_2}$, or $y_2 - y_1 = 2(y - y_1)$. Then, $y_2 - y_1 = 2y - 2y_1$

$$2y = y_1 + y_2$$
$$y = \frac{y_1 + y_2}{2}.$$

Exercise 2.5 (page 49)

1.

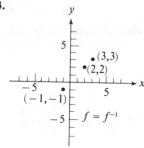

Points: (5,9), (4,7), (9,5), (7,4), (−2,3), (3,−2)

3.

Points: (3,3), (2,2), (−1,−1), $f = f^{-1}$

5.

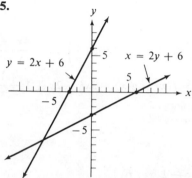

$y = 2x + 6$, $x = 2y + 6$

7.

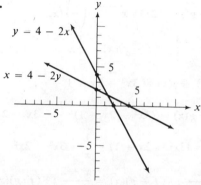

$y = 4 - 2x$, $x = 4 - 2y$

9.

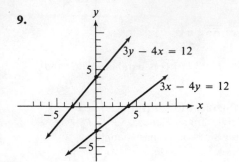

11.

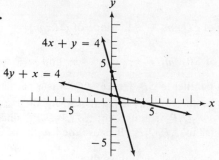

13. $F(x) = x$, $F^{-1}(x) = x$; $F[F^{-1}(x)] = F(x) = x$, $F^{-1}[F(x)] = F^{-1}(x) = x$.

15. $F(x) = 4 - 2x$, $F^{-1}(x) = \dfrac{4 - x}{2}$;

$$F[F^{-1}(x)] = F\left[\frac{4-x}{2}\right] = 4 - 2\left(\frac{4-x}{2}\right) = 4 - 4 + x = x,$$

$$F^{-1}[F(x)] = F^{-1}[4 - 2x] = \frac{4 - (4 - 2x)}{2} = \frac{4 - 4 + 2x}{2} = \frac{2x}{2} = x.$$

17. $F(x) = \dfrac{3x - 12}{4}$, $F^{-1}(x) = \dfrac{4x + 12}{3}$;

$$F[F^{-1}(x)] = F\left[\frac{4x + 12}{3}\right] = \frac{(4)\left(\dfrac{3x - 12}{4}\right) + 12}{3} = \frac{3x - 12 + 12}{3} = \frac{3x}{3} = x,$$

$$F^{-1}[F(x)] = F^{-1}\left[\frac{3x - 12}{4}\right] = \frac{(3)\left(\dfrac{4x + 12}{3}\right) - 12}{4} = \frac{4x + 12 - 12}{4} = \frac{4x}{4} = x.$$

Exercise 2.6 (page 51)

1. $(f + g)(x) = f(x) + g(x) = (3x - 2) + (2x + 1) = 5x - 1$;
$(f - g)(x) = f(x) - g(x) = (3x - 2) - (2x + 1) = x - 3$;
$(f \cdot g)(x) = f(x) \cdot g(x) = (3x - 2) \cdot (2x + 1) = 6x^2 - x - 2$;
$\left(\dfrac{f}{g}\right)(x) = \dfrac{f(x)}{g(x)} = \dfrac{3x - 2}{2x + 1}$

3. $(f + g)(x) = x^2 + 2x + 3$; $(f - g)(x) = -x^2 + 2x - 1$;
$(f \cdot g)(x) = 2x^3 + x^2 + 4x + 2$; $\left(\dfrac{f}{g}\right)(x) = \dfrac{2x + 1}{x^2 + 2}$

5. $(f + g)(x) = f(x) + g(x) = 2x^2 + \sqrt{x - 1}$; $(f - g)(x) = f(x) - g(x) = 2x^2 - \sqrt{x - 1}$;
$(f \cdot g)(x) = f(x) \cdot g(x) = 2x^2\sqrt{x - 1}$; $\left(\dfrac{f}{g}\right)(x) = \dfrac{2x^2}{\sqrt{x - 1}}$

7. $(f + g)(x) = \dfrac{5}{2}x - \dfrac{3}{2}$; $(f - g)(x) = \dfrac{3}{2}x - \dfrac{9}{2}$; $(f \cdot g)(x) = \dfrac{1}{2}(2x^2 + 3x - 9)$; $\left(\dfrac{f}{g}\right)(x) = \dfrac{4x - 6}{x + 3}$

9. $(f + g)(x) = f(x) + g(x) = 3x^2 - 2x + 1 + \dfrac{1}{x}$; $(f - g)(x) = f(x) - g(x) = 3x^2 - 2x + 1 - \dfrac{1}{x}$;
$(f \cdot g)(x) = f(x) \cdot g(x) = (3x^2 - 2x + 1) \cdot \dfrac{1}{x} = 3x - 2 + \dfrac{1}{x}$;
$\left(\dfrac{f}{g}\right)(x) = \dfrac{f(x)}{g(x)} = (3x^2 - 2x + 1) \div \dfrac{1}{x} = 3x^3 - 2x^2 + x$

11. $(f + g)(x) = \dfrac{x}{x - 1}$; $(f - g)(x) = \dfrac{x + 2}{x - 1}$; $(f \cdot g)(x) = \dfrac{-x - 1}{(x - 1)^2}$; $\left(\dfrac{f}{g}\right)(x) = -x - 1$

13. $(f \circ g)(x) = f[g(x)] = 3(2x + 1) - 2 = 6x + 1, \quad \{x \mid x \in R\}$ **15.** $2x^2 + 5, \quad \{x \mid x \in R\}$

17. $(f \circ g)(x) - 2(\sqrt{x - 1})^2 = 2x - 2$. The elements in the domain of $f \circ g$ must be in the domain of g, and each corresponding $g(x)$ must be in the domain of f. The domain of g is $\{x \mid x \in R, x \geq 1\}$ and the domain of f is $\{x \mid x \in R\}$; hence the domain of $f \circ g$ is $\{x \mid x \in R, x \geq 1\}$.

19. $f \circ g(x) = x, \quad \{x \mid x \in R\}$

21. $(f \circ g)(x) = f[g(x)] = 3\left(\dfrac{1}{x}\right)^2 - 2\left(\dfrac{1}{x}\right) + 1 = \dfrac{3}{x^2} - \dfrac{2}{x} + 1$. The domain of g is $\{x \mid x \in R, x \neq 0\}$

the domain of f is $\{x \mid x \in R\}$; hence the domain of $f \circ g$ is $\{x \mid x \in R, x \neq 0\}$.

23. $\dfrac{2 - x}{x}, \quad \{x \mid x \in R, x \neq 0, 1\}$ **25.** $(f + g)(2) = f(2) + g(2) = \dfrac{2 + 1}{2 - 1} + \dfrac{1}{2} = 3 + \dfrac{1}{2} = \dfrac{7}{2}$

27. Undefined.

29. $(f \circ g)(3) = f[g(3)]$ where $g(3) = \dfrac{1}{3}$;   hence $(f \circ g)(3) = f\left(\dfrac{1}{3}\right) = \dfrac{\frac{1}{3} + 1}{\frac{1}{3} - 1} = \dfrac{\frac{4}{3}}{-\frac{2}{3}} = -2$

31. $-\dfrac{25}{21}$ **33.** $(f + g)(x) = f(x) + g(x)$;   hence $(2x - 4) + g(x) = 5x + 2$, and $g(x) = 3x + 6$.

35. $g(x) = \dfrac{5x^2 + 2x}{2x - 4}$ **37.** $\left(\dfrac{g}{f}\right)(x) = \dfrac{g(x)}{f(x)} = \dfrac{g(x)}{2x - 4} = x + 1$;   hence $g(x) = 2x^2 - 2x - 4$.

39. Let $f(x) = 2x + 1$ and $f^{-1}(x) = \dfrac{x - 1}{2}$;   then $(f \circ f^{-1})(x) = 2\left(\dfrac{x - 1}{2}\right) + 1 = x - 1 + 1 = x$,

and $(f^{-1} \circ f)(x) = \dfrac{(2x + 1) - 1}{2} = \dfrac{2x}{2} = x$.

Chapter 2 Review (page 53)   **1.** $\{4, 2, 3\}$ **2.** $\{1\}$ **3.** $\{x \mid x \neq -4, x \in R\}$ **4.** $\{x \mid x \geq 6, x \in R\}$

5. 13   **6.** -5 **7.** 1   **8.** $x^2 + 2xh + h^2 + 4$

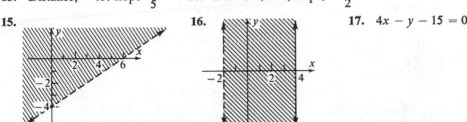

9. y-intercept $(0, 3)$, $(2, 0)$ **10.** $(0, 2)$, $(8, 0)$ **11.**   **12.**

13. Distance, $\sqrt{41}$; slope $\dfrac{-4}{5}$ **14.** Distance, $\sqrt{5}$; slope, $-\dfrac{1}{2}$

15.   **16.**   **17.** $4x - y - 15 = 0$

18. $x - 2y + 12 = 0$ **19.** $y = -4x + 6$;   slope, -4; y-intercept, 6

20. $y = \dfrac{3}{2}x - 8$;   slope, $\dfrac{3}{2}$; y-intercept, -8 **21.** $2x - 3y + 6 = 0$ **22.** $12x - y - 4 = 0$

23. **a.** $4x - 3y + 11 = 0$ **b.** $3x + 4y + 2 = 0$ **24.** **a.** $x + 6y + 21 = 0$ **b.** $6x - y - 22 = 0$

25. $3x - 2y + 10 = 0$ **26.** $14x + 4y - 27 = 0$

27.

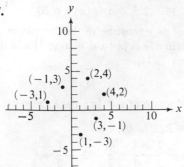

28.

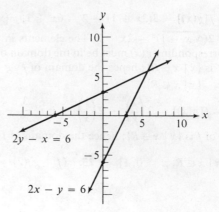

29. $(f + g)(x) = 2x^2 + x - 1$, $(f - g)(x) = 2x^2 - x - 3$, $(fg)(x) = 2x^3 + 2x^2 - 2x - 2$,

$\left(\dfrac{f}{g}\right)(x) = 2x - 2$, $(f \circ g)(x) = 2x^2 + 4x$

30. $(f + g)(x) = x^2 - x + 1 + \dfrac{1}{x - 1}$, $(f - g)(x) = x^2 - x + 1 - \dfrac{1}{x - 1}$, $(fg)(x) = \dfrac{x^2 - x + 1}{x - 1}$,

$\left(\dfrac{f}{g}\right)(x) = x^3 - 2x^2 + 2x - 1$, $(f \circ g)(x) = \dfrac{x^2 - 3x + 3}{(x - 1)^2}$

Exercise 3.1 (page 58)

1. $y, 4$; $x, 1$ and 4;

$\left(\dfrac{5}{2}, -\dfrac{9}{4}\right)$, minimum

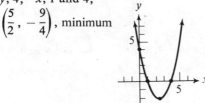

3. $y, -7$; $x, -1$ and 7;

$(3, -16)$, minimum

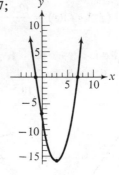

5. $y, -4$; $x, 1$ and 4;

$\left(\dfrac{5}{2}, \dfrac{9}{4}\right)$, maximum

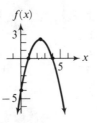

7. $y, 2$; no x;

$(0, 2)$, minimum

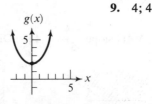

9. $4; 4$

11. Varying k has the effect of translating the graph along the y-axis.

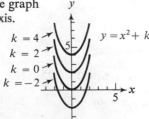

13. a. Parabola

b. No; there are two values of y associated with each value of x, except $x = 0$.

c. No

15.
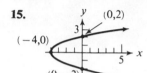

17.
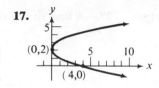

19.
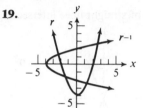

21.
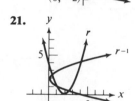

23.
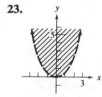

25.

27.

29.
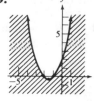

31.
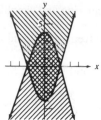

Exercise 3.2 (page 65)   1. a. $y = \pm\sqrt{4 - x^2}$ **b.** Domain $= \{x \mid -2 \le x \le 2\}$

3. a. $y = \pm 3\sqrt{4 - x^2}$

   b. Domain $= \{x \mid -2 \le x \le 2\}$

5. a. $y = \pm\dfrac{1}{2}\sqrt{16 - x^2}$ **b.** Domain $= \{x \mid -4 \le x < 4\}$

7. a. $y = \pm\dfrac{1}{\sqrt{3}}\sqrt{24 - 2x^2}$

   b. Domain $= \{x \mid -\sqrt{12} \le x \le \sqrt{12}\}$

9. a. $y = \pm\sqrt{x^2 - 1}$ **b.** Domain $= \{x \mid x \ge 1 \text{ or } x \le -1\}$

11. a. $y = \pm\sqrt{9 + x^2}$ **b.** Domain $= \{x \mid x \in R\}$

13. Circle; x-intercepts are 7 and   **15.** Ellipse; x-intercepts are 6 and   **17.** Two straight lines; inter-
    -7; y-intercepts are 7 and -7 -6; y-intercepts are 2 and -2 cepts at the origin

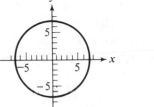

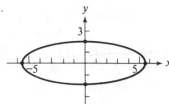

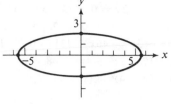

19. Hyperbola; x-intercepts are 3   **21.** Circle; x-intercepts are 1/2   **23.** Ellipse; x-intercepts are
    and -3; $y = \pm x$ and $-1/2$; y intercepts are 1/2   2 and -2; y-intercepts
          and $-1/2$ are $\sqrt{3}$ and $-\sqrt{3}$

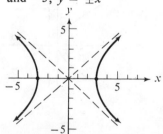

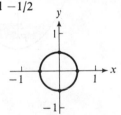

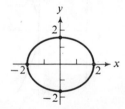

25. Pair of straight lines intersecting at the origin   **27.** A line, the *y*-axis

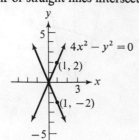

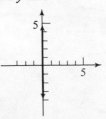

29. Since $x^2 + y^2 \geq 0$ for all $x, y \in R$, and $-1 < 0$, it follows that $x^2 + y^2 \neq -1$ for any $x, y \in R$.

31. Let $P(x, y)$ be a point at distance 4 from $(2, -3)$;   then by the distance formula
$$\sqrt{(x-2)^2 + (y+3)^2} = 4; \quad \text{squaring and simplifying yields } (x-2)^2 + (y+3)^2 = 16.$$

33. Given $Ax^2 + By^2 = C$ with intercepts $(a, 0)$ and $(0, b)$. If $y = 0$, $x = a$ and $x^2 = \dfrac{C}{A}$;

hence $a^2 = \dfrac{C}{A}$.   If $x = 0$, $y = b$ and $y^2 = \dfrac{C}{B}$; hence $b^2 = \dfrac{C}{B}$.

Because $Ax^2 + By^2 = C$, $\dfrac{x^2}{C/A} + \dfrac{y^2}{C/B} = 1$ and $\dfrac{x^2}{a^2} + \dfrac{y^2}{b^2} = 1$.

35. As $|x|$ increases, so do x^2 and Ax^2 of the fraction $\dfrac{C}{Ax^2}$;   the value of the fraction approaches zero,

and hence $\sqrt{1 - \dfrac{C}{Ax^2}}$ approaches 1, and $y = \pm\sqrt{\dfrac{A}{B}}\,|x|\left(\sqrt{1 - \dfrac{C}{Ax^2}}\right)$ approaches $y = \pm\sqrt{\dfrac{A}{B}}\,|x|$.

37.

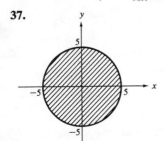

Exercise 3.3 (page 69)   **1.** $x^2 + y^2 - 2x = 0$ **3.** $x^2 + y^2 + 6x - 2y + 6 = 0$

5. $x^2 + y^2 - 2hx - 2ky + h^2 + k^2 - r^2 = 0$ **7.** $x^2 + y^2 - 6x - 10y + 9 = 0$

9. $x - y - 3 = 0$ **11.** $3x^2 + 3y^2 - 46x + 131 = 0$ **13.** $y^2 - 8x = 0$

15. $x^2 + 6x + 4y + 13 = 0$ **17.** $3x^2 + 4y^2 - 48 = 0$ **19.** $3x^2 - y^2 - 3 = 0$

21. $x^2 - 2py + p^2 = 0$ **23.** $(x + 4)^2 + (y - 1)^2 \geq 9$

Exercise 3.4 (page 72)

1. $y^2 = 16x$

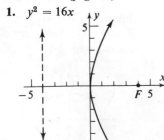

3. $x^2 = 4y$

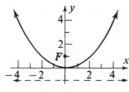

5. $y^2 = -3x$

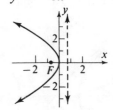

7. $x = -\dfrac{3}{2}; \left(\dfrac{3}{2}, 0\right)$ **9.** $y = \dfrac{5}{4}; \left(0, -\dfrac{5}{4}\right)$ **11.** $x = \dfrac{3}{16}; \left(-\dfrac{3}{16}, 0\right)$

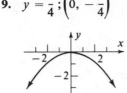

13. $y^2 = 20x$ **15.** $3x^2 = 16y$ **17.** $y^2 = \dfrac{4}{3}x$, or $3y^2 = 4x$.   **19.** $4x^2 = y$

21. $y^2 - 6x + 4y + 13 = 0$ **23.** $y^2 + 8x - 4y + 20 = 0$ **25.** $y^2 + 8x - 4y - 20 = 0$

Exercise 3.5 (page 74)   1. $\dfrac{x^2}{25} + \dfrac{y^2}{9} = 1$ **3.** $\dfrac{x^2}{4} + \dfrac{y^2}{9} = 1$ **5.** $\dfrac{x^2}{16} + \dfrac{y^2}{9} = 1$

7. $\dfrac{x^2}{25} + \dfrac{y^2}{75/7} = 1$ **9.** $\dfrac{x^2}{12} + \dfrac{y^2}{3} = 1$

11. Vertices $(3, 0)$ and $(-3, 0)$;   **13.** Vertices $(0, 6)$ and $(0, -6)$;

   foci $(\sqrt{5}, 0)$ and $(-\sqrt{5}, 0)$ foci $(0, 4\sqrt{2})$ and $(0, -4\sqrt{2})$

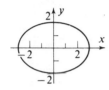   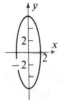

15. Vertices $(6, 0)$ and $(-6, 0)$;   **17.** Vertices $(0, \sqrt{10})$ and $(0, -\sqrt{10})$;

   foci $(\sqrt{11}, 0)$ and $(-\sqrt{11}, 0)$ foci $(0, \sqrt{5})$ and $(0, -\sqrt{5})$

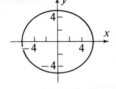   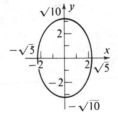

19. Vertices $(0, 3)$ and $(0, -3)$;   foci $\left(0, \dfrac{3\sqrt{3}}{2}\right)$ and $\left(0, -\dfrac{3\sqrt{3}}{2}\right)$

21. $8x^2 + 9y^2 + 32x - 72y + 104 = 0$ **23.** $9x^2 + 5y^2 - 90x + 10y + 50 = 0$

Exercise 3.6 (page 77)   **1.** $\dfrac{x^2}{9} - \dfrac{y^2}{7} = 1$ **3.** $\dfrac{y^2}{1} - \dfrac{x^2}{3} = 1$ **5.** $\dfrac{x^2}{4} - \dfrac{y^2}{21} = 1$ **7.** $\dfrac{y^2}{11} - \dfrac{x^2}{25} = 1$

9. $\dfrac{x^2}{3} - \dfrac{y^2}{2/9} = 1$ **11.** $\dfrac{x^2}{36} - \dfrac{y^2}{25} = 1$ or $\dfrac{y^2}{36} - \dfrac{x^2}{25} = 1$

13. a. Foci $(\sqrt{2}, 0)$ and $(-\sqrt{2}, 0)$;   vertices $(1, 0)$ and $(-1, 0)$

 b. $y = \pm \dfrac{b}{a} x$ or $y = x$ and $y = -x$

15. a. Foci $(\sqrt{10}, 0)$ and $(-\sqrt{10}, 0)$;   vertices $(1, 0)$ and $(-1, 0)$
 b. $y = 3x$ and $y = -3x$

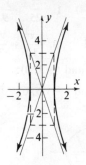

17. a. Foci $(0, \sqrt{2})$ and $(0, -\sqrt{2})$;   vertices $(0, 1)$ and $(0, -1)$
 b. $y = x$ and $y = -x$

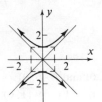

19. a. Foci $(0, \sqrt{13})$ and $(0, -\sqrt{13})$;   vertices $(0, 3)$ and $(0, -3)$

 b. $y = \dfrac{3}{2} x$ and $y = -\dfrac{3}{2} x$

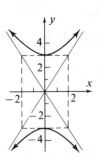

21. a. Foci $(2\sqrt{5}, 0)$ and $(-2\sqrt{5}, 0)$;   vertices $(4, 0)$ and $(-4, 0)$

 b. $y = \dfrac{1}{2} x$ and $y = -\dfrac{1}{2} x$

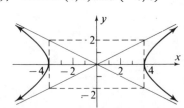

23. a. Foci $(0, 2\sqrt{10})$ and $(0, -2\sqrt{10})$;   vertices $(0, 2)$ and $(0, -2)$

   b. $y = \dfrac{1}{3}x$ and $y = -\dfrac{1}{3}x$

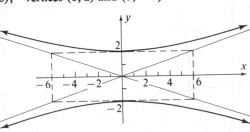

25. a. Foci $\left(\dfrac{3\sqrt{2}}{2}, 0\right)$ and $\left(-\dfrac{3\sqrt{2}}{2}, 0\right)$;   vertices $\left(\dfrac{3}{2}, 0\right)$ and $\left(-\dfrac{3}{2}, 0\right)$

   b. $y = \pm\dfrac{b}{a}x$, or $y = x$ and $y = -x$

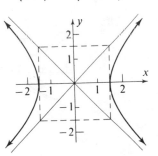

27. $x^2 - 3y^2 - 4x + 7 - 0$ **29.** $20x^2 - 16y^2 + 20x - 160y - 415 = 0$

Chapter 3 Review (page 78) **1.** x-intercepts $3, -2$;   axis of symmetry $x = \dfrac{1}{2}$;   minimum pt. $\left(\dfrac{1}{2}, \dfrac{-25}{4}\right)$

2. x-intercepts $5, 2$;   axis of symmetry $x = \dfrac{7}{2}$;   maximum pt. $\left(\dfrac{7}{2}, \dfrac{9}{4}\right)$

3.

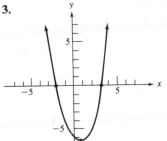

4.

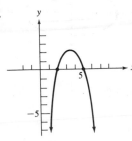

5.

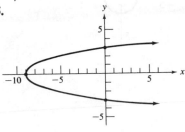

6.

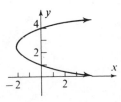

7.

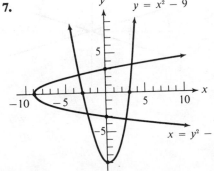

8.

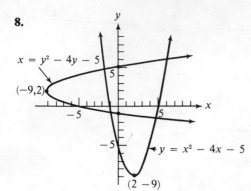

$x = y^2 - 4y - 5$

$(-9, 2)$

$y = x^2 - 4x - 5$

$(2, -9)$

9.

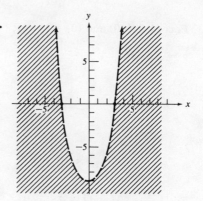

10.

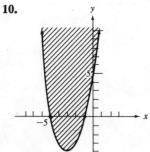

11.

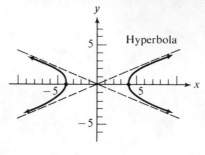

Hyperbola

12.

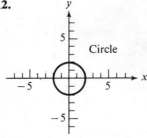

Circle

13.

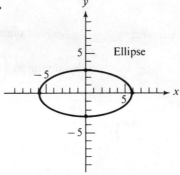

Ellipse

14.

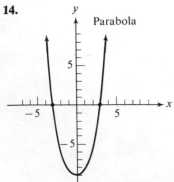

Parabola

15.

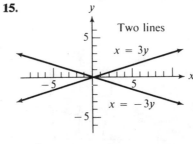

Two lines

$x = 3y$

$x = -3y$

16.

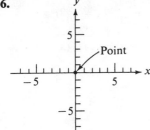

17. $(x - 2)^2 + (y - 3)^2 = 16$ **18.** $(x + 3)^2 + (y - 4)^2 = 4$

19. $3x^2 + 3y^2 + 4x - 34y + 59 = 0$ **20.** $15x^2 + 2xy + 15y^2 + 20x - 148y + 156 = 0$

21. $x^2 = 4(y - 3)$ **22.** $y^2 = 18\left(x - \dfrac{3}{2}\right)$ **23.** $y = 2, \ F(0, -2)$ **24.** $3x^2 = 2y$

25. $(x - 2)^2 = -8(y + 2)$ **26.** $(x - 2)^2 = 4(y - 5)$ **27.** $4x^2 + 9y^2 = 36$

28. $9x^2 + 4y^2 = 144$ **29.** Foci: $(0, 3), (0, -3)$;   vertices: $(0, 5), (0, -5)$

30. Foci: $(2\sqrt{7}, 0), (-2\sqrt{7}, 0)$;   vertices: $(4\sqrt{2}, 0), (-4\sqrt{2}, 0)$

31. $25(x + 6)^2 + 9(y - 6)^2 = 225$ **32.**

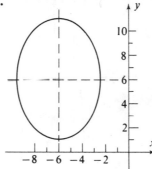

33. $15x^2 - y^2 = 15$ **34.** $4y^2 - 5x^2 = 80$

35. **a.** Foci: $(\sqrt{10}, 0), (-\sqrt{10}, 0)$;   vertices: $(3, 0), (-3, 0)$ **b.** $x + 3y = 0, x - 3y = 0$

    c.

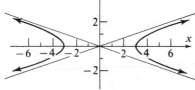

36. **a.** Foci: $(0, 2\sqrt{10}), (0, -2\sqrt{10})$;   vertices: $(0, 2), (0, -2)$ **b.** $x + 3y = 0, x - 3y = 0$

    c.

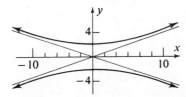

Exercise 4.1 (page 83) **1.** $2y^2 - y + \dfrac{13}{2} + \dfrac{-3/2}{2y + 1} \left(y \neq -\dfrac{1}{2}\right)$ **3.** $2y^3 - y^2 - \dfrac{1}{2}y - \dfrac{5}{4} + \dfrac{-13y/4 + 9/4}{2y^2 + y + 1}$

 5. $x^3 - x^2 - \dfrac{1}{x - 2}$ $(x \neq 2)$ **7.** $2x^2 - 2x + 3 + \dfrac{-8}{x + 1}$ $(x \neq -1)$

9. $2x^3 + 10x^2 + 50x + 249 + \dfrac{1251}{x - 5}$ $(x \neq 5)$ **11.** $x^2 + 2x - 3 + \dfrac{4}{x + 2}$ $(x \neq -2)$

13. $x^5 + x^4 + 2x^3 + 2x^2 + 2x + 1 + \dfrac{1}{x - 1}$ $(x \neq 1)$ **15.** $x^4 + x^3 + x^2 + x + 1$ $(x \neq 1)$

17. $3, 17, 47$ **19.** $56, 12, 326$ **21.** $-685, 95, 719$

Exercise 4.2 (page 87)

1. 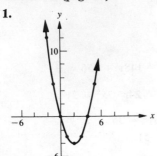   **3.** 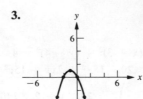   **5.**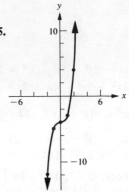

7. 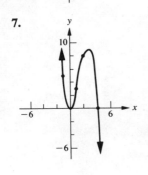   **9.** 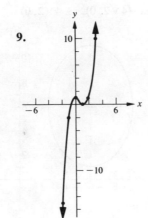   **11.**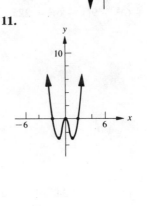

Exercise 4.3 (page 90)   **1.** 2 positive; 2 negative   **3.** No positive; 1 negative

5. No positive; no negative   **7.** Upper bound 3; lower bound -4

9. Upper bound 4; lower bound -3 **11.** Upper bound 2; lower bound -3

13. Upper bound 1; lower bound -2

21. 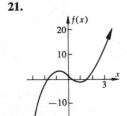   **23.** 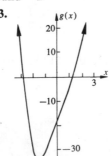   **25.**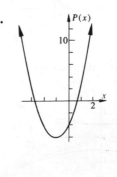

Exercise 4.4 (page 93)   **1.** $\{4\}$ **3.** $\{1, -2\}$ **5.** $\emptyset$ **7.** $\left\{-\dfrac{3}{2}\right\}$ **9.** $\left\{2, -\dfrac{7}{4}\right\}$ **11.** $\left\{\dfrac{3}{2}\right\}$

13. Consider $x^2 - 3 = 0$. If the equation has real zeros, then they are either rational or irrational. If rational, by Th. 4.9, the only possibilities are ± 1 and ± 3; however, direct substitution shows none of these satisfies the equation. Now, $\sqrt{3}$ is a real number, and since $(\sqrt{3})^2 - 3 = 3 - 3 = 0$, then $\sqrt{3}$ is a real zero of the equation, and since it is not among the rational zeros, it must be irrational.

Exercise 4.5 (page 95)   **1.** $\{i, -i\}$ **3.** $\left\{1, \dfrac{1}{2}\right\}$ **5.** $\left\{-\dfrac{1}{3}, 1 + \sqrt{5}, 1 - \sqrt{5}\right\}$

7. $\left\{-1, \dfrac{1}{2}, -\dfrac{1}{2} + i\dfrac{\sqrt{3}}{2}, -\dfrac{1}{2} - i\dfrac{\sqrt{3}}{2}\right\}$ **9.** $-2i$ **11.** $-i$ **13.** $i, -2i$

15. $3i, \dfrac{-3 + \sqrt{29}}{2}, \dfrac{-3 - \sqrt{29}}{2}$ **17.** $1 - i;\ x^3 - 2x + 4 = 0$

19. $\left\{2 + i, 2 - i, \dfrac{3}{2}\right\}$

Exercise 4.6 (page 100)   **1.** $x = 3$ **3.** $x = 3;\ x = -2$ **5.** $x = -1;\ x = -4$ **7.** $x = -1$

9.
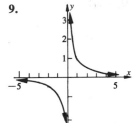

11.
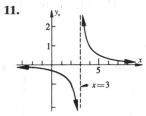

13.
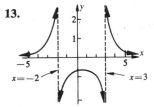

15.
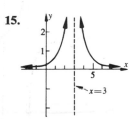

17. $x = 2;\ x = -2;\ y = 0$ **19.** $x = 4;\ y = x + 4$ **21.** $x = 4;\ x = -1;\ y = 1$

23.

25.
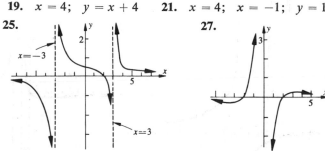

27.
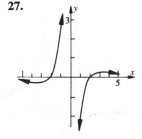

29.
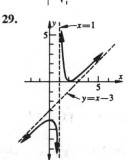

31.
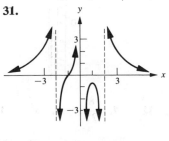

Chapter 4 Review (page 102)   **1.** $2n + 1$ **2.** $r + 1$ **3.** $2x^2 - 3x + 1$

4. $3y^2 - 2y + 3 + \dfrac{1}{y + 3}$ **5.** 1   **6.** 53   **7.**

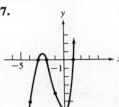

8.

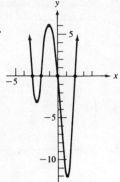

9. 3 positive; no negative zeros   **10.** 1 positive; 1 negative   **11.** Upper bound, 5; lower bound, -2

12. Upper bound, 2; lower bound, -5 **13.** $f(2) = -1 < 0, \quad f(3) = 9 > 0$

14. $g(0) = 1 > 0, \quad g(1) = -2 < 0$ **15.** $\{4\}$ **16.** $\left\{-\dfrac{1}{2}, 3\right\}$ **17.** $\{2, 2i, -2i\}$

18. $\left\{\dfrac{3}{2}, 2 + 2i, 2 - 2i\right\}$

19.

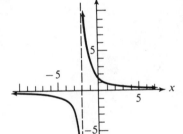

20.

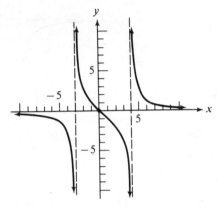

21.

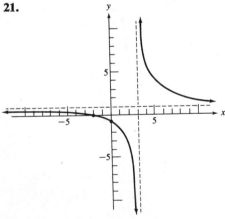

22.

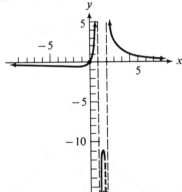

Exercise 5.1 (page 106)   **1.** 2   **3.** $\dfrac{1}{27}$ **5.** 32   **7.** $\dfrac{8}{27}$ **9.** x^2 **11.** $a^{17/12}$ **13.** $x^{1/3}$

15. $\dfrac{1}{x^{16/15}}$ **17.** $\dfrac{y^2}{x}$ **19.** $\left(\dfrac{x}{y}\right)^{5/4}$ **21.** $x^n \cdot y^4$ **23.** $x^{3n/2}$ **25.** $x^{5n/2} \cdot y^{3m/2-1}$ **27.** $x - x^{2/3}$

29. $x - 2x^{1/2}y^{-1/2} + y^{-1}$ **31.** $x + y - (x + y)^{3/2}$ **33.** $x + y$ **35.** 5   **37.** $2|x|$ or $-2|x|$

39. $\dfrac{2}{|x|(x+1)^{1/2}}$

Exercise 5.2 (page 109) **1.** $(0, 1), (1, 3), (2, 9)$ **3.** $(0, -1), (1, -5), (2, -25)$

5. $(-3, 8), (0, 1), \left(3, \dfrac{1}{8}\right)$ **7.** $\left(-2, \dfrac{1}{100}\right), (1, 10), (0, 1)$

9. 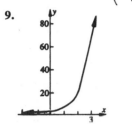   **11.** 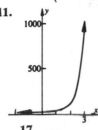   **13.**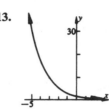

15. 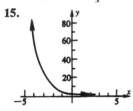   **17.** 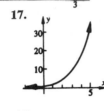   **19.** No; a constant function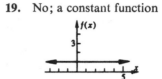

21. **a.** $\{-2\}$ **b.** $\{-4\}$ **c.** $\left\{\dfrac{3}{4}\right\}$

Exercise 5.3 (page 112) **1.** $\log_4 16 = 2$ **3.** $\log_3 27 = 3$ **5.** $\log_{1/2}\dfrac{1}{4} = 2$ **7.** $\log_8 \dfrac{1}{2} = -\dfrac{1}{3}$

9. $\log_{10} 100 = 2$ **11.** $\log_{10}(0.1) = -1$ **13.** $2^6 = 64$ **15.** $3^2 = 9$ **17.** $\left(\dfrac{1}{3}\right)^{-2} = 9$

19. $10^3 = 1000$ **21.** 2   **23.** 3   **25.** -1 **27.** 1   **29.** 2   **31.** -1 **33.** $\{2\}$
35. $\{2\}$ **37.** $\{64\}$ **39.** $\{-3\}$ **41.** $\{100\}$ **43.** $\{4\}$
45. By definition, $\log_b 1$ is a number such that $b^{\log_b 1} = 1$; therefore $\log_b 1 = 0$ **47.** $\log_b x + \log_b y$

49. $\log_b x - \log_b y$ **51.** $5\log_b x$ **53.** $\dfrac{1}{3}\log_b x$ **55.** $\dfrac{1}{2}(\log_b x - \log_b z)$

57. $\dfrac{1}{3}(\log_{10} x + 2\log_{10} y - \log_{10} z)$ **59.** $\log_b 2xy^3$ **61.** $\log_b x^{1/2}y^{2/3}$

63. $\log_b \dfrac{x^3 y}{z^2}$ **65.** $\log_{10} \dfrac{x(x-2)}{z^2}$

Exercise 5.4 (page 117) **1.** 2   **3.** -3 or $7 - 10$ **5.** -4 or $6 - 10$ **7.** 4   **9.** 0.8280
11. $9.9101 - 10$ **13.** $8.9031 - 10$ **15.** 2.3945   **17.** 4.10   **19.** 3.67   **21.** 0.0642
23. 5480   **25.** 0.000718   **27.** 0.6246   **29.** 3.1824   **31.** 4.5695   **33.** $9.7095 - 10$
35. 3.225   **37.** 10.52   **39.** 0.05076   **41.** 0.7495
43. $\log_{10} 3.751$; 0.751 is closer to a tabulated value   **45.** 9.1   **47.** 5000   **49.** 113
51. **a.** -1.2679 **b.** -3.5813

Exercise 5.5 (page 123) **1.** 4.014   **3.** 2.299   **5.** 0.000461   **7.** 64.34   **9.** 2.010
11. 3.435×10^{-10} **13.** 0.4582   **15.** 0.2776   **17.** 9.872   **19.** 4.745

21. $\left\{\dfrac{\log_{10} 7}{\log_{10} 2}\right\}$ **23.** $\left\{\dfrac{\log_{10} 8}{\log_{10} 3} - 1\right\}$ **25.** $\left\{\dfrac{1}{2}\left(\dfrac{\log_{10} 3}{\log_{10} 7} + 1\right)\right\}$ **27.** $n = \dfrac{\log_{10} y}{\log_{10} \dot{x}}$ **29.** $t = \dfrac{\log_{10} y}{k\log_{10} e}$

31. 1.343   **33.** 5%   **35.** 20 yrs   **37.** 12 yrs   **39.** $7396, $7430   **41.** 7   **43.** 7.7
45. 6.2   **47.** 1.0×10^{-3} **49.** 2.5×10^{-6} **51.** 6.3×10^{-8} **53.** 1.11 sec

Exercise 5.6 (page 128) **1.** 3.32   **3.** 3.41   **5.** 1.08   **7.** 0.79   **9.** 1.0986   **11.** 2.8332
13. 5.7900   **15.** 6.1093   **17.** 1.6487   **19.** 29.964   **21.** 1.260   **23.** 0.8607   **25.** 0.0821
27. 0.7600   **29.** 12.05 gr   **31.** 30.0 in., 16.1 in.

Chapter 5 Review (page 129) **1.** x^6y^{12} **2.** $\dfrac{x^3}{y^4}$ **3.** x^2 **4.** $\dfrac{x^2}{y^3}$ **5.** $x^{11/4}$ **6.** $\dfrac{y^{1/2}}{x^{1/3}}$

7. x^{2n-1} **8.** $x^{2n-2}y^4$ **9.** $\dfrac{1}{x^{n/2}y^n}$ **10.** xy **11.** $y^3 - y^2$ **12.** $y^2 - y$

13.

14. $x = -3$ **15.** $\log_{16}\frac{1}{4} = -\frac{1}{2}$ **16.** $\log_7 343 = 3$ **17.** $2^3 = 8$

18. $10^{-4} = 0.0001$ **19.** $y = 4$ **20.** 1,000   **21.** $\dfrac{1}{3}\log_{10} x + \dfrac{2}{3}\log_{10} y$

22. $\log_{10} 2 + 3\log_{10} R - \dfrac{1}{2}\log_{10} P - \dfrac{1}{2}\log_{10} Q$ **23.** $\log_b \dfrac{x^2}{\sqrt[3]{y}}$

24. $\log_{10}\dfrac{\sqrt[3]{x^2 y}}{z^3}$ **25.** 1.6232   **26.** $7.4969 - 10$ **27.** 2.8340

28. $8.6172 - 10$ **29.** 67.4   **30.** 0.0466   **31.** 2.655   **32.** 0.6643   **33.** 2.90   **34.** 895

35. 363.2   **36.** 49.65   **37.** 1449   **38.** 0.7711   **39.** $\dfrac{\log_{10} 2}{\log_{10} 3}$ **40.** $\dfrac{\log_{10} 80}{\log_{10} 3} - 1$ **41.** 2.18

42. 6.9   **43.** 4.8676   **44.** 121.51   **45.** 0.7866   **46.** 4.059

Exercise 6.1 (page 134) **1.** $\dfrac{\pi}{3}$, pos.   **3.** $\dfrac{3\pi}{7}$, pos.   **5.** $\dfrac{\pi}{5}$, pos.   **7.** $\dfrac{2\pi}{3}$, pos.   **9.** $\dfrac{\pi}{4}$, pos.

11. $\dfrac{4}{5}$, I   **13.** $-\dfrac{12}{13}$, IV   **15.** $\dfrac{\sqrt{5}}{3}$, IV   **17.** $-\dfrac{1}{2}$, III   **19.** $\dfrac{4}{5}$, IV   **21.** $-\sin\dfrac{\pi}{3}$

23. $\cos\dfrac{3\pi}{4}$ **25.** $-\sin\dfrac{13\pi}{6}$

27. Because the circle is symmetric to the horizontal axis,
it follows that if $\dfrac{\pi}{2} < s < \pi$, so that $(\cos s, \sin s) =$
$(-x, y)$, then $(\cos(-s), \sin(-s)) = (-x, -y)$, from
which the theorem follows.

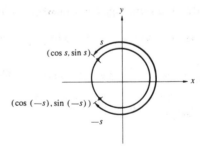

Exercise 6.2 (page 140) **1.** $\dfrac{1}{\sqrt{2}}$ **3.** $-\dfrac{1}{2}$ **5.** 1   **7.** -1 **9.** 0   **11.** 0   **13.** $\dfrac{1}{\sqrt{2}}$ **15.** $-\dfrac{1}{2}$

17. $\dfrac{\pi}{4}$ **19.** $\dfrac{5\pi}{6}$ **21.** $\dfrac{2\pi}{3}$ **23.** $\dfrac{4\pi}{3}$

25. Suppose that $0 < a < 2\pi$. Then if a is a period of the cosine function,

$$\cos(s + a) = \cos s, \quad \text{for all } s \in R.$$

Let $s = 0$; then

$$\cos(s + a) = \cos(0 + a) = \cos a = \cos 0 = 1.$$

But $\cos a$ must correspond to the first coordinate of a point on the unit circle, and the only point with first coordinate 1 is $(1, 0)$. This implies $a = 2k\pi$, contrary to hypothesis.

27. Let $f(s) = \cos s + \sin s$. Now we know that $\cos s = \cos (s + 2\pi)$ and $\sin s = \sin (s + 2\pi)$, so that
$$f(s) = \cos s + \sin s = \cos (s + 2\pi) + \sin (s + 2\pi) = f(s + 2\pi).$$
Thus $f(s) = f(s + 2\pi)$ and 2π is indeed a period of $f(s)$.

Exercise 6.3 (page 144) **1.** $-\cos \dfrac{2\pi}{5}$ **3.** $-\cos \dfrac{\pi}{8}$ **5.** $\cos \dfrac{\pi}{7}$ **7.** $-\cos \dfrac{\pi}{6}$ **9.** $-\cos \dfrac{3\pi}{7}$

11. $-\cos \dfrac{2\pi}{5}$ **13.** $-\cos \dfrac{2\pi}{11}$ **15.** $\cos \dfrac{2\pi}{9}$ **17.** 0.8309 **19.** 0.2482 **21.** 0.1403

23. 0.3474 **25.** -0.4267 **27.** 0.4801 **29.** -0.5319 **31.** -0.9950 **33.** 0.26

35. 0.79 **37.** 0.535

39. $\cos (x_1 - x_2) = \cos (x_1 + (-x_2))$
$$= \cos x_1 \cos (-x_2) - \sin x_1 \sin (-x_2) = \cos x_1 \cos x_2 + \sin x_1 \sin x_2$$

41. $\cos (2\pi - x) = \cos 2\pi \cos x + \sin 2\pi \sin x = 1 \cdot \cos x + 0 \cdot \sin x = \cos x$

43. $\cos \left(\dfrac{\pi}{2} + x\right) = \cos \dfrac{\pi}{2} \cos x - \sin \dfrac{\pi}{2} \sin x = 0 \cdot \cos x - 1 \cdot \sin x = -\sin x$

45. $\cos \left(\dfrac{3\pi}{2} + x\right) = \cos \dfrac{3\pi}{2} \cos x - \sin \dfrac{3\pi}{2} \sin x = 0 \cdot \cos x - (-1) \cdot \sin x = \sin x$

Exercise 6.4 (page 147) **1.** $-\sin \dfrac{3\pi}{5}$ **3.** $-\sin \dfrac{\pi}{8}$ **5.** $-\sin \dfrac{4\pi}{11}$ **7.** $\sin \dfrac{\pi}{7}$ **9.** $\sin \dfrac{\pi}{6}$

11. $-\sin \dfrac{5\pi}{11}$ **13.** $\sin \dfrac{3\pi}{5}$ **15.** $-\sin \dfrac{\pi}{5}$ **17.** 0.3051 **19.** 0.9356 **21.** 0.9975 **23.** 0.8634

25. 0.8016 **27.** -0.7243 **29.** -0.9391 **31.** 0.1889 **33.** 1.50 **35.** 1.03 **37.** 0.412

39. $\sin (x_1 - x_2) = \sin (x_1 + (-x_2))$
$$= \sin x_1 \cos (-x_2) + \cos x_1 \sin (-x_2) = \sin x_1 \cos x_2 - \cos x_1 \sin x_2$$

41. $\sin (\pi + x) = \sin \pi \cos x + \cos \pi \sin x = 0 \cdot \cos x + (-1) \cdot \sin x = -\sin x$

43. $\sin \left(\dfrac{\pi}{2} + x\right) = \sin \dfrac{\pi}{2} \cos x + \cos \dfrac{\pi}{2} \sin x = 1 \cdot \cos x + 0 \cdot \sin x = \cos x$

45. $\sin \left(\dfrac{3\pi}{2} + x\right) = \sin \dfrac{3\pi}{2} \cos x + \cos \dfrac{3\pi}{2} \sin x = (-1) \cdot \cos x + 0 \cdot \sin x = -\cos x$

Exercise 6.5 (page 155)

1.

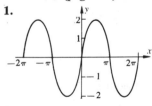

3.

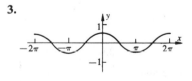

5.

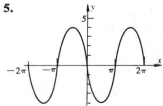

7.

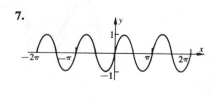

9.

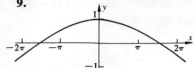

11.

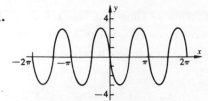

13.

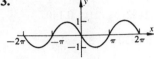

15.

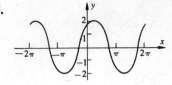

17.

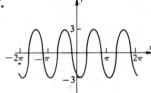

19.

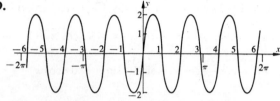

NOTE: LONG MARKS π UNITS; SHORT MARKS INTEGERS

21.

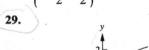

NOTE:
LONG MARKS π UNITS;
SHORT MARKS INTEGERS

23. $\left\{ -2\pi, -\dfrac{3\pi}{2}, -\pi, -\dfrac{\pi}{2}, 0, \dfrac{\pi}{2}, \pi, \dfrac{3\pi}{2}, 2\pi \right\}$

25. $\left\{ -\dfrac{3\pi}{2}, \dfrac{3\pi}{2} \right\}$

27. $\{-6, -5, -4, -3, -2, -1, 0, 1, 2, 3, 4, 5, 6\}$

29.

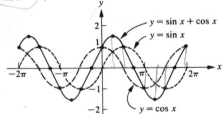

31.

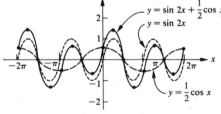

33.

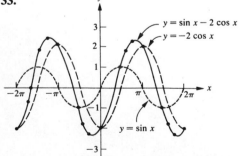

35.

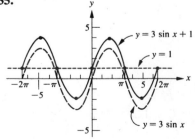

37.

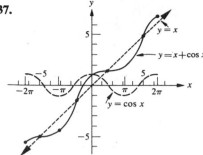

Exercise 6.6 (page 158)   **1.** $\frac{1}{2}\sqrt{2-\sqrt{2}}$ **3.** $\frac{1}{2}\sqrt{2+\sqrt{2}}$ **5.** $\frac{1}{2}\sqrt{2-\sqrt{2}}$

7. $\frac{1}{2}\sqrt{2-\sqrt{2-\sqrt{3}}}$ **9. a.** $-\frac{7}{25}$ **b.** $\frac{24}{25}$ **c.** $-\frac{2\sqrt{5}}{5}$ **d.** $-\frac{\sqrt{5}}{5}$ **11.** $\frac{1}{2}$ **13.** $-\frac{\sqrt{3}}{2}$

15. By Formula (3), $\cos 2x = 1 - 2\sin^2 x$, hence $\sin^2 x = \frac{1}{2}(1 - \cos 2x)$.

17. Use $\cos 3x = \cos(2x + x)$, Th. 6.3, Th. 6.9; then

$$\begin{aligned}
\cos 3x &= \cos(2x + x) = \cos 2x \cos x - \sin 2x \sin x \\
&= (\cos^2 x - \sin^2 x)\cos x - (2\sin x \cos x)\sin x \\
&= \cos^3 x - \sin^2 x \cos x - 2\sin^2 x \cos x \\
&= \cos^3 x - 3\sin^2 x \cos x.
\end{aligned}$$

From Th. 6.1, with $s = x$,

$$\cos^2 x + \sin^2 x = 1 \quad \text{and} \quad \sin^2 x = 1 - \cos^2 x;$$

hence

$$\begin{aligned}
\cos^3 x - 3\sin^2 x \cos x &= \cos^3 x - 3(1 - \cos^2 x)\cos x = \cos^3 x - 3\cos x + 3\cos^3 x \\
&= 4\cos^3 x - 3\cos x.
\end{aligned}$$

19. Let $x = x + \pi$; then use Th. 6.6 and substitution; $\cos 2\pi = 1$, $\sin 2\pi = 0$; then

$$\sin 2x = \sin[2(x + \pi)] = \sin(2x + 2\pi) = \sin 2x;$$

hence, π is a period of $\sin 2x$. To show π is fundamental period: suppose there exists a, $0 < a < \pi$, such that $\sin 2(x + a) = \sin 2x$. Then $\sin 2(x + a) = \sin(2x + 2a) = \sin 2x$ implies $a = 0$, but assumption was $0 < a < \pi$, so no a exists; hence π is the fundamental period of $\sin 2x$.

21. Let $x = x + 4\pi$; then use substitution $\cos 2\pi = 1$ and $\sin 2\pi = 0$; then

$$\sin \frac{x}{2} = \sin\left[\frac{1}{2}(x + 4\pi)\right] = \sin\left[\frac{x}{2} + 2\pi\right] = \sin \frac{x}{2};$$

hence 4π is a period for $\sin \frac{x}{2}$. To show π is the fundamental period, use procedure of Exercise 19 above, assuming $0 < a < 4\pi$.

23. By Formula (4) with $x = \frac{x}{2}$, we have $\cos 2\left(\frac{x}{2}\right) = 2\cos^2\left(\frac{x}{2}\right) - 1$.

Then $\cos x = 2\cos^2\left(\frac{x}{2}\right) - 1$ and $\cos^2\left(\frac{x}{2}\right) = \frac{1 + \cos x}{2}$; hence $\cos \frac{x}{2} = \pm\sqrt{\frac{1 + \cos x}{2}}$.

Exercise 6.7 (page 165)   **1.** $\cos x = -\frac{15}{17}$, $\sec x = -\frac{17}{15}$, $\csc x = -\frac{17}{8}$, $\tan x = \frac{8}{15}$, $\cot x = \frac{15}{8}$

3. $\sin x = \frac{5}{13}$, $\cos x = \frac{12}{13}$, $\sec x = \frac{13}{12}$, $\csc x = \frac{13}{5}$, $\cot x = \frac{12}{5}$

5. $\sin x = -\frac{1}{\sqrt{10}}$, $\cos x = \frac{3}{\sqrt{10}}$, $\sec x = \frac{\sqrt{10}}{3}$, $\csc x = -\sqrt{10}$, $\tan x = -\frac{1}{3}$

7. $\tan(\pi - x) = \dfrac{\tan \pi - \tan x}{1 + \tan \pi \tan x} = \dfrac{0 - \tan x}{1 + 0 \cdot \tan x} = -\tan x \quad \left(x \neq \dfrac{(2n+1)\pi}{2}, n \in J\right)$

9. $\tan\left(\dfrac{\pi}{2} - x\right) = \dfrac{\sin\left(\dfrac{\pi}{2} - x\right)}{\cos\left(\dfrac{\pi}{2} - x\right)} = \dfrac{\sin \dfrac{\pi}{2} \cos x - \cos \dfrac{\pi}{2} \sin x}{\cos \dfrac{\pi}{2} \cos x + \sin \dfrac{\pi}{2} \sin x}$

$= \dfrac{1 \cdot \cos x - 0 \cdot \sin x}{0 \cdot \cos x + 1 \cdot \sin x} = \dfrac{\cos x}{\sin x} = \cot x \quad (x \neq n\pi, n \in J)$

11. $\tan \dfrac{\pi}{7}$ **13.** $-\tan \dfrac{\pi}{11}$ **15.** $-\tan \dfrac{5\pi}{13}$ **17.** $\tan \dfrac{3\pi}{7}$ **19.** $-\sqrt{3}$ **21.** $-\dfrac{1}{\sqrt{3}}$

23. $-\sqrt{3}$ **25.** 5.177   **27.** 2.360   **29.** -0.0601 **31.** 0.1003   **33.** 1.247

35. -1.381 **37.** 0.97   **39.** 1.15   **41.** 1.55

Chapter 6 Review (page 168)   **1.** $\dfrac{\pi}{3}$, positive   **2.** $\dfrac{5\pi}{4}$, negative   **3.** $-\dfrac{5}{13}$ **4.** $-\dfrac{15}{17}$

5. $\cos \dfrac{2\pi}{3}$ **6.** $\sin \dfrac{5\pi}{4}$ **7.** $\dfrac{1}{2}$ **8.** $\dfrac{\sqrt{3}}{2}$ **9.** $-\dfrac{\sqrt{2}}{2}$ **10.** 0   **11.** $\dfrac{5\pi}{3}$ **12.** $\dfrac{7\pi}{4}$

13. $-\cos \dfrac{\pi}{5}$ **14.** $-\cos \dfrac{3\pi}{8}$ **15.** $-\cos \dfrac{5\pi}{7}$ **16.** $\cos \dfrac{3\pi}{11}$ **17.** 0.9090   **18.** 0.2141

19. -0.4447 **20.** -0.3058 **21.** 1.12   **22.** 0.701   **23.** $-\sin \dfrac{\pi}{8}$ **24.** $\sin \dfrac{\pi}{7}$

25. $-\sin \dfrac{\pi}{3}$ **26.** $\sin \dfrac{2\pi}{11}$ **27.** 0.9967   **28.** 0.3184   **29.** -0.4169 **30.** -0.6131

31. 0.44   **32.** 0.695   **33.**

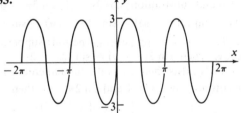

34.

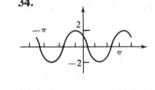

35.

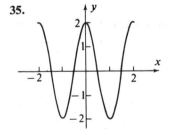

36.

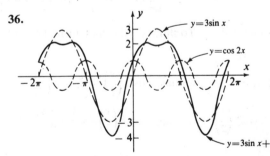

37. -0.92 **38.** 0.68   **39.** -0.74 **40.** 0.98

41. $\sin x = -\dfrac{3}{5}$, $\cos x = \dfrac{4}{5}$, $\cot x = -\dfrac{4}{3}$, $\sec x = \dfrac{5}{4}$, $\csc x = -\dfrac{5}{3}$

42. $\cos x = -\dfrac{12}{13}$, $\tan x = -\dfrac{5}{12}$, $\cot x = -\dfrac{12}{5}$, $\sec x = -\dfrac{13}{12}$, $\csc x = \dfrac{13}{5}$

43. -1.784 **44.** -2.176 **45.** 1.15   **46.** 0.237

Exercise 7.1 (page 173)　**1. a.** $0°$　**b.** $90°$　**c.** $180°$　**d.** $270°$　**e.** $360°$　**3.** $40°$　**5.** $252°$　**7.** $17.2°$
9. $207.5°$　**11.** 0.35　**13.** 2.27　**15.** 7.33　**17.** 13.08　**19.** $57.3°$
21. $390°, -330°; 30° + 360°k, k \in J$　**23.** $120°, 480°; -240° + 360°k, k \in J$
25. $60°, -300°; 420° + 360°k, k \in J$　**27.** $30°, 390°; -330° + 360°k, k \in J$
29. $s \approx 2.09$ cm　**31.** $s \approx 0.72$ m　**33.** $s \approx 4.71$ m

Exercise 7.2 (page 179)　**1.** $\sin \alpha = \dfrac{4}{5}, \quad \cos \alpha = \dfrac{3}{5}, \quad \tan \alpha = \dfrac{4}{3}, \quad \cot \alpha = \dfrac{3}{4}, \quad \sec \alpha = \dfrac{5}{3}, \quad \csc \alpha = \dfrac{5}{4}$

3. $\sin \alpha = -\dfrac{\sqrt{2}}{2}, \quad \cos \alpha = \dfrac{\sqrt{2}}{2}, \quad \tan \alpha = -1, \quad \cot \alpha = -1, \quad \sec \alpha = \sqrt{2}, \quad \csc \alpha = -\sqrt{2}$

5. $\sin \alpha = -\dfrac{4}{5}, \quad \cos \alpha = -\dfrac{3}{5}, \quad \tan \alpha = \dfrac{4}{3}, \quad \cot \alpha = \dfrac{3}{4}, \quad \sec \alpha = -\dfrac{5}{3}, \quad \csc \alpha = -\dfrac{5}{4}$

7. $\sin \alpha = 1, \quad \cos \alpha - 0, \quad \tan \alpha$ not defined, $\quad \cot \alpha = 0, \quad \sec \alpha$ not defined, $\quad \csc \alpha = 1$

9. IV　**11.** I　**13.** IV　**15.** $-\dfrac{1}{\sqrt{3}}$　**17.** $-\dfrac{1}{\sqrt{2}}$　**19.** $-\dfrac{1}{2}$　**21.** $-\dfrac{1}{2}$　**23.** $\dfrac{\sqrt{3}}{2}$

25. $\sqrt{3}$　**27.** 0.5299　**29.** 0.6494　**31.** 1.046　**33.** 0.808　**35.** 0.152　**37.** 5.911
39. 0.7431　**41.** -0.8391　**43.** -0.6428　**45.** 0.8772　**47.** 1.827　**49.** -0.0608
51. If the coordinates are associated with lengths of line segments as indicated in the
figure, then by the Pythagorean theorem, $OP_1 = \sqrt{x_1^2 + y_1^2}$ and $OP_2 = \sqrt{x_1^2 + y_1^2}$.
Since $\triangle OAP_1$ and $\triangle OBP_2$ are similar, their corresponding sides are proportional and the desired results follow by substituting in the following ratios:

$$\frac{OA}{AP_1} = \frac{OB}{BP_2}; \quad \frac{OA}{OP_1} = \frac{OB}{OP_2}; \quad \text{and} \quad \frac{AP_1}{OP_1} = \frac{BP_2}{OP_2}$$

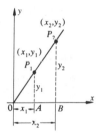

Exercise 7.3 (page 183)　**1.** $A = 36° 50', B = 53° 10', c = 10$　**3.** $A = 36°, a = 87.2, c = 148.3$
5. $B = 68°, a = 6.0, b = 14.8$　**7.** 16

9. $\cos \theta = \dfrac{-\sqrt{3}}{2}, \quad \tan \theta = \dfrac{1}{\sqrt{3}}, \quad \cot \theta = \sqrt{3}, \quad \csc \theta = -2, \quad \sec \theta = \dfrac{-2}{\sqrt{3}}$

11. $\sin \theta = \dfrac{2\sqrt{10}}{7}, \quad \tan \theta = -\dfrac{2\sqrt{10}}{3}, \quad \cot \theta = -\dfrac{3}{2\sqrt{10}}, \quad \sec \theta = -\dfrac{7}{3}, \quad \csc \theta = \dfrac{7}{2\sqrt{10}}$

13. $\cos \theta = \dfrac{-4}{5}, \quad \tan \theta = \dfrac{-3}{4}, \quad \cot \theta = \dfrac{-4}{3}, \quad \csc \theta = \dfrac{5}{3}, \quad \sec \theta = \dfrac{-5}{4}$

15. $\sin \theta = \dfrac{-5}{13}, \quad \cos \theta = \dfrac{-12}{13}, \quad \cot \theta = \dfrac{12}{5}, \quad \csc \theta = \dfrac{-13}{5}, \quad \sec \theta = \dfrac{-13}{12}$

17. $\sin \theta = \dfrac{-\sqrt{3}}{2}, \quad \cos \theta = \dfrac{-1}{2}, \quad \tan \theta = \sqrt{3}, \quad \cot \theta = \dfrac{1}{\sqrt{3}}, \quad \csc \theta = \dfrac{-2}{\sqrt{3}}$

19. $\dfrac{3 + \sqrt{2}}{2}$　**21.** 20.6　**23.** 394.8　**25.** 18.4 inches　**27.** $38° 40', 51° 20'$　**29.** 82.6 feet

Exercise 7.4 (page 188)　**1.** $C = 70°, a = 19.7, c = 18.8$　**3.** $B = 58°40', a = 74.4, b = 74.8$
5. $B = 47° 40', b = 66.2, c = 34.0$　**7.** One triangle possible　**9.** One triangle possible
11. Two triangles possible　**13.** A right triangle　**15.** $B = 18° 40', C = 48° 20', b = 1.7$
17. $A = 108° 40', C = 28° 40', a = 8.7$

19. $A = 107° 20'$, $B = 40° 10'$, $a = 2.7$; or $A = 7° 40'$, $B = 139° 50'$, $a = 0.4$

21. No possible solution   **23.** $B = 25°40'$, $C = 94°20'$, $c = 15.9$

25. $A = 93° 40'$, $C = 53° 00'$, $a = 763$; or $A = 19° 40'$, $C = 127° 00'$, $a = 257$

27. 21   **29.** 197.6 feet

31. From the law of sines and the properties of proportion, $\dfrac{a}{\sin \alpha} = \dfrac{b}{\sin \beta}$; $\dfrac{a}{b} = \dfrac{\sin \alpha}{\sin \beta}$;

$\dfrac{a}{b} + 1 = \dfrac{\sin \alpha}{\sin \beta} + 1$; then $\dfrac{a + b}{b} = \dfrac{\sin \alpha + \sin \beta}{\sin \beta}$.

33. Dividing result from Exercise 32 by the result of Exercise 31 gives

$\dfrac{a - b}{a + b} = \dfrac{\sin \alpha - \sin \beta}{\sin \alpha + \sin \beta}$; from Theorems 6.6 and 6.7 we see that

$$\sin x_1 \cos x_2 = \frac{1}{2}[\sin (x_1 + x_2) + \sin (x_1 - x_2)];$$

hence

$$\sin (x_1 + x_2) + \sin (x_1 - x_2) = 2 \sin x_1 \cos x_2.$$

Let $\alpha = x_1 + x_2$ and $\beta = x_1 - x_2$; then $2x_1 = \alpha + \beta$, $x_1 = \frac{1}{2}(\alpha + \beta)$; $2x_2 = \alpha - \beta$,

$x_2 = \frac{1}{2}(\alpha - \beta)$; so $\sin \alpha + \sin \beta = 2 \sin \frac{1}{2}(\alpha + \beta) \cos \frac{1}{2}(\alpha - \beta)$.

Using this result and replacing β by $-\beta$ yields

$$\sin \alpha - \sin \beta = 2 \sin \frac{1}{2}(\alpha - \beta) \cos \frac{1}{2}(\alpha + \beta).$$

Then

$\dfrac{a - b}{a + b} = \dfrac{\sin \alpha - \sin \beta}{\sin \alpha + \sin \beta} = \dfrac{2 \sin \frac{1}{2}(\alpha - \beta) \cos \frac{1}{2}(\alpha + \beta)}{2 \sin \frac{1}{2}(\alpha + \beta) \cos \frac{1}{2}(\alpha - \beta)} = \left[\dfrac{\sin \frac{1}{2}(\alpha - \beta)}{\cos \frac{1}{2}(\alpha - \beta)}\right] \cdot \left[\dfrac{\cos \frac{1}{2}(\alpha + \beta)}{\sin \frac{1}{2}(\alpha + \beta)}\right] =$

$\tan \frac{1}{2}(\alpha - \beta) \cdot \cot \frac{1}{2}(\alpha + \beta) = \dfrac{\tan \frac{1}{2}(\alpha - \beta)}{\tan \frac{1}{2}(\alpha + \beta)}$.

Exercise 7.5 (page 191) **1.** $c = 6.8$, $\alpha = 132° 50'$, $\beta = 17° 10'$ **3.** $b = 10.3$, $\alpha = 23° 40'$, $\gamma = 34°$

5. $a = 14.9$, $\beta = 75° 10'$, $\gamma = 24° 10'$.   **7.** $\alpha = 38° 10'$, $\beta = 81° 50'$, $\gamma = 60°$

9. $71° 40'$ **11.** 10.0

13. From the law of cosines, $a^2 = b^2 + c^2 - 2bc \cos \alpha$; $\cos \alpha = \dfrac{b^2 + c^2 - a^2}{2bc}$; then

$1 + \cos \alpha = 1 + \dfrac{b^2 + c^2 - a^2}{2bc} = \dfrac{(b^2 + 2bc + c^2) - a^2}{2bc} = \dfrac{(b + c)^2 - a^2}{2bc} = \dfrac{(b + c + a)(b + c - a)}{2bc}$.

15. Multiplying the results of Exercises 13 and 14 gives

$1 - \cos^2 \alpha = \dfrac{(b + c + a)(b + c - a)(a - b + c)(a + b - c)}{4b^2c^2}$;

since

$s = \dfrac{a + b + c}{2}$, $s - a = \dfrac{b + c - a}{2}$, $s - b = \dfrac{a - b + c}{2}$, and $s - c = \dfrac{a + b - c}{2}$.

Using the above results with the fact that $\mathscr{A} = \dfrac{1}{2} bc \sin \alpha$ and $\sin \alpha = \sqrt{1 - \cos^2 \alpha}$ yields

$\mathscr{A} = \dfrac{1}{2} bc \sqrt{\dfrac{(b + c + a)(b + c - a)(a - b + c)(a + b - c)}{4b^2c^2}}$

$= \sqrt{\left(\dfrac{b + c + a}{2}\right)\left(\dfrac{b + c - a}{2}\right)\left(\dfrac{a - b + c}{2}\right)\left(\dfrac{a + b - c}{2}\right)} = \sqrt{s(s - a)(s - b)(s - c)}$.

17. Let α be the right angle. Then a is the length of the hypotenuse in the right triangle; from the law of cosines and the fact that $\cos 90° = 0$,
$$a^2 = b^2 + c^2 - 2bc \cos 90° = b^2 + c^2.$$

19. $\mathscr{A} \approx 16$

Chapter 7 Review (page 192) **1.** $67.5°$ **2.** $499.9°$ **3.** 0.70^R **4.** 10.64^R **5.** $30°, 390°$

6. $-202°, 158°$ **7.** $5.0''$ **8.** $4.2'$ **9.** III **10.** III **11.** $-\dfrac{\sqrt{3}}{2}$ **12.** $-\dfrac{1}{2}$ **13.** 1

14. $-\dfrac{1}{2}$ **15.** $\dfrac{1}{2}$ **16.** $\dfrac{1}{\sqrt{2}}$ **17.** 0.9205 **18.** 2.912 **19.** 0.8660 **20.** -0.7071

21. -0.2079 **22.** 1.881 **23.** 0.8577 **24.** 0.9168 **25.** $B = 34°, a \approx 17.8, c \approx 21.5$

26. $A = 70°, a \approx 5.8, b \approx 2.1$ **27.** 7.5 **28.** $\mathscr{P} = 60$ in., $\mathscr{A} = 150\sqrt{3}$ sq in. ≈ 259.8 sq in.

29. $C = 78°, b \approx 2.0, c \approx 4.1$ **30.** $A = 67°10', a \approx 8.6, b \approx 5.0$ **31.** 2 **32.** None

33. $B \approx 26° 40', A \approx 131° 20', a \approx 10.0$ or $B \approx 153° 20', A \approx 4° 40', a \approx 1.1.$

34. For $B \approx 26° 40', \mathscr{A} \approx 11.2.$ For $B \approx 153° 20', \mathscr{A} \approx 1.2.$

35. $c \approx 3.5, \alpha \approx 44° 10', \beta \approx 111° 50'$ **36.** $a \approx 2.9, \beta \approx 62° 20', \gamma \approx 85° 50'$

37. Approximately $72°$ **38.** Approximately 7.7

Exercise 8.1 (page 197) **1.** $-\tan x$ **3.** $\tan x$ **5.** $\tan(x_1 + x_2)$ **7.** $\cos 2x$ **9.** $\sin(x_1 - x_2)$

11. $\tan 6x$ **13.** $\sin^2 x$ **15.** $\cos 2x_1$

17. By Identity 12, $\tan x = \dfrac{\sin x}{\cos x}$; hence $\cos x \tan x = \cos x \left(\dfrac{\sin x}{\cos x}\right) = \sin x.$

19. By Identity 18, $\cot x = \dfrac{\cos x}{\sin x}$; then $\cot^2 x = \dfrac{\cos^2 x}{\sin^2 x}$; hence $\sin^2 x \cot^2 x = \sin^2 x \left(\dfrac{\cos^2 x}{\sin^2 x}\right) = \cos^2 x.$

21. By Identity 14, $1 + \tan^2 \alpha = \sec^2 \alpha$; and by Identity 16, $\sec \alpha = \dfrac{1}{\cos \alpha}$. Then $\sec^2 \alpha = \dfrac{1}{\cos^2 \alpha}$;

hence $\cos^2 \alpha (1 + \tan^2 \alpha) = \cos^2 \alpha (\sec^2 \alpha) = \cos^2 \alpha \left(\dfrac{1}{\cos^2 \alpha}\right) = 1.$

23. By Identity 1, $\sin^2 \alpha + \cos^2 \alpha = 1$; hence $1 - \cos^2 \alpha = \sin^2 \alpha.$ Using Identities 16, 17, and 18 with this yields $\sec \alpha \csc \alpha - \cot \alpha = \left(\dfrac{1}{\cos \alpha}\right)\left(\dfrac{1}{\sin \alpha}\right) - \dfrac{\cos \alpha}{\sin \alpha} = \dfrac{1 - \cos^2 \alpha}{\cos \alpha \sin \alpha} = \dfrac{\sin^2 \alpha}{\cos \alpha \sin \alpha} =$

$\dfrac{\sin \alpha}{\cos \alpha} = \tan \alpha.$

25. By Identities 16 and 12, $\dfrac{\sin \theta \sec \theta}{\tan \theta} = \dfrac{(\sin \theta)\left(\dfrac{1}{\cos \theta}\right)}{\left(\dfrac{\sin \theta}{\cos \theta}\right)} = \dfrac{\sin \theta}{\sin \theta} = 1.$

27. By Identity 14, $\tan^2 \theta + 1 = \sec^2 \theta$; hence $\sec^2 \theta - 1 = \tan^2 \theta$; also, by Identity 1, $\sin^2 \theta + \cos^2 \theta = 1$, and if $\sin^2 \theta \neq 0$, then $\dfrac{\sin^2 \theta}{\sin^2 \theta} + \dfrac{\cos^2 \theta}{\sin^2 \theta} = \dfrac{1}{\sin^2 \theta}$, or $1 + \cot^2 \theta = \csc^2 \theta$; hence $\csc^2 \theta - 1 = \cot^2 \theta.$ Using these results with Identity 19 yields $(\sec^2 \theta - 1)(\csc^2 \theta - 1) = (\tan^2 \theta)(\cot^2 \theta) = (\tan^2 \theta)\left(\dfrac{1}{\tan^2 \theta}\right) = 1.$

29. From Identity 1, since $\sin^2 \theta + \cos^2 \theta = 1$, $1 - \sin^2 \theta = \cos^2 \theta.$ From this result and Identity 16, $\dfrac{1}{1 + \sin \theta} + \dfrac{1}{1 - \sin \theta} = \dfrac{1 - \sin \theta + 1 + \sin \theta}{(1 + \sin \theta)(1 - \sin \theta)} = \dfrac{2}{1 - \sin^2 \theta} = \dfrac{2}{\cos^2 \theta} = 2\left(\dfrac{1}{\cos \theta}\right)^2 = 2 \sec^2 \theta.$

31. From Identity 18, $\sin x \cot x = \sin x \left(\dfrac{\cos x}{\sin x}\right) = \cos x.$

33. From Identities 16 and 12,

$$\sec x - \cos x = \frac{1}{\cos x} - \cos x = \frac{1 - \cos^2 x}{\cos x} = \frac{\sin^2 x}{\cos x} = \sin x \left(\frac{\sin x}{\cos x}\right) = \sin x \tan x.$$

35. From Exercise 27 above, $1 + \cot^2 x = \csc^2 x$; also, from Identity 19, $\cot x = \dfrac{1}{\tan x}$;

hence $\tan x \cot x = 1$ and $\tan x = \dfrac{1}{\cot x}$, $\cot x \neq 0$. Then

$$\frac{1 + \tan^2 x}{\tan^2 x} = \frac{1}{\tan^2 x} + 1 = \cot^2 x + 1 = 1 + \cot^2 x = \csc^2 x.$$

37. From Identity 12 and the fact that $1 - \cos^2 \alpha = \sin^2 \alpha$ (from Exercise 23 above)

$$\tan^2 \alpha - \sin^2 \alpha = \frac{\sin^2 \alpha}{\cos^2 \alpha} - \sin^2 \alpha = \frac{\sin^2 \alpha - \sin^2 \alpha \cos^2 \alpha}{\cos^2 \alpha} = \frac{\sin^2 \alpha(1 - \cos^2 \alpha)}{\cos^2 \alpha}$$

$$= \left(\frac{\sin \alpha}{\cos \alpha}\right)^2 (1 - \cos^2 \alpha) = \tan^2 \alpha \sin^2 \alpha = \sin^2 \alpha \tan^2 \alpha.$$

39. From Identity 14,

$$\frac{1}{\sec \alpha - \tan \alpha} \cdot \frac{\sec \alpha + \tan \alpha}{\sec \alpha + \tan \alpha} = \frac{\sec \alpha + \tan \alpha}{\sec^2 \alpha - \tan^2 \alpha} = \frac{\sec \alpha + \tan \alpha}{(\tan^2 \alpha + 1) - \tan^2 \alpha} = \sec \alpha + \tan \alpha$$

41. From Identities 9, 12, 14, and 16, $\dfrac{2 \tan \alpha}{1 + \tan^2 \alpha} = \dfrac{2\left(\dfrac{\sin \alpha}{\cos \alpha}\right)}{\sec^2 \alpha} = \dfrac{2\left(\dfrac{\sin \alpha}{\cos \alpha}\right)}{\left(\dfrac{1}{\cos \alpha}\right)^2} = 2 \sin \alpha \cos \alpha = \sin 2\alpha.$

43. From Exercise 27 above, $1 + \cot^2 \theta = \csc^2 \theta$, and from Identities 19 and 22, $\cot 2\theta = \dfrac{\cot^2 \theta - 1}{2 \cot \theta}$.

   From these results and Identities 17 and 18,

$$\cot \theta - \cot 2\theta = \cot \theta - \frac{\cot \theta^2 - 1}{2 \cot \theta} = \frac{2 \cot^2 \theta - \cot^2 \theta + 1}{2 \cot \theta} = \frac{\cot^2 \theta + 1}{2 \cot \theta} = \frac{\csc^2 \theta}{2 \cot \theta} =$$

$$\frac{\csc x}{2}\left(\frac{\csc \theta}{\cot \theta}\right) = \frac{\csc \theta}{2}\left(\frac{\dfrac{1}{\sin \theta}}{\dfrac{\cos \theta}{\sin \theta}}\right) = \frac{\csc \theta}{2 \cos \theta} = \frac{1}{2 \cos \theta \sin \theta} = \frac{1}{\sin 2\theta} = \csc 2\theta.$$

45. From Identities 8c, 9, and the fact that $1 - \sin^2 \theta = \cos^2 \theta$ derived from Identity 1, followed by Identity 18,

$$\frac{1 + \cos 2\theta}{\sin 2\theta} = \frac{1 + (1 - 2 \sin^2 \theta)}{2 \sin \theta \cos \theta} = \frac{2(1 - \sin^2 \theta)}{2 \sin \theta \cos \theta} = \frac{2 \cos^2 \theta}{2 \sin \theta \cos \theta} = \frac{\cos \theta}{\sin \theta} = \cot \theta.$$

47. From Identity 8c, $\dfrac{1 - \cos 2x}{2} = \dfrac{1 - (1 - 2 \sin^2 x)}{2} = \dfrac{2 \sin^2 x}{2} = \sin^2 x.$

Exercise 8.2 (page 200)   1. $\left\{p \mid p = \dfrac{\pi}{6} + 2k\pi, k \in J\right\} \cup \left\{p \mid p = \dfrac{5\pi}{6} + 2k\pi, k \in J\right\}$ 3. $\emptyset$

5. $\left\{r \mid r = \dfrac{\pi}{4} + 2k\pi, k \in J\right\} \cup \left\{r \mid r = \dfrac{7\pi}{4} + 2k\pi, k \in J\right\}$

7. $\left\{x \mid x = \dfrac{4\pi}{3} + 2k\pi, k \in J\right\} \cup \left\{x \mid x = \dfrac{5\pi}{3} + 2k\pi, k \in J\right\}$

9. a. $\left\{\dfrac{\pi}{3} + 2k\pi\right\} \cup \left\{\dfrac{5\pi}{3} + 2k\pi\right\}, k \in J$

   b. $\left\{\left(\dfrac{\pi}{3} + 2k\pi\right)^R\right\} \cup \left\{\left(\dfrac{5\pi}{3} + 2k\pi\right)^R\right\}, k \in J;$ $\{(60 + 360k)^\circ\} \cup \{(300 + 360k)^\circ\}, k \in J$

11. a. $\left\{\dfrac{\pi}{3} + k\pi\right\}, k \in J$ **b.** $\left\{\left(\dfrac{\pi}{3} + k\pi\right)^R\right\}, k \in J$; $\{(60 + 180k)°\}, k \in J$

13. a. $\left\{\dfrac{\pi}{4} + 2k\pi\right\} \cup \left\{\dfrac{7\pi}{4} + 2k\pi\right\}, k \in J$

     b. $\left\{\left(\dfrac{\pi}{4} + 2k\pi\right)^R\right\} \cup \left\{\left(\dfrac{7\pi}{4} + 2k\pi\right)^R\right\}, k \in J$; $\{(45 + 360\,k)°\} \cup \{(315 + 360k)°\}, k \in J$

15. a. $\{0.25 + 2k\pi\} \cup \{2.89 + 2k\pi\}, k \in J$

     b. $\{(0.25 + 2k\pi)^R\} \cup \{2.89 + 2k\pi)^R\}, k \in J$; $\{14° \, 30' + 360°k\} \cup \{165° \, 30' + 360°k\}, k \in J$

17. a. $\{1.25 + k\pi\}, k \in J$ **b.** $\{(1.25 + k\pi)^R\}, k \in J$; $\{71° \, 30' + 180°k\}, k \in J$

19. a. $\{1.16 + 2k\pi\} \cup \{5.12 + 2k\pi\}, k \in J$

     b. $\{(1.16 + 2k\pi)^R\} \cup \{(5.12 + 2k\pi)^R\}, k \in J$; $\{66° \, 30' + 360°k\} \cup \{293° \, 30' + 360°k\}, k \in J$

21. $\left\{x \mid x - \dfrac{\pi}{3} + k\pi\right\}, k \in J$ **23.** $\left\{x \mid x = \dfrac{\pi}{4} + \dfrac{k\pi}{2}\right\}, k \in J$ **25.** $\left\{x \mid x - \dfrac{\pi}{2} + k\pi\right\}, k \in J$

27. $\{30°, 45°, 135°, 150°, 225°, 315°\}$ **29.** $\{0°, 120°, 180°, 240°\}$ **31.** $\left\{\dfrac{\pi^R}{4}, \dfrac{5\pi^R}{4}\right\}$

33. $\{0^R\}$ **35.** $\{1.11^R, 1.77^R, 4.25^R, 4.91^R\}$ **37.** $\{0.67^R, 2.48^R\}$ **39.** $\{1.9, 0, -1.9\}$

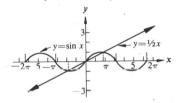

Exercise 8.3 (page 204)   **1.** $\{22° \, 30', 157° \, 30', 202° \, 30', 337° \, 30'\}$ **3.** $\{60°, 300°\}$

 5. $\{0°, 60°, 120°, 180°, 240°, 300°\}$ **7.** $\{45°, 225°\}$

 9. $\{67° \, 30', 157° \, 30', 247° \, 30', 337° \, 30'\}$ **11.** $\{30°, 90°, 150°, 210°, 270°, 330°\}$

13. $\left\{\dfrac{\pi}{2} + k\pi, \dfrac{\pi}{6} + 2k\pi, \dfrac{5\pi}{6} + 2k\pi\right\}, k \in J$ **15.** $\left\{\dfrac{\pi}{2} + k\pi, \dfrac{\pi}{3} + 2k\pi, \dfrac{5\pi}{3} + 2k\pi\right\}, k \in J$

17. $\left\{k\pi, \dfrac{\pi}{2} + k\pi\right\}, k \in J$ **19.** $\left\{\dfrac{k\pi}{2}\right\}, k \in J$

Exercise 8.4 (page 208)   **1.** $\dfrac{\pi}{6}$ **3.** $\dfrac{\pi}{4}$ **5.** Does not exist.   **7.** $\dfrac{-\pi}{6}$ **9.** 0.10   **11.** 0.39

13. -1.04 **15.** 1.29   **17.** $\dfrac{\pi}{4}$ **19.** $\dfrac{\pi}{3}$ **21.** $\dfrac{\sqrt{3}}{2}$ **23.** $\dfrac{-\sqrt{3}}{2}$ **25.** $\dfrac{\sqrt{3}}{2}$ **27.** $\dfrac{2}{3}$

29. $\dfrac{3 + 4\sqrt{3}}{10}$ **31.** $\dfrac{3\pi}{4}$ **33.** $\sqrt{1 - x^2}$ **35.** $\dfrac{y}{\sqrt{1 - y^2}}$ **37.** $\sqrt{\dfrac{1 + x}{2}}$

39. $x = \dfrac{1}{2}\cos\dfrac{y}{3}$ **41.** $x = \tan 2y - \pi$ **43.** Since $-\dfrac{\pi}{2} \leq \text{Arcsin}\dfrac{2}{5} \leq \dfrac{\pi}{2}$,

     $\tan\left(\text{Arcsin}\dfrac{2}{5}\right) = \dfrac{2}{\sqrt{21}}$; hence $\text{Arcsin}\dfrac{2}{5} = \text{Arctan}\dfrac{2}{\sqrt{21}}$.

45. Not always. Let Arccos (cos x) $= \alpha$. Then $0 \leq \alpha \leq \pi$ and $\cos \alpha = \cos x$; so $\alpha = x$ if and only if $0 \leq x \leq \pi$.

Exercise 8.5 (page 213)

1. 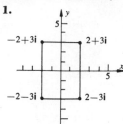 **3.** 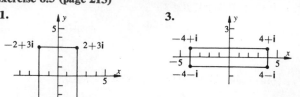 **5.**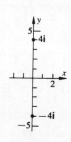

7. 4   **9.** $\sqrt{13}$ **11.** 2   **13.** $\sqrt{5}$ **15.** $3\sqrt{2}$ cis 45°   **17.** 5 cis 0°   **19.** 4 cis 330°

21. $-2 - 2\sqrt{3}\,i$ **23.** $3\sqrt{3} - 3i$ **25.** $6 + 6\sqrt{3}i$ **27. a.** $-3 + 3i$ **b.** $\dfrac{3}{2} + \dfrac{3}{2}i$

29. a. 108 **b.** $\dfrac{1}{6} - \dfrac{\sqrt{3}}{6}i$ **31. a.** $10 + 10i$ **b.** $\dfrac{1}{10} - \dfrac{7}{10}i$

33. Let $z = a + bi$; then $\bar{z} = a - bi$. $(a + bi) + (a - bi) = (a + a) + (b - b)i = 2a + 0i = 2a \in R$; $(a + bi) \cdot (a - bi) = (a^2 + b^2) + (ab - ab)i = (a^2 + b^2) + 0i = (a^2 + b^2) \in R$.

35. From the result of Exercise 34, and from Th. 8.1-I
$(a + bi)^3 = (a + bi)^2(a + bi) = (r^2 \text{ cis } 2\theta)(r \text{ cis } \theta) = r^3 \text{ cis } (2\theta + \theta) = r^3 \text{ cis } 3\theta.$

Exercise 8.6 (page 217)   **1.** $-64\sqrt{3} + 64i$ **3.** $1 + 0i$ **5.** $\dfrac{729}{2} + \dfrac{729\sqrt{3}}{2}i$ **7.** $\dfrac{-\sqrt{3}}{64} + \dfrac{1}{64}i$

9. $\dfrac{\sqrt{3} - 1}{64} - \dfrac{\sqrt{3} + 1}{64}i$ **11.** $-\dfrac{1}{2} - \dfrac{1}{2}i$

13. 2 cis 9°; 2 cis 81°; 2 cis 153°; 2 cis 225°; 2 cis 297°

15. 2 cis 30°; 2 cis 102°; 2 cis 174°; 2 cis 246°; 2 cis 318°

17. cis 45°; cis 105°; cis 165°; cis 225°; cis 285°; cis 345°

19. 2 cis 60°; 2 cis 132°; 2 cis 204°; 2 cis 276°; 2 cis 348°

21. cis $25\frac{5}{7}$°; cis $77\frac{1}{7}$°; cis $128\frac{4}{7}$°; cis 180°; cis $231\frac{3}{7}$°; cis $282\frac{6}{7}$°; cis $334\frac{2}{7}$°

23. $(x - 2 \text{ cis } 45°)(x - 2 \text{ cis } 135°)(x - 2 \text{ cis } 225°)(x - 2 \text{ cis } 315°)$

25. The four roots are $-1, 1, i, -i$; hence their sum is $0 + 0i$.

Exercise 8.7 (page 220)   **1.** $(6, 125°), (6, -235°), (-6, 305°), (-6, -55°)$

3. $(-2, -90°), (-2, 270°), (2, 90°), (2, -270°)$ **5.** $(6, -240°), (6, 120°), (-6, -60°), (-6, 300°)$

7. $\left(\dfrac{5}{\sqrt{2}}, \dfrac{5}{\sqrt{2}}\right)$ **9.** $\left(\dfrac{\sqrt{3}}{4}, -\dfrac{1}{4}\right)$ **11.** $\left(\dfrac{-10}{\sqrt{2}}, \dfrac{-10}{\sqrt{2}}\right)$ **13.** $(6, 45°), (6, -315°)$

15. $(2, 240°), (2, -120°)$ **17.** $(0, \theta°), (0, -\theta°)$, for any $\theta > 0$ **19.** $r = 5$ **21.** $r \sin \theta = -4$

23. $r^2(\cos^2 \theta + 9 \sin^2 \theta) = 9$ **25.** $x^2 + y^2 = 25$ **27.** $x^2 + y^2 - 9x = 0$ **29.** $y^2 = 4x + 4$

31. $\sec \theta/2 = \dfrac{1}{\cos (\theta/2)}$ and $\cos^2 (\theta/2) = \dfrac{1 + \cos \theta}{2}$; hence $r = \sec^2 (\theta/2) = \dfrac{1}{\cos^2 (\theta/2)} = \dfrac{2}{1 + \cos \theta}$.

Now $\cos \theta = \dfrac{x}{r}$ and $r^2 = x^2 + y^2$; so $r = \dfrac{2}{1 + \dfrac{x}{r}}$ and $r\left(1 + \dfrac{x}{r}\right) = 2$;

hence $r + x = 2$, or $r = 2 - x$. Squaring each member gives $r^2 = 4 - 4x + x^2$ and substituting for r^2 gives $x = 1 - \dfrac{1}{4}y^2$, whose graph is a parabola.

33.

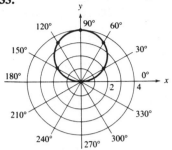

35.

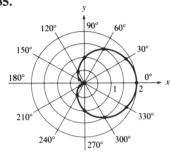

37.

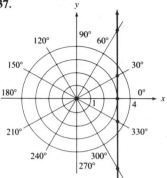

39.

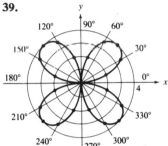

41.

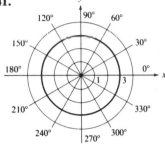

Chapter 8 Review (page 222) **1.** $\tan x = \sin x \cdot \sec x = \sin x \cdot \dfrac{1}{\cos x} = \dfrac{\sin x}{\cos x} = \tan x$

2. $\cos x - \sin x = \left(1 - \dfrac{\sin x}{\cos x}\right)\cos x = (1 - \tan x)\cos x$

3. $\dfrac{1 - \tan^2 \alpha}{\tan \alpha} = \dfrac{1}{\tan \alpha} - \tan \alpha = \cot \alpha - \tan \alpha$

4. $\dfrac{\cos^2 \alpha}{1 - \sin \alpha} = \dfrac{\cos \alpha}{\dfrac{1}{\cos \alpha} - \dfrac{\sin \alpha}{\cos \alpha}} = \dfrac{\cos \alpha}{\sec \alpha - \tan \alpha}$

5. $\dfrac{\sin 2\theta}{1 + \cos 2\theta} = \dfrac{2 \sin \theta \cos \theta}{1 + (2 \cos^2 \theta - 1)} = \dfrac{2 \sin \theta \cos \theta}{2 \cos^2 \theta} = \dfrac{\sin \theta}{\cos \theta} = \tan \theta$

6. $(\cos^2 \theta - \sin^2 \theta)^2 + \sin^2 2\theta = (\cos 2\theta)^2 + \sin^2 2\theta = 1$

7. $\{x \mid x = \dfrac{\pi}{6} + 2\pi k\} \cup \left\{x \mid x = -\dfrac{\pi}{6} + 2\pi k\right\}, k \in J$

8. $\left\{y \mid y = \dfrac{2\pi}{3} + k\pi, k \in J\right\}$ **9.** $\left\{x \mid x - \dfrac{\pi}{6} + k\pi, k \in J\right\}$

10. $\{0°, 180°, 360°\}$ **11.** $\{30°, 90°, 150°, 270°\}$ **12.** $\left\{\dfrac{\pi}{2}\right\}$

13. $\{20°, 40°, 140°, 160°, 260°, 280°\}$ **14.** $\{90°, 270°\}$ **15.** $\left\{x \mid x = \dfrac{\pi}{8} + k\dfrac{\pi}{2}, k \in J\right\}$

16. $\left\{x \mid x = \dfrac{\pi}{8} + k\dfrac{\pi}{2}\right\} \cup \left\{x \mid x \approx 0.554 + k\dfrac{\pi}{2}\right\}, k \in J$ **17.** $\dfrac{\pi}{6}$ **18.** $\dfrac{\pi}{6}$ **19.** $55° \ 00'$

20. $59° \ 00'$ **21.** $\dfrac{\pi}{6}$ **22.** $\dfrac{\pi}{4}$ **23.** 0 **24.** $-\dfrac{1}{2}$ **25.** $\dfrac{1}{\sqrt{x^2 + 1}}$ **26.** $\dfrac{\sqrt{1 - x^2}}{x}$

27. $x = \dfrac{1}{2}\cos y$ **28.** $x = \pi + \tan \dfrac{1}{2}y$ **29.** $2\sqrt{2} \text{ cis } 45°$ **30.** $10 \text{ cis } 300°$

31. $-\dfrac{3\sqrt{3}}{2}+\dfrac{3}{2}i$ **32.** $2-2\sqrt{3}i$ **33. a.** $-9\sqrt{3}+9i$ **b.** $0-\dfrac{2}{9}i$ **34. a.** $4+19i$

b. $-\dfrac{16}{29}-\dfrac{11}{29}i$ **35.** $-\dfrac{243\sqrt{2}}{2}-\dfrac{243\sqrt{2}}{2}i$

36. $-1-\sqrt{3}i$ **37.** $2^{1/6}\operatorname{cis}15°,\quad 2^{1/6}\operatorname{cis}135°,\quad 2^{1/6}\operatorname{cis}255°$

38. $6^{1/4}\operatorname{cis}82.5°,\quad 6^{1/4}\operatorname{cis}172.5°,\quad 6^{1/4}\operatorname{cis}262.5°,\quad 6^{1/4}\operatorname{cis}352.5°$

39. $\{\operatorname{cis}45°,\operatorname{cis}135°,\operatorname{cis}225°,\operatorname{cis}315°\}$

40. $\{\operatorname{cis}0°,\operatorname{cis}72°,\operatorname{cis}144°,\operatorname{cis}216°,\operatorname{cis}288°\}$

41. $(-4, 120°), (4, 300°), (4, -60°), (-4, -240°)$ **42.** $(-3\sqrt{2}, 3\sqrt{2})$ **43.** $r=9\cos\theta$

44. $x^2+y^2-4x=0$ **45.**

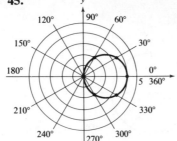

46.

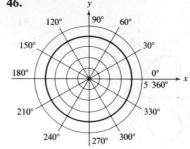

Exercise 9.1 (page 227)   **1. a.** For $n=1:\dfrac{n}{2}=\dfrac{1}{2};\dfrac{n(n+1)}{4}=\dfrac{1(1+1)}{4}=\dfrac{1}{2}$

b. For $n=k:\dfrac{1}{2}+\dfrac{2}{2}+\dfrac{3}{2}+\cdots+\dfrac{k}{2}=\dfrac{k(k+1)}{4}$ and $(k+1)$th term $=\dfrac{k+1}{2}$;   hence

$$\dfrac{1}{2}+\dfrac{2}{2}+\dfrac{3}{2}+\cdots+\dfrac{k}{2}+\dfrac{k+1}{2}=\dfrac{k(k+1)}{4}+\dfrac{k+1}{2}=\dfrac{k^2+k+2k+2}{4}$$

$$=\dfrac{k^2+3k+2}{4}=\dfrac{(k+1)(k+2)}{4}.$$

3. a. For $n=1: 2n=2(1)=2; n(n+1)=1(1+1)=2.$

b. For $n=k: 2+4+6+\cdots+2k=k(k+1)$ and $(k+1)$th term is $2(k+1)$;

hence $2+4+6+\cdots+2k+2(k+1)=k(k+1)+2(k+1)=(k+1)(k+2).$

5. a. For $n=1: n^2-1^2=1;\dfrac{n(n+1)(2n+1)}{6}=\dfrac{1(2)(3)}{6}=1.$

b. For $n=k: 1^2+2^2+3^2+\cdots+k^2=\dfrac{k(k+1)(2k+1)}{6}$ and $(k+1)$th term is $(k+1)^2$;

hence $1^2+2^2+3^2+\cdots+k^2+(k+1)^2=\dfrac{k(k+1)(2k+1)}{6}+(k+1)^2$

$$=\dfrac{k(k+1)(2k+1)+6(k+1)^2}{6}=\dfrac{(k+1)[k(2k+1)+6(k+1)]}{6}$$

$$=\dfrac{(k+1)(2k^2+7k+6)}{6}=\dfrac{(k+1)(k+2)(2k+3)}{6}=\dfrac{(k+1)[(k+1)+1][2(k+1)+1]}{6}.$$

7. a. For $n=1: (2n-1)^3=(2-1)^3=1^3=1; n^2(2n^2-1)=1(2-1)=1(1)=1.$

b. For $n=k: 1^3+3^3+5^3+\cdots+(2k-1)^3=k^2(2k^2-1)$ and the $(k+1)$th term is

$[2(k+1)-1]^3=(2k+1)^3;$

hence $1^3+3^3+5^3+\cdots+(2k-1)^3+(2k+1)^3=k^2(2k^2-1)+(2k+1)^3$

$$=2k^4+8k^3+11k^2+6k+1;$$

by use of the factor theorem and synthetic division,

$2k^4+8k^3+11k^2+6k+1=(k+1)(k+1)(2k^2+4k+1);$

also

$$2k^2 + 4k + 1 = 2(k^2 + 2k + 1) - 2 + 1 = 2(k + 1)^2 - 1;$$

hence

$$2k^4 + 8k^3 + 11k^2 + 6k + 1 = (k + 1)^2[2(k + 1)^2 - 1].$$

9. **a.** For $n = 1$: $n(n + 1) = 1(2) = 2$; $\dfrac{n(n + 1)(n + 2)}{3} = \dfrac{1(2)(3)}{3} = 2.$

   b. For $n = k$: $1 \cdot 2 + 2 \cdot 3 + 3 \cdot 4 + \cdots + k(k + 1) = \dfrac{k(k + 1)(k + 2)}{3}$

   and the $(k + 1)$th term is

   $(k + 1)[(k + 1) + 1] = (k + 1)(k + 2);$ hence

   $$1 \cdot 2 + 2 \cdot 3 + 3 \cdot 4 + \cdots + k(k + 1) + [(k + 1)(k + 2)] = \frac{k(k + 1)(k + 2)}{3} + (k + 1)(k + 2)$$

   $$= \frac{[k(k + 1)(k + 2)] + [3(k + 1)(k + 2)]}{3} = \frac{(k + 1)(k + 2)(k + 3)}{3}$$

   $$= \frac{(k + 1)[(k + 1) + 1][(k + 1) + 2]}{3}.$$

11. For $n = k$: $2 + 4 + 6 + \cdots + 2k = k(k + 1) + 2$ and $(k + 1)$th term is $2(k + 1)$; hence
     $$2 + 4 + 6 + \cdots + 2k + 2(k + 1) = k(k + 1) + 2 + 2(k + 1) = (k^2 + 3k + 2) + 2$$
     $$= (k + 1)(k + 2) + 2 = (k + 1)[(k + 1) + 1] + 2.$$
     However, for $n = 1$: $2n = 2(1) = 2$; $n(n + 1) + 2 = 1(2) + 2 = 4$. Hence not true for every $n \in N$.

Exercise 9.2 (page 230) **1.** $-4, -3, -2, -1$ **3.** $-\dfrac{1}{2}, 1, \dfrac{7}{2}, 7$ **5.** $2, \dfrac{3}{2}, \dfrac{4}{3}, \dfrac{5}{4}$ **7.** $0, 1, 3, 6$

9. $-1, 1, -1, 1$ **11.** $1, 0, -\dfrac{1}{3}, \dfrac{1}{2}$ **13.** $11, 15, 19$ **15.** $x + 2, x + 3, x + 4$

17. $2x + 7, 2x + 10, 2x + 13$ **19.** $32, 128, 512, 2048$ **21.** $\dfrac{8}{3}, \dfrac{16}{3}, \dfrac{32}{3}, \dfrac{64}{3}$

23. $\dfrac{x}{a}, -\dfrac{x^2}{a^2}, \dfrac{x^3}{a^3}, -\dfrac{x^4}{a^4}$ **25.** $4n + 3, 31$ **27.** $-5n + 8, -92$ **29.** $48(2)^{n-1}, 1536$

31. $-\dfrac{1}{3}(-3)^{n-1}, -243$ **33.** $2; 3; 41$ **35.** 28th   **37.** 3

39. **a.** For $n = 1$: $s_n = s_1 = a$; $ar^{n-1} = a(r)^0 = a(1) = a.$

   b. For $n = k$: $s_k = ar^{k-1}$; to obtain s_{k+1}, multiply s_k by r; hence, from multiplying both sides by r, $s_{k(r)} = s_k(r) = (ar^{k-1})(r) = ar^{k-1+1} = ar^{(k+1)-1}.$

Exercise 9.3 (page 235) **1.** $1 + 4 + 9 + 16$ **3.** $-\dfrac{1}{2} + \dfrac{1}{4} - \dfrac{1}{8}$ **5.** $1 + \dfrac{1}{2} + \dfrac{1}{4} + \cdots$

7. $\displaystyle\sum_{j=1}^{4} x^{2j-1}$ **9.** $\displaystyle\sum_{j=1}^{5} j^2$ **11.** $\displaystyle\sum_{j=1}^{\infty} j(j + 1)$ **13.** $\displaystyle\sum_{j=1}^{\infty} \dfrac{j + 1}{j}$ **15.** 63   **17.** 806   **19.** -6

21. 1092   **23.** $\dfrac{31}{32}$ **25.** $\dfrac{364}{729}$ **27.** 168   **29.** 196

31. Since $s_n = a + (n - 1)d$ and $S_{n+1} = S_n + s_{n+1}$, it follows that

   a. For $n = 1$: $S_n = s_n = s_1 = a$; $\dfrac{n}{2}(a + s_n) = \dfrac{1}{2}(2a) = a.$

   b. For $n = k$: $S_k = \dfrac{k}{2}(a + s_k)$, $s_k = a + (k - 1)d$; now $s_{k+1} = a + [(k + 1) - 1]d = a + kd,$

adding produces

$$S_k + s_{k+1} = S_{k+1} = \frac{k}{2}(a + s_k) + (a + kd) = \frac{k}{2}[a + a + (k-1)d] + \frac{2}{2}(a + kd)$$

$$= \frac{2ka + k^2d - kd + 2a + 2kd}{2} = \frac{2ka + 2a + k^2d + kd}{2} = \frac{(k+1)2a + k(k+1)d}{2}$$

$$= \frac{(k+1)}{2}[2a + kd] = \frac{k+1}{2}[a + (a + kd)] = \frac{k+1}{2}[a + s_{k+1}].$$

33. Let $s_1, s_2, s_3, \ldots, s_n$ and $t_1, t_2, t_3, \ldots, t_n$ be two sequences with terms in arithmetic progression; then, by Exercise 32, $(s_1 + t_1), (s_2 + t_2), \ldots, (s_n + t_n)$ is also a sequence with terms in arithmetic progression.

$$S_1 = s_1 + s_2 + s_3 + \cdots + s_n = \sum_{j=1}^{n} s_j, \quad S_2 = t_1 + t_2 + t_3 + \cdots + t_n = \sum_{j=1}^{n} t_j, \quad \text{and}$$

$$S_1 + S_2 = \sum_{j=1}^{n} s_j + \sum_{j=1}^{n} t_j;$$

however, $(S_1 + S_2) = (s_1 + t_1) + (s_2 + t_2) + \cdots + (s_n + t_n)$, and $S_1 + S_2 = \sum_{j=1}^{n}(s_j + t_j);$

hence $\sum_{j=1}^{n} s_j + \sum_{j=1}^{n} t_j = \sum_{j=1}^{n}(s_j + t_j).$

Exercise 9.4 (page 240) **1.** $8 \cdot 7 \cdot 6 \cdot 5 \cdot 4 \cdot 3 \cdot 2 \cdot 1$ **3.** $6 \cdot 5 \cdot 4 \cdot 3 \cdot 2 \cdot 1$ **5.** $5 \cdot 4 \cdot 3 \cdot 2 \cdot 1 = 120$

7. $\dfrac{9 \cdot 8 \cdot 7!}{7!} = 72$ **9.** $\dfrac{5 \cdot 4 \cdot 3 \cdot 2 \cdot 1 \cdot 7!}{8 \cdot 7!} = 15.$ **11.** $\dfrac{8 \cdot 7 \cdot 6!}{2 \cdot 1 \cdot 6!} = 28$ **13.** $3!$ **15.** $\dfrac{6!}{2!}$ **17.** $\dfrac{8!}{5!}$

19. $\dfrac{6!}{5!1!} = 6$ **21.** $\dfrac{3!}{3!0!} = 1$ **23.** $\dfrac{7!}{0!7!} = 1$ **25.** $\dfrac{5!}{2!3!} = 10$

27. $(n)(n-1)(n-2) \cdots \cdots 3 \cdot 2 \cdot 1$ **29.** $(3n)(3n-1)(3n-2) \cdots \cdots 3 \cdot 2 \cdot 1$

31. $(n-2)(n-3)(n-4) \cdots \cdots 3 \cdot 2 \cdot 1$ **33.** $(n+2)(n+1)$ **35.** $\dfrac{n+1}{n+3}$ **37.** $\dfrac{2n-1}{2n-2}$

39. $x^5 + 15x^4 + 90x^3 + 270x^2 + 405x + 243$ **41.** $x^4 - 12x^3 + 54x^2 - 108x + 81$

43. $8x^3 - 6x^2y + \dfrac{3}{2}xy^2 - \dfrac{1}{8}y^3$ **45.** $\dfrac{1}{64}x^6 + \dfrac{3}{8}x^5 + \dfrac{15}{4}x^4 + 20x^3 + 60x^2 + 96x + 64$

47. $x^{20} + 20x^{19}y + \dfrac{20 \cdot 19}{2!}x^{18}y^2 + \dfrac{20 \cdot 19 \cdot 18}{3!}x^{17}y^3$,   or

$$\binom{20}{0}x^{20} + \binom{20}{1}x^{19}y + \binom{20}{2}x^{18}y^2 + \binom{20}{3}x^{17}y^3$$

49. $a^{12} + 12a^{11}(-2b) + \dfrac{12 \cdot 11}{2!}a^{10}(-2b)^2 + \dfrac{12 \cdot 11 \cdot 10}{3!}a^9(-2b)^3$,   or

$$\binom{12}{0}a^{12} + \binom{12}{1}a^{11}(-2b) + \binom{12}{2}a^{10}(-2b)^2 + \binom{12}{3}a^9(-2b)^3$$

51. $x^{10} + 10x^9(-\sqrt{2}) + \dfrac{10 \cdot 9}{2!}x^8(-\sqrt{2})^2 + \dfrac{10 \cdot 9 \cdot 8}{3!}x^7(-\sqrt{2})^3$,   or

$$\binom{10}{0}x^{10} + \binom{10}{1}x^9(-\sqrt{2}) + \binom{10}{2}x^8(-\sqrt{2})^2 + \binom{10}{3}x^7(-\sqrt{2})^3$$

53. 1.22 **55.** $\$1480$ **57.** $-3003a^{10}b^5$ **59.** $3360x^6y^4$

61. **a.** $1 - x + x^2 - x^3 + \cdots$ **b.** $1 - x + x^2 - x^3 + \cdots$

Exercise 9.5 (page 248) **1.** $1, 5, 8$ **3.** $2, 6, 16$ **5.** $2, 2, 4$ **7.** 4 **9.** 24 **11.** 16 **13.** 64

15. 24 **17.** 216 **19.** 375 **21.** 30 **23.** 10 **25.** 48 **27.** $\dfrac{5!}{2!}$, or 60

29. $\dfrac{8!}{3!}$, or 6720   **31.** $P_{5,3} = \dfrac{5!}{2!} = \dfrac{5 \cdot 4!}{2!} = 5\left(\dfrac{4!}{2!}\right) = 5(P_{4,2})$

33. $P_{n,3} = \dfrac{n!}{(n-3)!} = \dfrac{n(n-1)!}{(n-3)!} = n\left[\dfrac{(n-1)!}{(n-3)!}\right] = n(P_{n-1,2})$ **35.** 9   **37.** 24   **39.** 48

Exercise 9.6 (page 252)   **1.** 7   **3.** 15   **5.** $\dbinom{52}{5}$ **7.** $\dbinom{13}{5} \cdot \dbinom{13}{5} \cdot \dbinom{13}{3}$ **9.** $4 \cdot \dbinom{13}{5}$

11. 164   **13.** 10   **15.** 210

Chapter 9 Review (page 253)   **1.** Formula holds for 1. Assume formula holds for k; test for $k + 1$:

$$3 + 6 + 9 + \cdots + 3k + 3(k+1) = \frac{3k(k+1)}{2} + 3(k+1).$$

Right-hand member is equivalent to:

$$\frac{3k^2 + 3k}{2} + \frac{6(k+1)}{2} = \frac{3k^2 + 9k + 6}{2} = \frac{3(k^2 + 3k + 2)}{2} = \frac{3(k+1)(k+2)}{2} = \frac{3(k+1)((k+1)+1)}{2}$$

2. Formula holds for 1. Assume formula holds for k; test for $k + 1$:

$$\frac{1}{2} + \frac{1}{4} + \frac{1}{8} + \cdots + \frac{1}{2^k} + \frac{1}{2^{k+1}} = 1 - \frac{1}{2^k} + \frac{1}{2^{k+1}}.$$

Right-hand member is equivalent to: $\dfrac{2^{k+1} - 2 + 1}{2^{k+1}} = \dfrac{2^{k+1} - 1}{2^{k+1}} = 1 - \dfrac{1}{2^{k+1}}.$

3. 13, 16, 19   **4.** $u - 4,\ a - 6,\ a - 8$ **5.** $-18,\ 54,\ -162$ **6.** $\dfrac{3}{2}, \dfrac{9}{4}, \dfrac{27}{8}$

7. $s_n = 5n - 8;\ s_7 = 27$ **8.** $s_n = (-2)\left(\dfrac{-1}{3}\right)^{n-1};\ s_5 = \dfrac{-2}{81}$ **9.** 25   **10.** 6th term

11. $2 + 6 + 12 + 20$ **12.** $\displaystyle\sum_{k=1}^{\infty} x^{k+1}$ **13.** 119   **14.** $\dfrac{121}{243}$ **15.** $5 \cdot (2 \cdot 1)$ **16.** 48

17. 21   **18.** $\dfrac{1}{n(n+1)!}$ **19.** $x^{10} - 20x^9y + 180x^8y^2 - 960x^7y^3$ **20.** $-15{,}360x^3y^7$ **21.** 16

22. 64   **23.** 128   **24.** 360   **25.** 792   **26.** 1,033,885,600   **27.** 200   **28.** 84

Exercise 10.1 (page 259)   **1. a.** $(2, 1, 1)$ **b.** $(-1, 0, 3)$ **c.** $(6, 2, -2)$

3. a. $(4, 0, -16)$ **b.** $(0, \sqrt{2}, -4), (0, -\sqrt{2}, -4)$ **c.** No solution exists.

5. a. -1 **b.** -4 **c.** -2 **7. a.** $(0, 0, 2)$ **b.** $(0, 3, 0)$ **c.** $(6, 0, 0)$

9. a. $(0, 0, 2), (0, 0, -2)$ **b.** $(0, 2, 0), (0, -2, 0)$ **c.** $(\sqrt{2}, 0, 0), (-\sqrt{2}, 0, 0)$

11. a. No solution exists.   **b.** $(0, -9, 0)$ **c.** $(3, 0, 0), (-3, 0, 0)$

13. a. $(0, 0, 0)$ **b.** $(0, 0, 0)$ **c.** $(0, 0, 0)$ **15.** $(2, -2, z)$

17. $(\sqrt{13}, y, 2), (-\sqrt{13}, y, 2)$ **19.** $(x, 3, z)$ **21.** $(-2, y, z)$

23.

25.

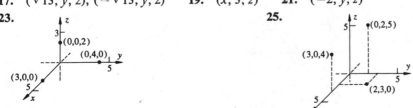

27.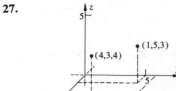

29.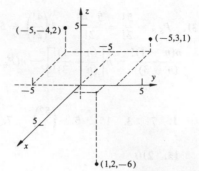

31. $\sqrt{19}$ **33.** $\sqrt{14}$ **35.** $\sqrt{22}$ **37.** $-1, 0, 0$ **39.** $0, 1, 0$ and $0, -1, 0$

Exercise 10.2 (page 262)

1.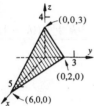

3.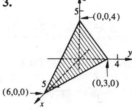

5.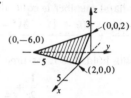

7.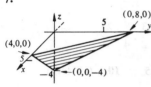

9.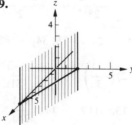

11.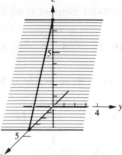

13.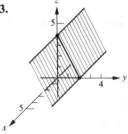

15.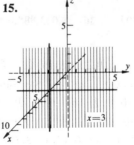

17.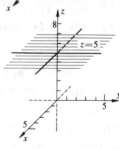

19.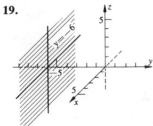

21.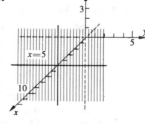

23. a.

b.

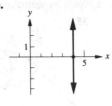

c.

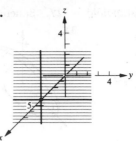

25. $\dfrac{83}{\sqrt{122}}$ **27.** $\dfrac{21}{\sqrt{30}}$

Exercise 10.3 (page 268)

1.

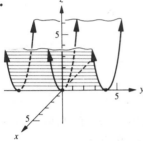

3.

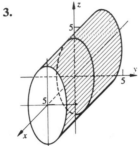

5.

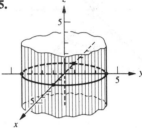

7.

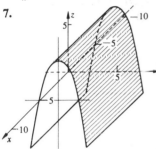

9.

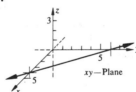

11.

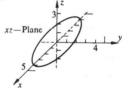

13.

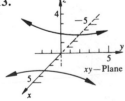

15.   No trace exists.

17.

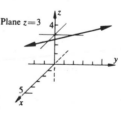

19.

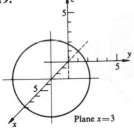

21.

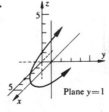

23.

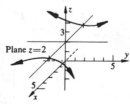

25. Circular paraboloid   **27.** Sphere   **29.** Elliptic hyperboloid of two sheets

31. 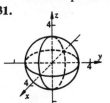   **33.** 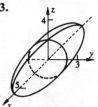   **35.**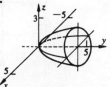

37.

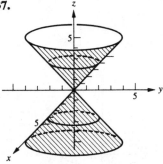

Chapter 10 Review (page 269)   **1.** $(0, 0, 3\sqrt{2}), (0, 0, -3\sqrt{2})$ **2.** $(0, -6, 0)$ **3.** $(3, 0, 0), (-3, 0, 0)$
4. $(1, 1, \sqrt{19}), (1, 1, -\sqrt{19})$ **5.** 4   **6.** -6 **7.** $(1, -12, z)$ **8.** $(x, 0, 2), (x, 0, -2)$
9.   **10.**   **11.** $\sqrt{26}$ **12.** $\sqrt{113}$

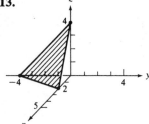   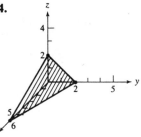

13.   **14.**   **15.**

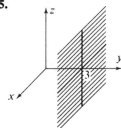

16.   **17.**   **18.**

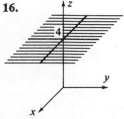   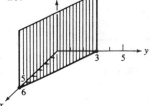   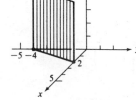

19. 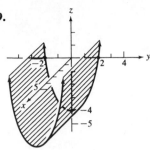   **20.** 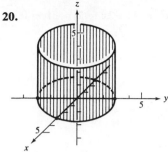   **21.**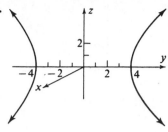

22. 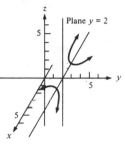   **23.** Elliptic paraboloid   **24.** Sphere

25. Elliptic hyperboloid of one sheet   **26.** Elliptic hyperboloid of two sheets.

27. 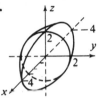   **28.**

Exercise A.1 (page 273)

1. a. 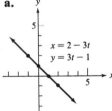   **b.** $x + y - 1 = 0$ **3. a.** 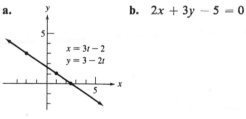   **b.** $2x + 3y - 5 = 0$

5. a. 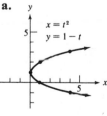   **b.** $x = 1 - 2y + y^2$ **7. a.** 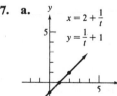   **b.** $x - y - 1 = 0$

9. a. 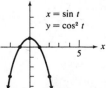   **b.** $x^2 + y = 1$ **11. a.** 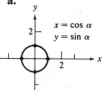   **b.** $x^2 + y^2 = 1$

13. a.

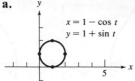

$x = 1 - \cos t$
$y = 1 + \sin t$

b. $x^2 + y^2 - 2x - 2y + 1 = 0$

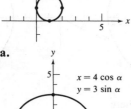

15. a.

$x = 4 \cos \alpha$
$y = 3 \sin \alpha$

b. $\dfrac{x^2}{16} + \dfrac{y^2}{9} = 1$

17. $x = \dfrac{4}{3 - t}$; $y = tx = \dfrac{4t}{3 - t}$ **19.** $x = \dfrac{-1}{2 + 5t}$; $y = \dfrac{-t}{2 + 5t}$

21. $x = \dfrac{4}{\pm \sqrt{1 + t^2}}$; $y = \dfrac{4t}{\pm \sqrt{1 + t^2}}$

Exercise A.2 (page 278)   1. $(-7, -3)$ **3.** $(5, -1)$ **5.** $x' - 2y' = 13$
7. $x'^2 - 2y'^2 + 4x' + 20y' - 52 = 0$ **9.** $(x + 5)^2 + (y - 1)^2 = 9$

11. $(x - 3)^2 = 8(y - 3)$

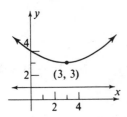

(3, 3)

(-5, 1)

13. $\dfrac{(x + 2)^2}{25} + \dfrac{(y - 2)^2}{9} = 1.$

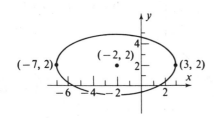

(-7, 2)   (-2, 2)   (3, 2)

15. $\dfrac{(x - 3)^2}{9} - \dfrac{(y - 2)^2}{16} = 1$

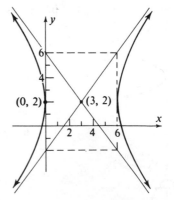

(0, 2)   (3, 2)

17. $y'^2 = 12x'$ **19.** $x'^2 + y'^2 = \dfrac{25}{4}$ **21.** $\dfrac{x'^2}{4} + \dfrac{y'^2}{6} = 1$ **23.** $\dfrac{x'^2}{16} - \dfrac{y'^2}{9} = 1$

25.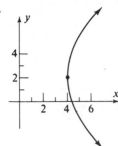

27.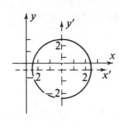

29.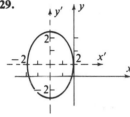

31.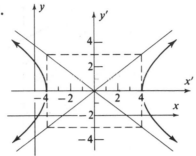

Exercise A.3 (page 281) 1. $x' = \dfrac{-3\sqrt{3}+5}{2}, y' = \dfrac{3+5\sqrt{3}}{2}$ **3.** $x' = 4 + 3\sqrt{3}, \; y' = -4\sqrt{3} + 3$

5. $x' = 3 + 4\sqrt{3}, \; y' = 3\sqrt{3} - 4$ **7.** $x' = -3\sqrt{3} - \frac{5}{2}, \; y' = -3 + \frac{5}{2}\sqrt{3}$

9. $-x'^2 + 4y'^2 = 16$; hyperbola **11.** $5x'^2 + y'^2 = 36$; ellipse

13. $11x'^2 + y'^2 = 44$; ellipse

15. $8x'^2 - 2y'^2 - 8\sqrt{2}x - 6\sqrt{2}y = 15$; eliminating the first-degree term gives $4x''^2 - y''^2 = 5$; hyperbola

17. $2x'^2 - \sqrt{2}x' - 3\sqrt{2}y' + 5 = 0$; eliminating the first-degree terms gives

$2x''^2 - 3\sqrt{2}y'' + \dfrac{19}{4} = 0$; parabola

19. Using the equations on page 280,

$A' + C' = A\cos^2\alpha + B\sin\alpha\cos\alpha + C\sin^2\alpha + A\sin^2\alpha - B\sin\alpha\cos\alpha + C\cos^2\alpha$
$= A(\cos^2\alpha + \sin^2\alpha) + C(\sin^2\alpha + \cos^2\alpha)$

and, because $\cos^2\alpha + \sin^2\alpha = 1$, $A' + C' = A + C$.

Appendix Review (page 282) 1. a. 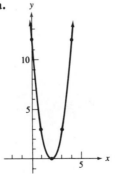 **b.** $y = 3x^2 - 12x + 12$

2. **a.** 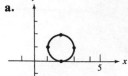   **b.** $x^2 + y^2 - 4x - 2y + 4 = 0$

3. $x = \dfrac{3}{1 - 4t}$; $y = -\dfrac{3}{4} + \dfrac{3}{4(1 - 4t)}$ **4.** $x = \pm\sqrt{\dfrac{6}{2 + 3t^2}}$; $y = \pm\sqrt{2 - \dfrac{2}{2 + 3t^2}}$

5. $(-7, 4)$ **6.** $2x'^2 - 12x' - y' + 18 = 0$

7. $(x - 1)^2 + (y + 3)^2 = 16$ **8.** $y^2 = -8(x - 7)$

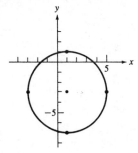   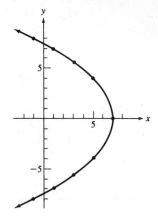

9. $x'^2 = 6y'$ **10.** $2x'^2 + y'^2 = \dfrac{73}{8}$

11. 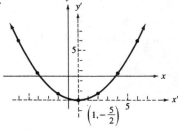   **12.**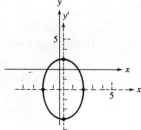

13. $x' = -\sqrt{3} + \frac{1}{2}$, $y' = 1 + \dfrac{\sqrt{3}}{2}$ **14.** $x' = -2\sqrt{2}$, $y' = 2\sqrt{2}$

15. $3x'^2 - y'^2 = 20$; hyperbola   **16.** $7x''^2 + 11y''^2 = \dfrac{158}{11}$; ellipse

Index